Statistics with JMP: Graphs, Descriptive Statistics, and Probability

Statistics with JMP: Graphs, Descriptive Statistics, and Probability

Peter Goos

University of Leuven and University of Antwerp, Belgium

David Meintrup

University of Applied Sciences Ingolstadt, Germany

Library of Congress Cataloging-in-Publication Data applied for.

A catalogue record for this book is available from the British Library.

ISBN: 9781119035701

Set in 10/12pt Times by Laserwords Private Limited, Chennai, India

1 2015

To Marijke, Bas, Loes and Mien
To Béatrice and Werner

Contents

Preface

This book is the result of a thorough revision of the lecture notes "Descriptive Statistics and Probability" that were developed by Peter Goos for the course "Statistics for business and economics 1" at the Faculty of Applied Economics of the University of Antwerp in Belgium. Encouraged by the success of the Dutch version of this book (entitled *Beschrijvende Statistiek en Kansrekenen*, published in 2013 by Acco Leuven/Den Haag), we joined forces to create an English version. The book provides a detailed treatment of basic probability theory, descriptive statistics, and graphical representations of data. We pay equal attention to mathematical aspects, the interpretation of all the statistical concepts that are introduced, and their practical application. In order to facilitate the understanding of the methods and to appreciate their usefulness, the book contains many examples involving real-life data. To demonstrate the broad applicability of statistics and probability, these examples have been taken from various fields of application, including business, economics, sport, engineering, and natural sciences.

We had two motivations in writing this book. First, we wanted to provide students and teachers with a resource that goes beyond other textbooks of similar scope in its technical and mathematical content. It has become increasingly fashionable for authors and statistics teachers to sweep technicalities and mathematical derivations under the carpet. We decided against this, because we feel that students should be encouraged to apply their mathematical knowledge and that doing so deepens their understanding of statistical methods. Reading this book requires some knowledge of mathematics, including the use of derivatives, integrals, and some matrix algebra. In most countries, these topics are taught in secondary or high school. Moreover, these topics are often revisited in introductory mathematics courses at university. Therefore, we are convinced that many university students have a sufficiently strong mathematical background to appreciate and benefit from the more thorough nature of this book. A particular strength is that all mathematical derivations are shown in detail. We included all intermediate steps, even those that might be trivial for mathematicians. We hope that this keeps the book readable for less mathematically gifted readers and also shows that the mathematical derivations are actually not as difficult as these readers might first imagine.

Our second motivation was to ensure that the concepts introduced in the book can be successfully put into practice. To this end, we show how to generate graphs, calculate descriptive statistics and compute probabilities using the statistical software package JMP (pronounced "jump"). We chose JMP as supporting software because it is powerful yet easy to use, and suitable for a wide range of statistically oriented courses (including descriptive statistics, hypothesis testing, regression, analysis of variance, design of experiments, reliability, multivariate methods, and statistical and predictive modeling). We believe that introductory courses in statistics and probability should use such software so that the enthusiasm of students is not nipped in the bud. Indeed, we find that, because of the way students can easily interact with JMP, it can actually spark enthusiasm for statistics and probability in class.

In summary, our approach to teaching descriptive statistics and probability theory combines theoretical and mathematical depth, detailed and clear explanations, numerous practical examples, and the use of a user-friendly and yet very powerful statistical package. Our companion book *Statistics with JMP: Hypothesis Tests, ANOVA and Regression*, (based on *Verklarende Statistiek: Schatten en Toetsen*, Acco Leuven/Den Haag, 2014), follows the same philosophy.

Software

As mentioned, we use JMP as enabling software. With the purchase of a hard copy of this book, you receive a one-year license for JMP's Student Edition. The license period starts when you activate your copy of the software using the code included with this book (see inside front cover). To download JMP's Student Edition, visit http://www.jmp.com/wiley. For students accessing a digital version of the book, your lecturer may contact Wiley in order to procure unique codes with which to download the free software. For more information about JMP, go to http://www.jmp.com. JMP is available for Windows and Mac operating systems. This book is based on JMP version 12 for Windows.

In our examples, we do not assume any familiarity with JMP: the step-by-step instructions are detailed and accompanied by screenshots. For more explanations and descriptions, www.jmp.com offers a substantial amount of free material, including many video demonstrations. In addition, there is a JMP Academic User Community where you can access content, discuss questions and collaborate with other JMP users worldwide: instructors can share teaching resources and best practices, students can ask questions, and everyone can access the latest resources provided by the JMP Academic Team. To join the community, go to http://community.jmp.com/academic.

Data files

Throughout the book, various data sets are used. We strongly encourage everybody who wants to learn statistics to actively try things out using data. JMP files containing the data sets as well as JMP scripts to reproduce figures, tables, and analyses, can be downloaded from the publisher's companion website to this book:

http://www.wiley.com/go/goosandmeintrup/JMP

There, we also provide some additional supporting files to generate maps, or visualize probability distributions and densities. For instructors who would like to use the book in their courses, there are slides available that cover the material presented. The information on how to access these teaching resources can also be found on the companion website.

Peter Goos
peter.goos@biw.kuleuven.be
David Meintrup
david.meintrup@thi.de

Acknowledgments

We would like to thank numerous people who have made the publication of this book possible. The first author, Peter Goos, is very grateful to Professor Willy Gochet from the University of Leuven, who introduced him to the topics of statistics and probability. Professor Gochet allowed Peter to use his lecture notes as a backbone for his own course material, which later developed into this book. The second author, David Meintrup, would like to thank Antonio Sáez for providing a perfect working environment during his sabbatical at the University of Jaén, Spain.

The authors are very grateful for the support and advice offered by several people from the JMP Division of SAS: Brady Brady, Ian Cox, Bradley Jones, Volker Kraft, John Sall, and Mia Stephens. It is Volker who brought the two authors together and encouraged them to work on a series of English books on statistics with JMP (the second book is entitled *Statistics with JMP: Hypothesis Tests, ANOVA and Regression*). A very special thank you goes to Ian, whose suggestions substantially improved this book. The authors would also like to thank Leonids Aleksandrovs, Kris Annaert, Stefan Becuwe, Filip De Baerdemaeker, Roselinde Kessels, Ida Ruts, Bagus Sartono, Evelien Stoffels, Anja Struyf, Utami Syafitri, Peter Thijssen, Anil Haydar Topal, Katrien Van Driessen, Ellen Vandervieren, Kristel Van Rompaey, Diane Verbiest, Sara Weyns, and Simone Willis for their detailed comments and constructive suggestions, and technical assistance in creating figures.

Finally, we thank Debbie Jupe, Heather Kay, Sangeetha Parthasarathy and Prachi Sinha Sahay at Wiley.

1

What is statistics?

The world is ready for the truth; the modern age is here; every year another report appears that examines poverty by means of statistical research rather than romantic claptrap.

(from *The Crimson Petal and the White,* Michael Faber, p. 334)

In this introductory chapter, we give a general description of the topics of statistics and probability theory. Some examples illustrate the purpose and applications of both disciplines, as well as the differences between them. As statistics has more applications in science, industry, and economics than probability theory, statistics is typically given far more attention in degree subjects like business, industrial and bio-science engineering, applied economics, and natural or social sciences. Nevertheless, one should pay some attention to probability theory as well. In fact, both disciplines are strongly connected to each other: it is impossible to understand the working of statistical inference without a sound knowledge of probability theory. Therefore, in this book, we discuss both probability theory and statistics.

1.1 Why statistics?

For many years, statistics has been a subject, often a dreaded one, in several fields of study at universities and colleges. The reason is that quite a few people will, sooner or later, be confronted with problems of data analysis during their professional activities. A sound statistical background not only allows us to analyze the data and to make concrete decisions based on the analysis, but it also provides an advantage in the data collection process.

Nevertheless, statistics is not immediately perceived as useful by most students. This is mainly due to the fact that, during a statistics course, they are still unfamiliar with the sorts of practical decision problems managers, economists, engineers, and

Statistics with JMP: Graphs, Descriptive Statistics, and Probability, First Edition. Peter Goos and David Meintrup.
© 2015 John Wiley & Sons, Ltd. Published 2015 by John Wiley & Sons, Ltd.
Companion Website: wiley.com/go/goosandmeintrup

researchers face on a daily basis. Many students will start realizing the usefulness of statistics when they start to work on their bachelor's or master's thesis. The many examples in this basic course are intended to advance this awareness by several years.

In an introductory statistics course, one often finds a whole series of quotes as an attempt to motivate students. A classic example is "Statistical thinking will one day be as necessary for efficient citizenship as the ability to read and write." from the British writer Herbert George Wells (1866–1946). More recent is the judgment by the US quality guru W. Edwards Deming, to whom a large part of the downright spectacular economic recovery in Japan after World War II is attributed. He claimed that "Statistics is too important to be left to statisticians. The goal is to have many statistically-skilled workers: engineers, scientists, managers..." Hal Varian, chief economist at Google says the following: "I keep saying that the most sexy job in the next 10 years will be statistician. And I'm not kidding." In Europe, Willy Buysse, former CEO at SN Brussels Airlines, states that too few decisions are made based on data. Only recently, his many years of diligence establishing a research department, where statistical and other quantitative methods are used to address all sorts of problems, has been rewarded.

Another justification for a thorough training in statistical methods can be found in the so-called Six Sigma improvement program. The purpose of this program is to solve concrete problems with a large financial impact both in service and industrial companies, and to reduce the number of faults and defects to 3.4 per million opportunities. The approach is based on statistical methods, as presented in Figure 1.1. The figure shows that the traditional method to solve a practical problem is to immediately search for practical solutions. This approach is typically based on guessing and trial-and-error, so that it will often take a long time to find a final solution to the problem. The Six Sigma improvement program promotes a more thoughtful, scientific approach to problems. First, data is collected in the so-called measurement phase. Then, using statistical methods, the data is carefully examined. This often leads to interesting insights and recommendations to improve existing products, services, or processes. The Six Sigma approach also relies on the use of statistical process control and statistically designed experiments. Hence, statistics helps to find the best possible solution for all kinds of practical problems.

To achieve a successful cooperation between practitioners, on one hand, and statisticians, on the other, some openness is required on both sides. Engineers, economists, or scientists need a solid knowledge of the basic principles and techniques of statistics. Statistics is thus an indispensable skill in the repertoire of

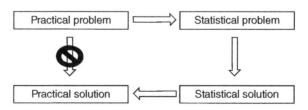

Figure 1.1   Using statistical methods to solve problems.

an effective employee. This explains why statistics is taught not only in the first and second years of many bachelor's degrees in engineering, sciences, and economics, but also later, for example in master's programs.

Finally, a thorough training in statistics is also a prerequisite for students of political and social sciences. They will also be confronted with numerous data sets in their professional careers that are impossible to interpret without a statistical background. For them, statistics is a stepping stone to econometric research methods.

1.2 Definition of statistics

The word statistics may sound familiar to anyone. A statistic usually refers to numerical information, for example, information about

- the population of a country: birth and death rates, immigration and emigration, ... (such statistics are called population statistics),

- the economy: employment and unemployment rates, investments, prices, gross national products (GNP), ... (these statistics are called economic statistics), or

- a company or sector: sales figures, income statements, growth, acquisitions, layoffs, ... (these figures are called business statistics).

More formally, statistics can be defined as the set of methodologies for collecting, representing, analyzing, and interpreting data. This shows that the statistical science is a very general auxiliary science, which plays an important role in almost any environment. Applications of statistics are countless in engineering, medicine, economics, natural sciences, and business management, but statistics is also used in literature, history, political science, criminology, and even musicology.

In our modern society, data is massively present:

- computer files in companies contain sales data, cost data, and customer data (such as addresses, ordered quantities, and order frequencies),

- the financial pages of newspapers contain stock prices, commodity prices, and exchange rates,

- federal and regional authorities regularly publish data on population, trade, and industry, and

- the Internet is a source of numerous data sets.

Companies collect data naturally and actively. Among other things, this takes place by carrying out experiments (e.g., to design new products), in the context of statistical process control, or by measuring all kinds of properties of products, services, and processes. By continuously analyzing data, quality departments of companies attempt to deliver products or services with as few defects as possible and with the highest reliability. In addition, business processes are organized in such a way that waste is minimized, inspections of finished products are reduced to the minimum, and customer requirements are satisfied with minimal costs.

Research agencies collect data via surveys by phone, by post, via the Internet or by street interviews. Such surveys are designed to gather information about the shopping behavior of consumers, about the voting behavior of the population, or public opinion on social issues.

Statistics allows us to turn data into usable information. The role that statistics plays herein may be best illustrated based on some examples.

1.3   Examples

Example 1.3.1 *An airline conducted a study on the behavior of its passengers on intercontinental flights and recorded*

- *the number of passengers with reservations that do not show up (the so-called no-shows),*

- *the weight of the luggage of passengers (often there is a limit of 20 kilograms), and*

- *the time the passengers arrive before the official departure time of the flight (for intercontinental flights, the passengers are asked to be at the airport at least two hours prior to departure).*

The company recorded this data over several months and then made a distinction between passengers in economy class and passengers in business class. The data is analyzed with the aim of instituting appropriate policies. An example may be to allow overbooking, that is, to take more reservations than there are seats on the plane, or to apply more stringent action against passengers carrying too much luggage.

Example 1.3.2 *In the production of coffee, the humidity during production is of crucial importance for the quality of the final product. The humidity is kept under control by a system that does not work flawlessly. Therefore, several measurements of the humidity are taken daily to determine whether it remains within appropriate limits. This approach is referred to as statistical process control.*

Example 1.3.3 *A filling machine for bottles usually has several filling heads, so that many bottles can be filled in parallel. In such a filling process, operators typically weigh a certain number of bottles every hour, to verify that each filling head delivers the desired amount of liquid into the bottles. Another interesting question in this context is whether differences occur between measurements that have been carried out by different operators.*

Example 1.3.4 *Thanks to loyalty cards, supermarkets collect massive data sets. Data that is typically recorded includes*

- *the amount spent per visit at the store, maybe broken down into categories (food, clothing, …),*

- *the number of items sold,*

- *the payment method (cash, debit card, credit card, or voucher).*

Researchers use statistical methods to summarize this huge amount of information and to present it in a way suitable for decision making. Supermarkets exploit this information to send out personalized promotional materials.

Example 1.3.5   *Financial analysts are interested in the degree of risk of investing in a particular stock. To this end, they keep track of the monthly return rates of stocks over many years. They take into account price changes, but the dividends as well. Moreover, monthly return rates of the global market, for example the Euro Stoxx 50 index, are tracked. If the return rate of a stock rises or falls to a larger extent than the market, then the share is called risky. In the opposite case, one speaks of a share with little risk. Using statistical methods, one can investigate relations between the return rate of the stock and of the overall market.*

1.4   The subject of statistics

In each of the examples in the previous section, the interest is in one or more questions concerning a **population** of objects or elements, or concerning a **process** that generates objects or elements.

The data of the population or process is obtained by recording one or more properties or characteristics of their elements. These properties or characteristics are called **variables**. The name indicates that the value of the property varies from element to element. Therefore, statistics is sometimes referred to as the study of variability.

Usually, it is impossible to include all elements of a population or process in a study. Therefore, one works with a subset of the elements: the **sample**. It is not always easy to collect sample data in a correct way. In any statistical survey, one should pay a lot of attention to the data collection process. In this context, the abbreviation GIGO[1] is often used. This stands for garbage in, garbage out and refers to the fact that any statistical methods can only extract little reliable information from data of poor quality.

Example 1.4.1   *For a study of the electoral behavior in European elections, the population can be described easily: all citizens of Europe who are entitled to vote. Variables that could be registered in this context are gender, occupation, political beliefs, age, and so on.*

Example 1.4.2   *Tossing a die is a process that generates data. A possible sample involves throwing the die 50 times. Variables that could be registered are the number of dots or whether or not the number of dots is even.*

In Examples 1.3.2 and 1.3.3, we can consider all times at which the production process is in operation to be the population. At a limited or finite number of points in time,

---

[1] This abbreviation is a parody of the abbreviations FIFO (first in first out) and LIFO (last in first out), used in accounting for booking items in stock.

measurements or observations are made, for example, measurements of the humidity (Example 1.3.2) or weight (Example 1.3.3). All measurements together form the sample. For the financial analyst in Example 1.3.5, the sample is formed by a finite set of return rates and market indices. In Example 1.3.4, the population of interest for the researcher is the set of all customers of the supermarket. One possible sample consists of all customers that have visited the store during one month and that made use of their loyalty card.

The data collected in a sample can be represented in many ways using tables and graphs. In addition, one can calculate characteristic values or statistics, such as the mean, to generate a clear idea of the collected data. The different ways of presenting sample data are summarized under the term **descriptive statistics**. This topic is covered in Chapters 2 and 3.

In many cases, describing the sample data is only a first step in an investigation. A second phase involves analyzing and interpreting the sample. Analysis and interpretation is necessary in order to find answers to questions about the population or process that were set in advance, to test hypotheses, or to assess the quality of a proposed statistical model. The answers and conclusions obtained from the statistical analysis are generalized to the population or the process. This generalization is called **inference**, which explains the term **inferential statistics**.

The generalization of conclusions from sample data to an entire population or to a process immediately discloses the weakness of statistics: based on sample data, one can never make statements with certainty about the population or process in question. These statements may be considered reliable if statistically valid methods were used for the collection of the sample data. The degree of confidence in a particular statement is expressed by means of a probability, so that a basic knowledge of probability theory is required to be able to understand and apply statistical methods.

1.5 Probability

The words chance and probability sound even more familiar than the term statistics. Intuitively, everyone has a good idea of the meaning of a probability of 1/4 when participating in a gambling game. Such a probability can be used by virtually everyone to decide whether or not to participate in the game. However, the calculation of such a probability can already raise difficulties.

Probability theory studies processes or experiments in which the outcome is uncertain. Here, the terms process and experiment should be interpreted in their broadest sense. Examples are throwing a die, the price of a share when the stock exchange closes, a mortgage interest rate, the demand for laptop computers of a particular brand, the percentage of defective products in a production line during a certain period, the number of visitors to a website, or drawing a winner from all the participants in a lottery.

The difference between probability and statistics is that, in probability theory, populations and processes are studied directly, while statistics does this through sample data. Probability theory always starts with a set of assumptions about the population or the process. Some examples will illustrate this.

Example 1.5.1 *If the process is throwing a die, then, with the help of probability theory, we can try to figure out the probability of obtaining a six* 20 *or more times when we toss the die* 100 *times. This calculation is only possible if we make an important assumption about the die used: the die is fair, or, in other words, the die is completely homogeneous and symmetrical, so that it is equally likely to obtain a one as it is to obtain a two, a three, a four, a five, or a six.*

A statistical question about the die could be to investigate the fairness of the die. The die may be thrown a (large) number of times to collect the required sample data. Based on these data, one can draw a statistical conclusion about the hypothesis that the die is fair.

Example 1.5.2 *In an industrial filling process, one can calculate, based on some assumptions concerning the settings and the accuracy of the filling machine, the probability that a bottle will not be full enough. Another possibility is to calculate the probability that, in a lot with* 1,000 *bottles, at most* 5% *of the bottles will not have been filled enough.*

A statistical analysis of the same filling process typically may involve regular weighings of a number of bottles (the sample), in order to verify whether the average content of the bottles is too large or too little, and whether or not the content of the bottles varies too much.

Example 1.5.3 *Using probability theory, one could study the electoral behavior of the European population assuming that* 30% *will vote for party A,* 25% *for party B,* 20% *for party C, and* 25% *for smaller parties. Probability theory can then calculate that, for every* 500 *voters, on average* 150 *will opt for party A,* 125 *for Party B,* 100 *for Party C, and* 125 *will choose other parties.*

Statistics, however, will make a statistical prediction based on a sample of, for example, 2,000 *voters. This prediction can also be given with a margin of error.*

It is important to realize that statistics works with a limited amount of information obtained from a sample. Therefore, statements about populations and processes can be false. This is the weakness of statistics. Ideally, the probabilities of error are small. The probability for errors can be reduced by collecting a lot of high quality data in a sensible manner.

Probability theory also has a weakness: the assumptions about the studied process or population may be wrong, so that its conclusions are invalid.

1.6 Software

In probability and statistics, a lot of calculations are needed. It is important to create summary tables of all the data in a sample, or to represent the data graphically. This makes the use of a computer and of specialized statistical software necessary. As mentioned in the Preface, in this book, we use the statistical software package JMP®.

2

Data and its representation

A microphone in the sidewalk would provide an eavesdropper with a cacophony of clocks, seemingly random like the noise from a Geiger counter. But the right kind of person could abstract signal from noise and count the pedestrians, provide a male/female breakdown and a leg-length histogram …

(from *Cryptonomicon*, Neal Stephenson, p. 147)

Data is a set of measurements of one or more characteristics or variables of some elements of a population, or of a number of objects generated by a process. Different types of variables can be measured.

2.1 Types of data and measurement scales

Variables are classified according to the measurement scale on which they are measured. Categorical or qualitative variables are measured on a nominal scale or on an ordinal scale. Quantitative variables are either measured on an interval scale or on a ratio scale.

2.1.1 Categorical or qualitative variables

2.1.1.1 Nominal variables

Elements of a sample or a population can be classified using a nominal variable: the value of the variable places an element in a certain class or category. Examples of such variables are

- gender (male/female),
- nationality (Belgian, German, and so on),

Statistics with JMP: Graphs, Descriptive Statistics, and Probability, First Edition. Peter Goos and David Meintrup.
© 2015 John Wiley & Sons, Ltd. Published 2015 by John Wiley & Sons, Ltd.
Companion Website: wiley.com/go/goosandmeintrup

- religion (Catholic, Protestant, and so on), and

- whether or not one owns a car (yes/no).

Sometimes it can be useful to assign labels, code numbers, or code letters, to the different classes or categories. For example, a Belgian person may be assigned the code "1", a Dutch person the code "2", a French person the code "3", and a German person the code "4". It is important to note that these figures do not imply any order and/or quantity. Therefore, except for calculations of frequencies and percentages, most arithmetic operations on nominal variables are meaningless.

2.1.1.2 Ordinal variables

If a nominal variable implies a logical order between the elements of a sample, then the variable is ordinal. Typical examples of ordinal variables can be found in all kinds of surveys. There, respondents are typically asked whether they consider the quality of a product or service as "1: very good", "2: good", "3: moderate", "4: bad", or "5: very bad". In other surveys, the respondents are asked if they "1: strongly disagree", "2: rather disagree", "3: neither agree nor disagree", "4: rather agree", or "5: strongly agree" with a particular statement. Other examples of ordinal variables include the number of Michelin stars of restaurants and the number of stars of hotels.

An ordinal scale has no fixed measurement unit. This means that the difference between two levels cannot be expressed as a number of units on the measuring scale. For example, the difference between a hotel with three stars and one with two stars is not necessarily the same as the difference between a hotel with two stars and one with only one star. It is obvious that it is also not very useful to perform arithmetic operations with ordinal variables.

2.1.2 Quantitative variables

A variable that is measured on a quantitative scale can be expressed as a fixed number of measurement units. Examples are length, area, volume, weight, duration, number of bits per unit of time, price, income, waiting time, number of ordered goods, and so on. For quantitative variables, almost all arithmetic operations make sense. This is due to the fact that the difference between two levels of a quantitative variable can be expressed as a number of units in contrast to differences between two levels of an ordinal variable. Within the class of quantitative variables, a distinction is made between variables that are measured on an interval scale and variables measured on a ratio scale.

2.1.2.1 Interval scale

An interval scale has no natural zero point, that is, no natural lower limit. For variables measured on an interval scale, calculating ratios is not meaningful. Well-known examples of interval variables are the time read on a clock or the temperature expressed in degrees Celsius or Fahrenheit. The difference between

2 o'clock and 4 o'clock is the same as the difference between 21:00 and 23:00, but it's not like 4 o'clock is twice as late as 2 o'clock. This is due to the fact that time read on a clock has no absolute zero. The same applies to the temperature measured in degrees Celsius: 20°C is not four times as hot as 5°C.

2.1.2.2   Ratio scale

A ratio scale does have an absolute zero. Therefore, for variables measured on a ratio scale, ratios can be calculated. A length of 6 cm is twice as much as a length of 3 cm, as the length scale has an absolute zero point. Analogously, an order of six products is twice as large as an order of three products. The temperature measured in Kelvin does have an absolute minimum, so that temperature is sometimes measured on a ratio scale. Zero Kelvin (−273.15°C) is the coldest possible temperature, and therefore an absolute lower limit for the temperature.

2.1.2.3   Discrete versus continuous variables

A discrete variable can only take a finite or infinite countable number of different values, while a continuous variable can take a continuum of values. Examples of discrete variables are the number of passengers on a flight, the number of children in a family, or the number of insurances that a family contracted. Examples of continuous variables are length, duration, weight, and body mass index.

In practice, all observations of a continuous variable are discrete: a continuous length is measured up to a certain accuracy (e.g., one millimeter), thus turned into a discrete number. Nevertheless, we will consider length as a continuous variable.

2.1.3   Hierarchy of scales

It is clear that there is a hierarchy in the measurement scales. The highest or most informative measurement scale is the ratio scale, followed by the interval scale, the ordinal, and the nominal scale. Data that has been measured on a certain scale can be transformed into data of a lower measurement scale. Data measured on a ratio scale (e.g., length) are naturally interval scaled (the difference between 6 and 3 cm is the same as the difference between 15 and 12 cm), ordinal (ordering lengths is meaningful), and nominal (lengths can be divided into classes). Conversely, nominal data can never be transformed into ordinal or quantitative data. Therefore, all techniques that are applicable to nominal data are automatically also applicable to ordinal and quantitative data. All techniques that are applicable to ordinal data can be useful for quantitative data. One rarely makes a distinction between data measured on an interval scale and data measured on a ratio scale.

2.1.4   Measurement scales in JMP

JMP distinguishes between nominal, ordinal, and quantitative variables. The software refers to measurement scale as "Modeling type", and uses "Nominal", "Ordinal", and "Continuous" for nominal, ordinal, and quantitative variables, respectively.

2.2 The data matrix

Data is often presented in a matrix, with a row for each element or observation of a sample, and a column for every measured variable. A complete row in a data matrix is sometimes referred to as an **observation vector**.

Example 2.2.1 *Figure 2.1 contains data from a survey on a number of characteristics of Spanish red wines. The sample contains* 70 *wines. Figure 2.2 shows the symbols that JMP is using to indicate the different measurement scales, "Nominal", "Ordinal", and "Continuous". The variable "Name" is a nominal variable. The variables "Rating" and "Price category" are ordinal variables. The other variables are quantitative. The measurement scale of a variable can be changed in JMP by a right-click on the name of a column, and then selecting "Column info".*

In this chapter, we will mainly treat so-called univariate and bivariate representations of variables. A univariate representation refers to one variable, while a bivariate representation refers to two variables simultaneously. Likewise, multivariate data is nothing but data consisting of several variables. In the remainder of the chapter,

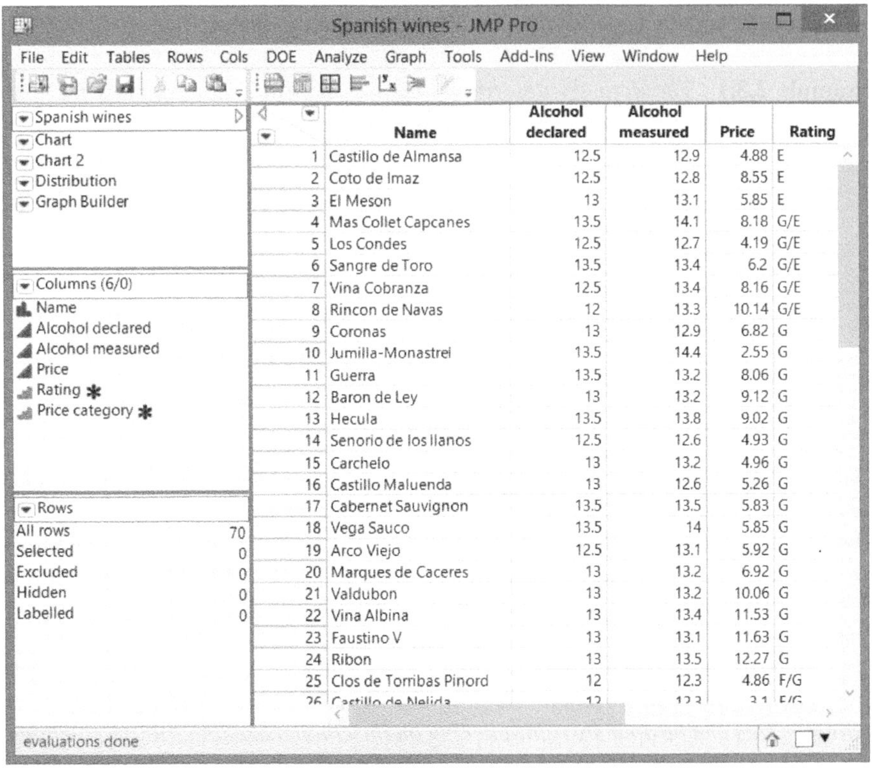

	Name	Alcohol declared	Alcohol measured	Price	Rating
1	Castillo de Almansa	12.5	12.9	4.88	E
2	Coto de Imaz	12.5	12.8	8.55	E
3	El Meson	13	13.1	5.85	E
4	Mas Collet Capcanes	13.5	14.1	8.18	G/E
5	Los Condes	12.5	12.7	4.19	G/E
6	Sangre de Toro	13.5	13.4	6.2	G/E
7	Vina Cobranza	12.5	13.4	8.16	G/E
8	Rincon de Navas	12	13.3	10.14	G/E
9	Coronas	13	12.9	6.82	G
10	Jumilla-Monastrel	13.5	14.4	2.55	G
11	Guerra	13.5	13.2	8.06	G
12	Baron de Ley	13	13.2	9.12	G
13	Hecula	13.5	13.8	9.02	G
14	Senorio de los llanos	12.5	12.6	4.93	G
15	Carchelo	13	13.2	4.96	G
16	Castillo Maluenda	13	12.6	5.26	G
17	Cabernet Sauvignon	13.5	13.5	5.83	G
18	Vega Sauco	13.5	14	5.85	G
19	Arco Viejo	12.5	13.1	5.92	G
20	Marques de Caceres	13	13.2	6.92	G
21	Valdubon	13	13.2	10.06	G
22	Vina Albina	13	13.4	11.53	G
23	Faustino V	13	13.1	11.63	G
24	Ribon	13	13.5	12.27	G
25	Clos de Torribas Pinord	12	12.3	4.86	F/G

Figure 2.1 Part of the data matrix on Spanish red wines.

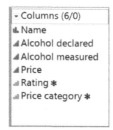

Figure 2.2   Symbols used by JMP for the different measurement scales.

we assume that we have a data sample. However, the various representations that we will address may also be used for data of entire populations.

2.3   Representing univariate qualitative variables

Categorical or qualitative variables allow us to put data into categories or classes. The absolute frequency, or simply the frequency, of a class is the number of elements of the sample that belong to that class. The relative frequency of a class is the ratio of the frequency and the total number of observations in the sample.

Example 2.3.1 *The data set described here on Spanish wines contains the final rating of the wines. The following coding is used:*

- *E: excellent,*

- *G/E: good to excellent,*

- *G: good,*

- *F/G: fair to good,*

- *F: fair, and*

- *P/F: poor to fair.*

The final rating is clearly a qualitative, ordinal variable. The absolute and relative frequencies for each class are shown in Table 2.1, which is called a frequency table. The same information can also be presented using a bar chart. Figure 2.3 shows two versions of a bar chart, which have exactly the same shape. The bar chart in Figure 2.3a shows the absolute frequencies, while that in Figure 2.3b displays the relative frequencies.

It is useful to let JMP know that a rating "Excellent" is better than a rating "Good to excellent", and that a rating "Good to excellent" is in turn better than a rating "Good". This can be done by right-clicking on the column heading "Rating", choosing "Column Properties" in the resulting pop-up menu, and selecting the option "Value Ordering". To create a bar chart in JMP, one can use the "Chart" option

Table 2.1   Frequency table for the final rating of Spanish red wines.

Rating	E	G/E	G	F/G	F	P/F	Sum
Abs. frequency	3	5	16	35	9	2	70
Rel. frequency	.043	.071	.229	.500	.129	.029	1

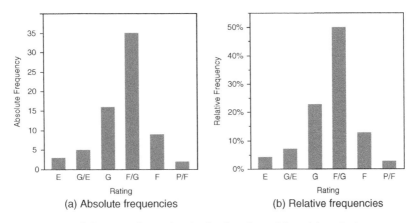

(a) Absolute frequencies   (b) Relative frequencies

Figure 2.3   Bar charts for the final rating of Spanish red wines.

*in the "Graph" menu. After choosing that option, the variable "Rating" has to be selected as well as the desired type of chart, "Bar Chart". For a bar chart showing absolute frequencies, the option "N" has to be chosen under "Statistics". In order to show relative frequencies instead, the option "% of Total" has to be picked. A frequency table can be obtained in JMP using the option "Tabulate" within the "Analyze" menu. If you want to display the result in a separate data table, you need to select the option "Make Into Data Table" in the pop-up menu that appears when clicking on the red triangle icon next to the word "Tabulate". This is illustrated in Figure 2.4. Such a red triangle is called a **hotspot** in JMP. Hotspots appear in practically all reports and data tables. Clicking a hotspot always opens a menu containing additional options that are specific to the graphical or statistical analysis you are doing.*

If the classes are arranged in decreasing order of their frequency and the cumulative frequencies are plotted, the result is called a Pareto chart, a Pareto diagram, or a Pareto plot. The purpose of a Pareto chart is to draw attention to the classes with the highest frequencies[1]. A cumulative representation of the frequencies means that the frequencies of the different classes are summed. This is clarified in the following example.

---

[1] In quality control, the classes with the highest frequencies are called the "vital few", while the classes with the lowest frequencies are called the "trivial many". A commonly used rule of thumb says that 80% of the quality problems can be attributed to 20% of the causes.

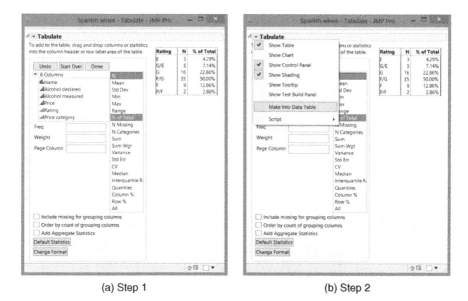

(a) Step 1   (b) Step 2

Figure 2.4   Creating a frequency table in JMP.

Example 2.3.2   *The quality department of a manufacturer of mobile phones inspected* 2530 *devices. During the inspection the employees found* 115 *faulty phones. Devices with scratched surfaces or cracks, deformed devices, and devices with missing parts (incomplete) were labeled as defective. The data, a bar chart, and the corresponding Pareto chart are shown in Figure 2.5.*

In the Pareto chart in Figure 2.5c, the left vertical axis is for the bars, while the right vertical axis is for the cumulative frequencies shown by means of the black line. It can easily be seen in the Pareto chart that the most common problem is missing parts. This problem has a relative frequency of 41.74%. *The second most common problem is the occurrence of scratches, with a relative frequency of* 27.83%. *The relative frequency of the two most common problems together is* 41.74% + 27.83% = 69.57%. *If we add the relative frequency of devices with cracks to this, we obtain a cumulative frequency of* 41.74% + 27.83% + 20% = 89.57%.

To create a Pareto chart in JMP, one can use the "Analyze" menu. In this menu, the option "Quality and Process" has to be chosen first. The next step is to select the option "Pareto Plot". Figure 2.6 shows the resulting dialog window, in which the variable "Type of Defect" has to be entered in the field "Y, Cause", and the variable "Absolute Frequency" has to be entered in the field "Freq".

Another graphical representation of absolute and relative frequencies for a qualitative variable is the pie chart.

Type of Defect	Absolute Frequency	Relative Frequency	Cumulative Frequency
Incomplete	48	41.74%	41.74%
Scratched	32	27.83%	69.57%
Cracks	23	20.00%	89.57%
Other	8	6.96%	96.52%
Deformed	4	3.48%	100.00%

(a) Data

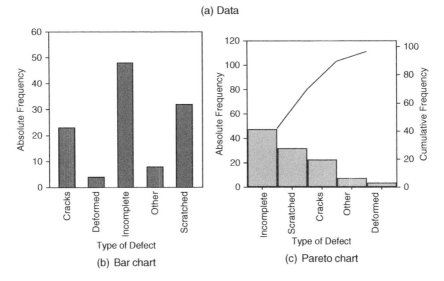

(b) Bar chart   (c) Pareto chart

Figure 2.5   Causes of defective mobile phones in Example 2.3.2.

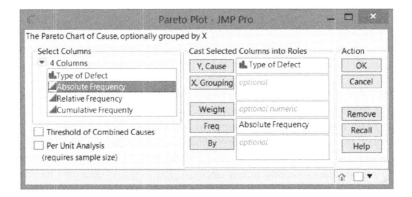

Figure 2.6   Dialog window for creating a Pareto chart in JMP.

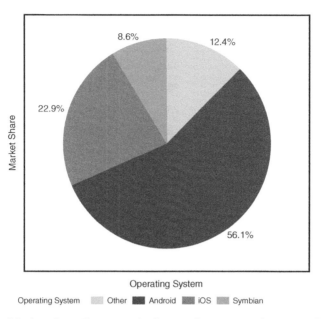

Figure 2.7   Market share (in percent) of operating systems for smartphones in the first quarter of 2012.

Example 2.3.3   *Figure 2.7 shows the market share (in percent) of various operating systems on smartphones in the first quarter of* 2012. *One possible way to make a pie chart in JMP is via the menu "Graph", by using the option "Chart", and selecting "Pie Chart".*

2.4   Representing univariate quantitative variables

2.4.1   Stem and leaf diagram

The stem and leaf diagram is an interesting representation of quantitative data because it does not only give a picture of the frequencies of the various kinds of values for the variable under study, it also preserves every individual observation.

Example 2.4.1   *Figure 2.8 shows a stem and leaf diagram of the price variable in the data set of Spanish red wines (see Example 2.2.1). Note that prices are unavailable for* 11 *wines in the data set, so that the stem and leaf diagram only contains information on* 59 *wines. Here, the stem shows the whole part of the price (the number before the decimal point), while the leaves represent the first digit after the decimal point of the* 59 *prices, after rounding to one decimal. The diagram indicates that the four cheapest wines cost €2.2, 2.5, 2.6, and 2.7. The most expensive wine costs €13.6. Most wines cost between €4 and 6.*

Creating a stem and leaf diagram in JMP can be done via the option "Distribution" in the "Analyze" menu. In the resulting dialog window, shown in Figure 2.9, the

Stem and Leaf

Stem	Leaf	Count
13	56	2
12	134	3
11	56	2
10	0115	4
9	01	2
8	1224667	7
7	2229	4
6	2389	4
5	023345789999	12
4	02236788999	11
3	0126	4
2	2567	4

2|2 represents 2.2

Figure 2.8   Stem and leaf diagram of the prices of Spanish red wines.

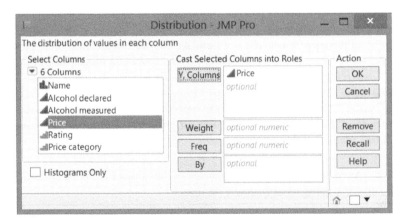

Figure 2.9   Creating a stem and leaf diagram in JMP: Step 1.

variable "Price" has to be entered in the field "Y, Columns". This results in an output involving a histogram and a lot of statistics. To obtain the stem and leaf diagram, one then has to click on the hotspot (red triangle icon) next to the word "Price". In the pop-up menu that appears after doing so, the option "Stem and Leaf" has to be selected. This step is shown in Figure 2.10.

2.4.2   Needle charts for univariate discrete quantitative variables

A needle chart, just like a bar chart, displays absolute or relative frequencies of the values of a variable. Therefore, the names "needle chart" and "bar chart" are often used interchangeably.

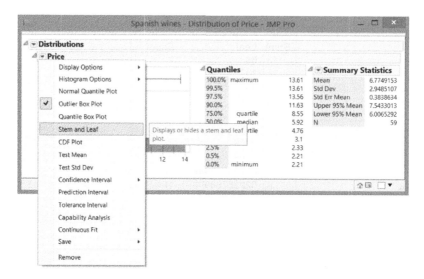

Figure 2.10   Creating a stem and leaf diagram in JMP: Step 2.

Example 2.4.2 *For 100 flights from Brussels to London, Brussels Airlines registered the number of passengers who did not show up, despite the fact that they reserved a seat in business class. In the professional jargon, one calls these "no-shows". The absolute and relative frequencies are shown in Table 2.2. The relative frequencies are displayed in Figure 2.11. The representation in Figure 2.11a was created in JMP with the option "Needle Chart", while the representation in Figure 2.11b was made with the option "Bar Chart". Both of these options become available after selecting the "Chart" platform in the "Graph" menu.*

Table 2.2   Absolute and relative frequencies of the numbers of passengers not showing up for 100 flights of Brussels Airlines.

Number of no-shows	0	1	2	3	4	5	6
Abs. frequency	11	38	32	9	6	3	1
Rel. frequency	11%	38%	32%	9%	6%	3%	1%

Example 2.4.3 *The first lottery drawing with 42 numbers in Belgium happened on April 30, 1984. When considering all drawings, some numbers were drawn more often than others, as shown in Table 2.3. For each integer from 1 to 42, the table contains the frequency, the relative frequency and the date on which it was drawn for the last time. A bar chart for the relative frequencies is shown in Figure 2.12.*

  It would be a good exercise to compare the relative frequencies in Table 2.3 and Figure 2.12 with the theoretical probability for drawing a certain number using a statistical hypothesis test. This topic is discussed in the book Statistics with JMP: Hypothesis Tests, ANOVA and Regression.

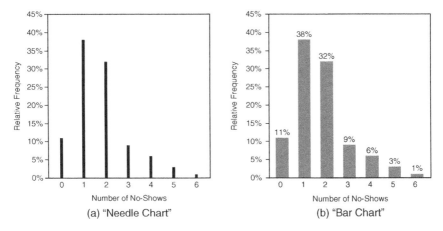

Figure 2.11   Graphical representations of the numbers of passengers who did not show up.

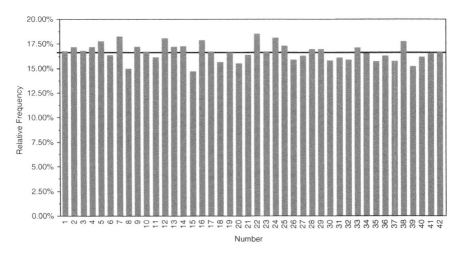

Figure 2.12   Bar chart of the relative frequencies of the 42 lottery numbers. The horizontal reference line represents the theoretical probability of $7/42 = 1/6$ that a specific number is drawn at any lottery drawing.

Example 2.4.4   *Two students organize a game night and want to test that the two dice they use are fair. The first student throws the first die 20 times and calculates the relative frequencies of the numbers of dots. The second student is more diligent and throws the second die 100 times. Using a needle diagram, each student compares his results for every number of dots with the theoretical probability of 1/6. The corresponding needle diagrams are shown in Figure 2.13. The results of the samples are shown in gray, while the theoretical probabilities are shown in black. In this context, one can introduce sampling frequencies (i.e., the observed relative frequencies)*

Table 2.3   Data for the lottery drawings in Belgium.
(source: http://www.nationale-loterij.be/, 04/01/2012)

Number	Number of drawings	Relative frequency	Date of most recent drawing
1	406	16.76%	28/09/2011
2	416	17.18%	27/08/2011
3	407	16.80%	24/09/2011
4	416	17.18%	21/09/2011
5	430	17.75%	28/09/2011
6	396	16.35%	24/09/2011
7	442	18.25%	21/09/2011
8	363	14.99%	17/09/2011
9	417	17.22%	14/09/2011
10	405	16.72%	03/09/2011
11	391	16.14%	20/08/2011
12	438	18.08%	20/08/2011
13	417	17.22%	10/09/2011
14	418	17.26%	24/09/2011
15	356	14.70%	16/07/2011
16	433	17.88%	24/08/2011
17	405	16.72%	28/09/2011
18	379	15.65%	28/09/2011
19	403	16.64%	17/09/2011
20	376	15.52%	10/09/2011
21	397	16.39%	31/08/2011
22	449	18.54%	17/09/2011
23	405	16.72%	28/09/2011
24	439	18.13%	14/09/2011
25	419	17.30%	14/09/2011
26	385	15.90%	28/09/2011
27	395	16.31%	24/09/2011
28	411	16.97%	24/09/2011
29	411	16.97%	06/08/2011
30	383	15.81%	21/09/2011
31	390	16.10%	14/09/2011
32	385	15.90%	07/09/2011
33	415	17.13%	24/09/2011
34	401	16.56%	17/09/2011
35	381	15.73%	24/08/2011
36	395	16.31%	17/08/2011
37	382	15.77%	07/09/2011
38	430	17.75%	03/09/2011
39	369	15.24%	24/09/2011
40	392	16.18%	24/08/2011
41	402	16.60%	28/09/2011
42	404	16.68%	24/08/2011

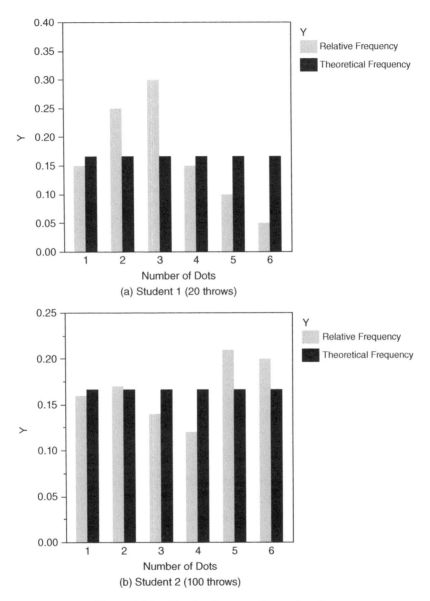

Figure 2.13   Needle diagrams for testing dice.

and population frequencies (i.e., the theoretical relative frequencies). The relative frequencies of the first student (with only 20 throws) deviate quite strongly from the theoretical probabilities, while the relative frequencies of the second student (who did 100 throws) are fairly close to the theoretical probabilities. Based on these needle diagrams, one may want to perform a statistical hypothesis test to determine whether the dice are fair or not. Hypothesis tests are not covered here, but in the book Statistics with JMP: Hypothesis Tests, ANOVA and Regression.

2.4.3   Histograms and frequency polygons for continuous variables

2.4.3.1   Histograms

Undoubtedly, the most popular way to visualize the values of a continuous quantitative variable is a histogram. A histogram involves several bars, the heights of which are absolute or relative frequencies. Each bar corresponds to an interval of values of the variable under study. These intervals are obtained by dividing the range of the sample values (i.e., the smallest interval covering all values measured for the quantitative variable) into a number of smaller intervals or classes. Typically, but not always, the same width is used for all these smaller intervals or classes. In a histogram showing relative frequencies, the sum of the heights of all bars is equal to 1. In a histogram showing absolute frequencies, the sum of all heights equals the number of observations.

Example 2.4.5   *Figure 2.14 shows a histogram of 50 breaking strengths (expressed in kg), each measured for a bundle of 20 woolen fibers. The minimum and maximum breaking strengths are 3.16 and 162.39 kg, respectively. The histogram involves 6 classes with a width of 28 kg. These choices ensure that the histogram covers all values of the variable breaking strength between 0 kg and 6 × 28 kg = 168 kg.*

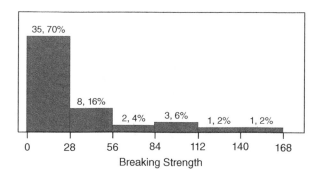

Figure 2.14   Histogram of the 50 breaking strengths in Example 2.4.5.

Note that the rectangles of a histogram are placed right next to each other. This emphasizes the continuous nature of the depicted variable and distinguishes histograms from bar charts for qualitative variables and needle charts for discrete quantitative variables.

Later, we will learn that we do not always use the original sample data in a statistical analysis. Instead, we will sometimes use transformed data. For example, instead of using the original data for a histogram, we could first apply a mathematical operation. A transformation that is frequently used is the logarithmic transformation. Sometimes, this transformation ensures that we obtain a more or less symmetrical histogram with one peak. A histogram for the natural logarithm of the breaking strengths

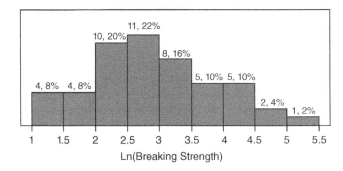

Figure 2.15   Histogram of 50 values of ln(*breaking strength*).

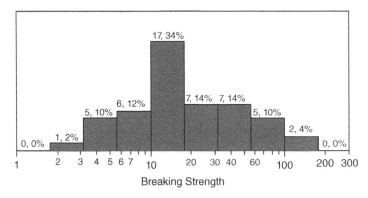

Figure 2.16   Histogram of 50 breaking strengths with logarithmic scale.

is shown in Figure 2.15. Note that this histogram displays the absolute frequencies and the relative frequencies, separated by a comma, on top of each bar.

Figure 2.16 shows a histogram similar to that in Figure 2.15. The difference between the two histograms is that the histogram in Figure 2.16 shows the original breaking strengths with a logarithmic scale on the horizontal axis, while the histogram in Figure 2.15 shows the natural logarithm of the breaking strengths on a linear scale. The linear scale in Figure 2.15 can be identified by the fact that the distance between 1 and 2 is the same as the distance between 3 and 4. On the logarithmic scale in Figure 2.16, this is not the case, but the distance between 1 (= 10^0) and 10 (= 10^1) is the same as the distance between 10 (= 10^1) and 100 (= 10^2).

2.4.3.2   Construction of histograms

A disadvantage of histograms and frequency polygons is that their ultimate form strongly depends on the number of intervals or classes chosen. The final aim of a histogram should be to give a clear picture of the location of the data. Too many classes provide too detailed an image, while too few classes in a histogram display insufficient details. Typically, we work with 5–20 classes. A classic rule of thumb is to set the number of classes to the square root of the number of observations. For a sample of 50 observations, one should use $\sqrt{50} \approx 7$ classes according to this rule of thumb.

Creating a histogram in JMP is extremely easy via the "Analyze" menu, in which you have to select the "Distribution" option. You will then obtain the dialog window shown in Figure 2.17. The next step is to indicate the variable whose distribution you wish to plot using the histogram. By default, JMP displays the histogram vertically, but it is easy to switch to a horizontal display. To do so, you need to click on the hotspot (red triangle) next to the name of the variable at the top of the output, and uncheck the option "Vertical" under "Histogram Options". Under "Histogram Options", you can also adjust the width of the intervals or classes ("Set Bin Width") and chose to display the absolute and/or relative frequencies ("Show counts" and/or "Show percents") at the top of each of the histogram's bars. All of these options are shown in Figure 2.18. The "Grabber" tool allows you to change the bin width

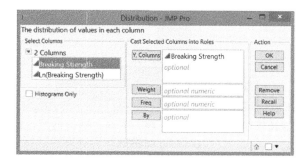

Figure 2.17   Dialog window for creating a histogram.

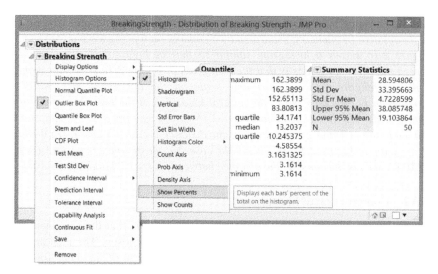

Figure 2.18   Options for a histogram in JMP.

dynamically. To do so, select the little hand symbol in the "Tools" menu, place your cursor anywhere in the histogram, and click and drag the histogram bars. Depending on the direction of your movement, you will dynamically increase or decrease the width of the histogram bars.

If you would like to add a title on the histogram's axis, or switch from a linear to a logarithmic scale, you can right-click on the axis. You will then get various options to adjust the axis according to your taste. These options are shown in Figure 2.19.

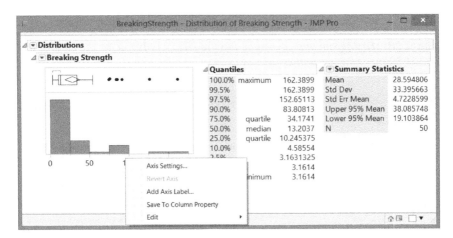

Figure 2.19   Options for the axis in a histogram in JMP.

Another interesting feature of histograms in JMP is that you can click and double-click on their bars. Clicking on a bar in a histogram automatically selects the corresponding rows in the data table. Double-clicking on a bar in a histogram creates a new data table, containing only the corresponding data. So, double-clicking on a bar in a histogram is a fast way to create a subset of the original data set. If you want to select several histogram bars, hold down the "Shift" key while you select the bars. Holding down the "Shift" key while double-clicking creates a data table with the data corresponding to all selected histogram bars.

2.4.3.3   Frequency polygons

In a frequency polygon, the bars of a histogram are replaced by straight lines that connect the tops of the adjacent bars. An example of a frequency polygon, along with the corresponding histogram, is shown in Figure 2.20.

2.4.3.4   Construction of frequency polygons

To construct a frequency polygon in JMP, we start by creating a histogram, as described previously. In the hotspot menu (red triangle icon), we then have to press "Save" and select the option "Level Midpoints". This step is shown in Figure 2.21. JMP has now created a new column in your data table, containing the midpoints

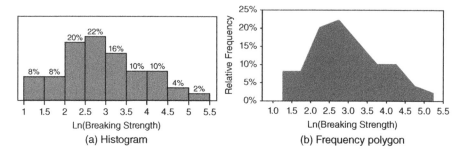

(a) Histogram   (b) Frequency polygon

Figure 2.20   Histogram and corresponding frequency polygon for the natural logarithm of 50 breaking strengths.

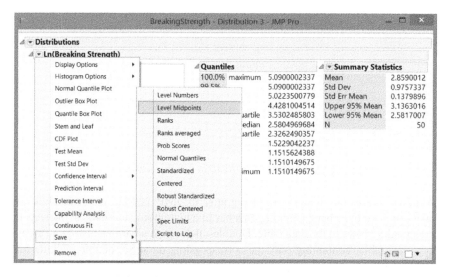

Figure 2.21   Constructing a frequency polygon: Step 1.

of the histogram bars. Next, we need to select the "Summary" option from the "Tables" menu. In the resulting dialog window, we have to choose "% of Total" from the "Statistics" drop-down menu, and drag the new variable containing the midpoints to the "Group" field. This second step is shown in Figure 2.22. Clicking "OK" will create a new data table, shown in Figure 2.23. Working with this new data table, we then need to select the "Graph Builder" in the "Graph" menu. This is a highly flexible platform for the creation of a wide range of graphics that we will use frequently. We will cover more details on the use of the Graph Builder in Section 2.5.1. For the purpose of creating a frequency polygon, we should drag the variable "% of Total" from the list of columns displayed at the top left to the drop zone called "Y", and the variable containing the midpoints to the drop zone called "X". Finally, we need to click the "Area" button from the toolbar on top of the window to get the desired frequency polygon. This is illustrated in Figure 2.24.

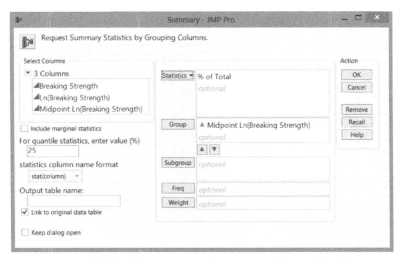

Figure 2.22   Constructing a frequency polygon: Step 2.

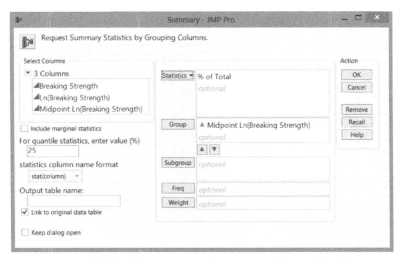

Figure 2.23   Constructing a frequency polygon: Intermediate data table.

By clicking on the button named "Done", renaming the axes by clicking on their labels and scaling the graph by dragging the corners, you can produce a graph that looks exactly as the frequency polygon shown in Figure 2.20b.

2.4.4   Empirical cumulative distribution functions

Empirical cumulative distribution functions can be constructed both for discrete and continuous quantitative variables. Graphical representations of such functions are used frequently, because they allow one to determine quantiles, such as the quartiles and the median of a data set (see Sections 3.1.1 and 3.2.2), in a single glance.

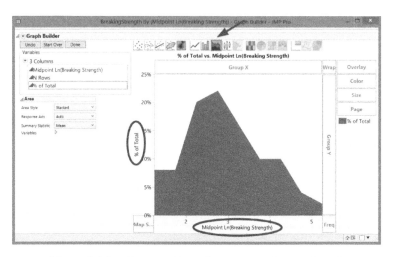

Figure 2.24   Constructing a frequency polygon: Step 3.

Also, to test whether sample data originated from a normally distributed population, the empirical cumulative distribution function is often used (e.g., in the Lilliefors test and the Kolmogorov–Smirnov test, see the book *Statistics with JMP: Hypothesis Tests, ANOVA and Regression*). The construction of an empirical cumulative distribution function can best be explained using an example.

Example 2.4.6 *Imagine that, in a small sample, we obtained the observations 6, 4, 3, 1, 7, 6, and 10. Ranking these seven observations from small to large, we get 1, 3, 4, 6, 6, 7, 10. In this sample, every value occurs once, except for the value 6, which occurs twice. These different values and the corresponding observed frequencies are shown in the first two rows of Table 2.4. The relative frequencies are calculated by dividing the observed frequencies by the number of observations, 7. Finally, the last row of the table shows the cumulative relative frequencies. The cumulative relative frequency of a sample value is simply the sum of its relative frequency and the relative frequencies of all the smaller observations in the sample. For instance, the cumulative relative frequency of the observation 4 is equal to the sum of the relative frequencies of the observations 1, 3, and 4. This yields the value 3/7. A graphical representation*

Table 2.4   Calculating the empirical cumulative distribution function for the sample in Example 2.4.6.

	Observations					
	1	3	4	6	7	10
Frequency	1	1	1	2	1	1
Rel. frequency	1/7	1/7	1/7	2/7	1/7	1/7
Cum. rel. frequency	1/7	2/7	3/7	5/7	6/7	1

of the cumulative relative frequencies for this example, all of which are given in the last row of Table 2.4, is given in Figure 2.25.

Example 2.4.7   *Figure 2.26 contains the graphical representations of the empirical cumulative distribution functions of the numbers of no-shows in Table 2.2 and of the breaking strengths of Example 2.4.5. It is a useful exercise to reconstruct the function in Figure 2.26a by yourself.*

Creating an empirical cumulative distribution function using JMP is quite easy. In the "Analyze" menu, choose the option "Distribution". Next, click on "CDF Plot" in the hotspot (red triangle) menu next to the name of the variable under study (in the figure, "Breaking strength"). This final step is shown in Figure 2.27. Note that CDF is the abbreviation of cumulative distribution function.

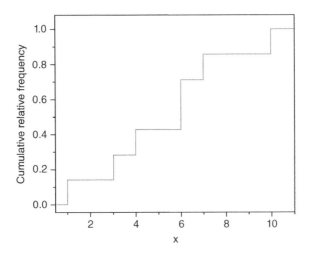

Figure 2.25   Graphical representation of the empirical cumulative distribution function for the sample in Example 2.4.6.

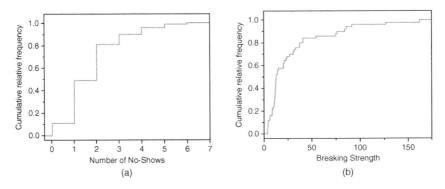

Figure 2.26   Empirical cumulative distribution functions of the numbers of no-shows in Table 2.2 and of the breaking strengths of Example 2.4.5.

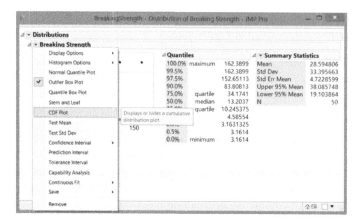

Figure 2.27   Creating an empirical cumulative distribution function in JMP.

2.5   Representing bivariate data

2.5.1   Qualitative variables

A **cross tabulation**, also known as a **contingency table**, is a convenient way to represent bivariate data in tabular form. A cross tabulation is designed for nominal and ordinal data, but it can also be used for quantitative variables provided their values are put into categories or classes.

Example 2.5.1 *Based on the Spanish red wine data described in Example 2.2.1, a cross tabulation can be made for the variables rating and price. The variable rating is an ordinal variable, but the price is a quantitative variable. Therefore, for that variable, we need to define several classes. Suppose that we use three classes or price categories: cheap (< €6), moderately priced and expensive (≥ €10). The resulting cross tabulation is displayed in Table 2.5.*

In JMP, we create a cross tabulation using the "Analyze" menu, with the "Fit Y by X" platform. The corresponding dialog window is shown in Figure 2.28. In this

Table 2.5   Cross tabulation for the data set of Spanish red wines.

Price category	Rating				Sum
	F/G	G	G/E	E	
Cheap (< €6)	2	1	7	21	31
Moderately priced	1	3	5	9	18
Expensive (≥ €10)	0	1	4	5	10
Sum	3	5	16	35	59

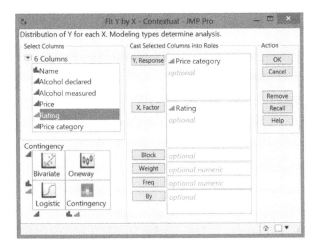

Figure 2.28   Creating a cross tabulation and mosaic plot in JMP.

dialog window, you need to enter the variable "Price category" as the y variable, and the variable "Rating" as the x variable. At first, this produces the output in Figure 2.29. Each cell in this table contains four numbers: the absolute frequency for each cell, and three relative frequencies. The number 2 in the first cell of the table tells us that there are two cheap wines with rating excellent (E). The number 3.39 tells us that 3.39% of all 59 wines are both cheap and excellent. The number 6.45 tells us that 6.45% of all 31 cheap wines are excellent. Finally, the number 66.67 tells us that 66.67% of all three excellent wines are cheap. The last row and the last column of the cross tabulation contain the column totals and the row totals, and the relative frequency of each price category and of each rating, respectively.

The initial cross tabulation produced by JMP can be simplified by unchecking some of the options in the hotspot (red triangle) menu next to the word "Contingency Table" at the top of the output.

A graphical alternative to a cross tabulation is called a **mosaic plot**. This graphical representation is produced together with a cross tabulation using the "Fit Y by X" platform. The mosaic plot corresponding to the cross tabulation in Table 2.5 and Figure 2.29 is shown in Figure 2.30. The interpretation of the mosaic plot is as follows:

- In the mosaic plot, every price category has its own color. This way, we see immediately that the cheap wines are the most numerous and expensive wines the least numerous.

- Each rectangle in the mosaic plot corresponds to a cell in the cross tabulation. The larger the surface area of a rectangle, the more observations correspond to that cell. The largest rectangle in the mosaic plot in Figure 2.30 is located at the lower right corner. This cell refers to the cheap wines with a rating of fair to good (F/G).

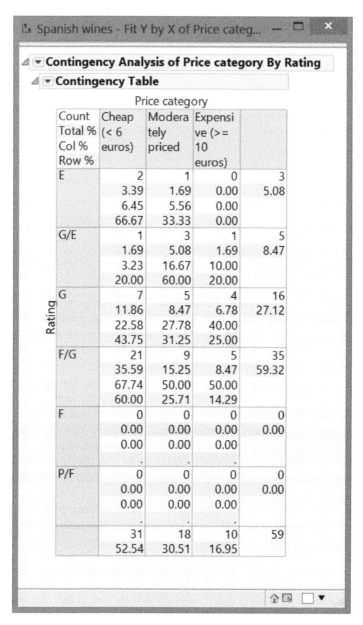

Figure 2.29   Initial cross tabulation produced by JMP.

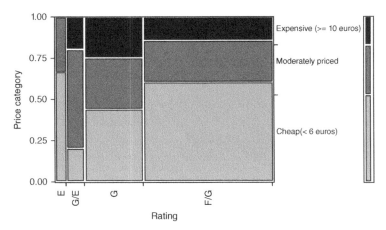

Figure 2.30   Mosaic plot corresponding to the cross tabulation in Table 2.5 and Figure 2.29.

- *The widest rectangles in the mosaic plot are for wines with rating fair to good (F/G). This means that the fair to good wines are the most numerous. The narrowest surfaces are for excellent (E) wines, which are the least numerous.*

- *The heights of the rectangles indicate how numerous the wines are in the different price categories for each of the ratings separately.*

- *Finally, the horizontal marks on the right vertical axis indicate the overall proportions of cheap, moderately priced, and expensive wines.*

If we switch the roles of the variables "Price category" and "Rating" in the dialog window in Figure 2.28, we obtain an alternative mosaic plot with the price categories on the horizontal axis instead of the vertical axis. This mosaic plot is shown in Figure 2.31.

In a mosaic plot in JMP, it is possible to click on a rectangle so that all observations in the data table associated with this area are highlighted. If you have created a histogram for the same data, then all parts of the histogram corresponding to the same observations are also highlighted.

As an alternative to the mosaic plot, a **multiple bar chart** *can be used to graphically display the information contained within a cross tabulation. In Figure 2.32, two multiple bar charts are shown for the variables "Price category" and "Rating".*

The creation of a multiple bar chart in JMP requires the use of the option "Graph Builder" in the "Graph" menu. This is a highly flexible platform for the creation of a wide range of graphics. The start screen of the "Graph Builder" is shown in Figure 2.33. On the left, the screen shows all variables in the data set of Spanish red wines. At the top of the start screen, a range of buttons is visible, each corresponding to a type of graph that can be created. Finally, in the center, the screen involves several drop zones for variables, named "X", "Y", "Group X", "Group Y", "Overlay", "Color", and "Size".

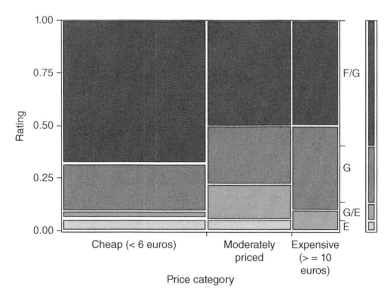

Figure 2.31   Alternative mosaic plot corresponding to the cross tabulation in Table 2.5 and Figure 2.29.

By dragging variable names to the various drop zones and choosing a chart type from the top, we can create a large number of graphical representations of data. For example, in order to get the multiple bar chart in Figure 2.32a, we first need to drag the variable "Price category" to the "X" zone, and then click the seventh button at the top of the screen to obtain a bar chart. This is illustrated in Figure 2.34.

Next, we need to drag the variable "Rating" to the "Overlay" zone. This is illustrated in Figure 2.35. Finally, clicking on the "Done" button completes the construction of the multiple bar chart. Figure 2.32b is obtained by using the "Stacked" bar option, obtained by right-clicking in the graphics area of the previous figure.

2.5.2   Quantitative variables

Data concerning two quantitative variables can be represented graphically using a so-called **scatter plot**. This is a two-dimensional figure, in which each dimension corresponds to a variable under study and each point corresponds to an observation. The first coordinate of any point is the value of the corresponding observation for the first variable, whereas its second coordinate is the value for the second variable. A scatter plot shows the relation or association between the two variables (see Section 3.9.2).

Example 2.5.2   *Figure 2.36 shows the scatter plot for the variables "Alcohol measured" (displayed on the horizontal axis) and "Price" (displayed on the vertical axis) for 59 Spanish red wines (see Example 2.2.1). In the figure, it is clearly visible that a high alcohol content is frequently associated with a high price, and a low alcohol content often corresponds to a low price.*

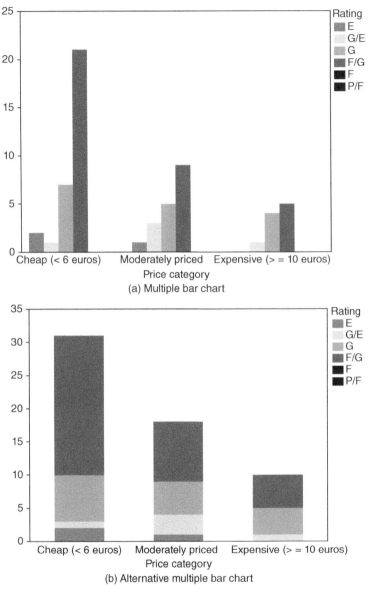

Figure 2.32   Alternative graphical representations of the cross tabulation in Table 2.5 and Figure 2.29.

Figure 2.33   Start screen of the "Graph Builder" in JMP.

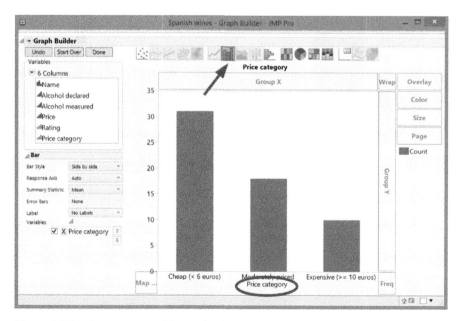

Figure 2.34   Construction of the multiple bar chart in Figure 2.32a with the "Graph Builder" in JMP: Step 1.

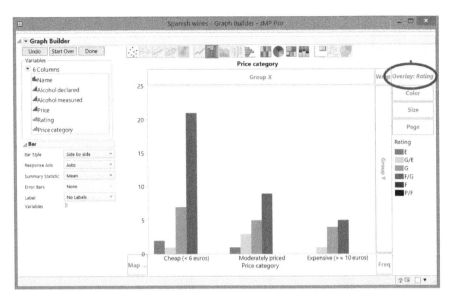

Figure 2.35   Construction of the multiple bar chart in Figure 2.32a with the "Graph Builder" in JMP: Step 2.

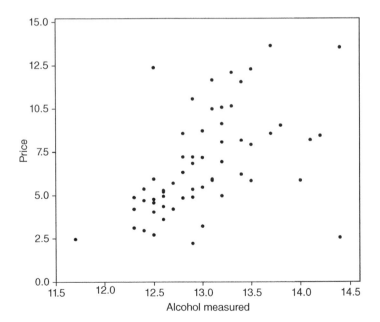

Figure 2.36   Scatter plot for the variables price and measured alcohol content for the data set of Spanish red wines.

There are different ways to create a scatter plot in JMP. One option is to make use of the "Graph Builder". If you wish to use this option, you have to drag the variable "Price" to the "Y" zone, and the variable "Alcohol measured" to the "X" zone. Finally, you need to make sure that, at the top of the "Graph Builder", only the button for a scatter plot has been activated. This is illustrated in Figure 2.37. An alternative method is to make use of the option "Scatterplot Matrix" in the "Graph" menu. With this option, you can create a matrix of scatter plots for data tables with more than two quantitative variables. This option can also be used for nominal or ordinal variables. Figure 2.38 shows a scatter plot matrix for "Rating", "Alcohol measured", "Alcohol declared" (on the bottle), and "Price" for the data set of Spanish red wines.

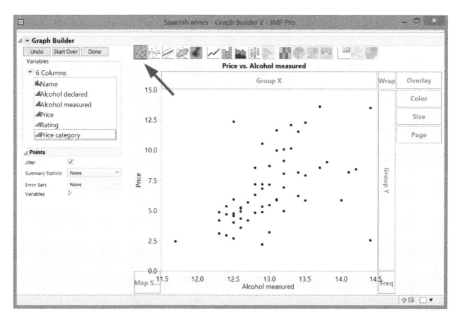

Figure 2.37   The construction of a scatter plot with the "Graph Builder" in JMP.

An interesting feature of any scatter plot in JMP is that clicking on a point in the scatter plot will highlight the corresponding row in the data table. Conversely, selecting a row in a data table will highlight the corresponding point in the scatter plot. The same thing holds for the selection of several points or rows.

2.6   Representing time series

If a variable is measured at successive time points, it is common to plot that variable on the vertical axis, put the time on the horizontal axis, and connect the successive data points by means of a straight line.

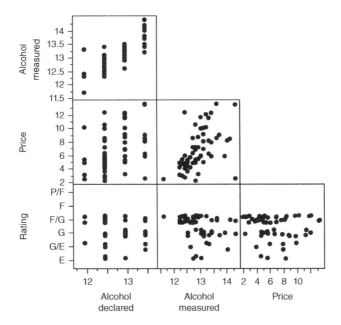

Figure 2.38   Scatterplot matrix.

Example 2.6.1 *On a dark Tuesday night in November* 2013, *John, George, Adam, Peter, and Frank, all members of the international research staff at the Department of Applied Statistics at the University of Cardiff, went to the local go-kart track. The initiative for the evening out came from Frank, who thought that the conventional snooker or bowling evenings were not exciting enough. The lap times in Figure 2.39 clarify why Frank insisted on a go-kart event. He invariably drove the fastest laps. The four others were significantly slower, especially in the first lap. Later they improved their performance, without really getting close to Frank's lap times.*

The construction of the graph in Figure 2.39 starts in the same way as the creation of a scatter plot. The only additional step required is that an extra button at the top of the "Graph Builder" is clicked. This button is shown in Figure 2.40. Clicking it ensures that successive points are connected.

2.7   The use of maps

In newspapers and on television, statistical information is often displayed using maps. This is also possible using JMP. The only requirement is that JMP recognizes the names of the geographical regions. This is no problem for the names of the various countries of the world, and for US states. By default, however, JMP does not recognize, for instance, the names of the Belgian or Dutch provinces and municipalities.

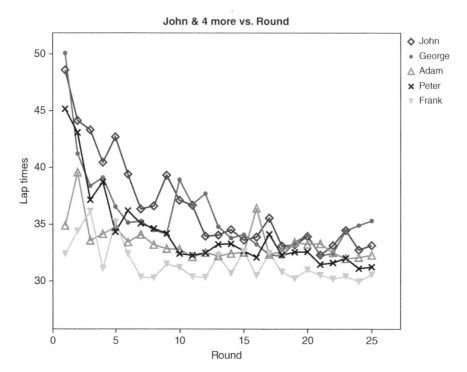

Figure 2.39   Lap times of five members of the research staff of the University of Cardiff on a go-kart track.

This can be resolved by loading two special files into JMP. When, for example, you are interested in the Belgian municipalities, you will need the names of the municipalities and a file that delimits the geographical area of these municipalities. The creation of these name and shape files is not easy, but they conform to the ESRI standard and can often be downloaded. For the Belgian municipalities, the files "Belgium-Cities-Names.jmp" and "Belgium-Cities-XY.jmp" were created.

Figure 2.41 presents a picture of the production of wind energy in the different European countries. Every country in Europe has a certain color tone in the figure. The darker the tone, the more energy the country produces using windmills. Figure 2.42 contains a similar graph for four different years. The starting point for the construction of both figures is the data table in Figure 2.43.

The data table contains the amount of wind energy (expressed in megawatts: MW) for each European country for each year from 1998 to 2010. The table also contains a column with the decimal logarithm of the amount produced. It is this logarithm that was used in Figures 2.41 and 2.42. Before explaining step by step how the figures can be reproduced, it is helpful to note that not all rows in the data table are used in the creation of the graphics (and any calculations). Indeed, some rows in the table have a small red prohibition sign. This prohibition sign indicates rows that are excluded from all calculations. In Figure 2.43, only the observations for the years 2001, 2004,

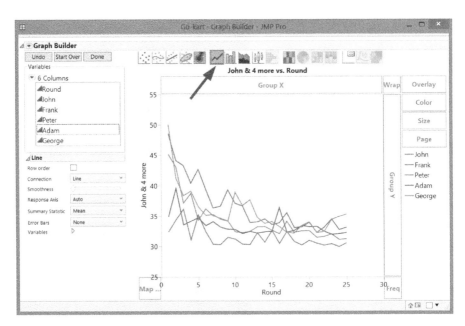

Figure 2.40   Graphical representation of a time series with the "Graph Builder" in JMP.

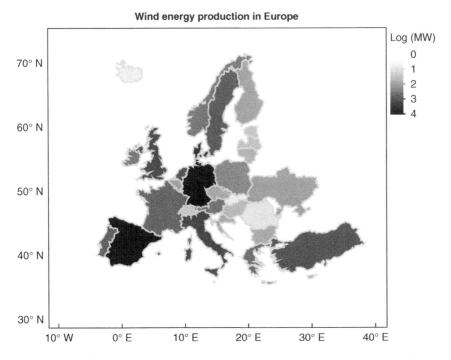

Figure 2.41   Graphical representation of the production of wind energy in Europe.

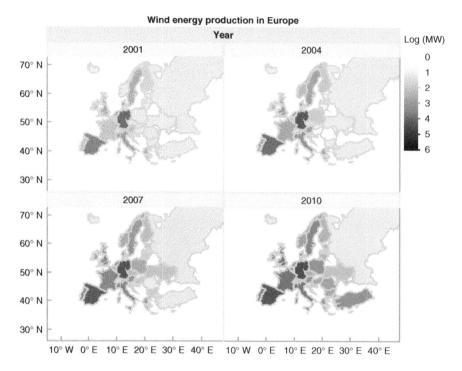

Figure 2.42   Graphical representation of the evolution of the production of wind energy in Europe.

2007, and 2010 are used. The fastest way to achieve this is by using a histogram of the variable "Year" and right-clicking on the bars for years that should be excluded. In the menu that appears, you need to choose "Row Exclude". If you want to undo the exclusion of these data points later on, you can select "Clear Row States" in the "Rows" menu.

Both Figures 2.41 and 2.42 can be created with the "Graph Builder". The first step that is required is to drag the variable "Country" to the zone named "Map Shape". You will immediately see a non-colored map of Europe, as shown in Figure 2.44. To display a color corresponding to the average production of wind power in each country, you need to drag the variable "Log (MW)" to the "Color" zone. JMP automatically chooses a color pattern that can be seen in the legend at the right of the figure (see Figure 2.45). If you prefer a different color pattern or if you would like to adjust the legend, you can right-click on the legend, select the "Gradient" option, and change whatever you like in the menu shown in Figure 2.46. Finally, if you want to get separate figures for the years 2001, 2004, 2007, and 2010, you have to drag the variable "Year" to the "Wrap" zone.

An alternative way to select a subset of your data for an analysis or a graph involves the use of **data filters**. JMP has a "Data Filter" in the "Rows" menu, and a local data filter embedded in each report window. In contrast with the data filter in the "Rows"

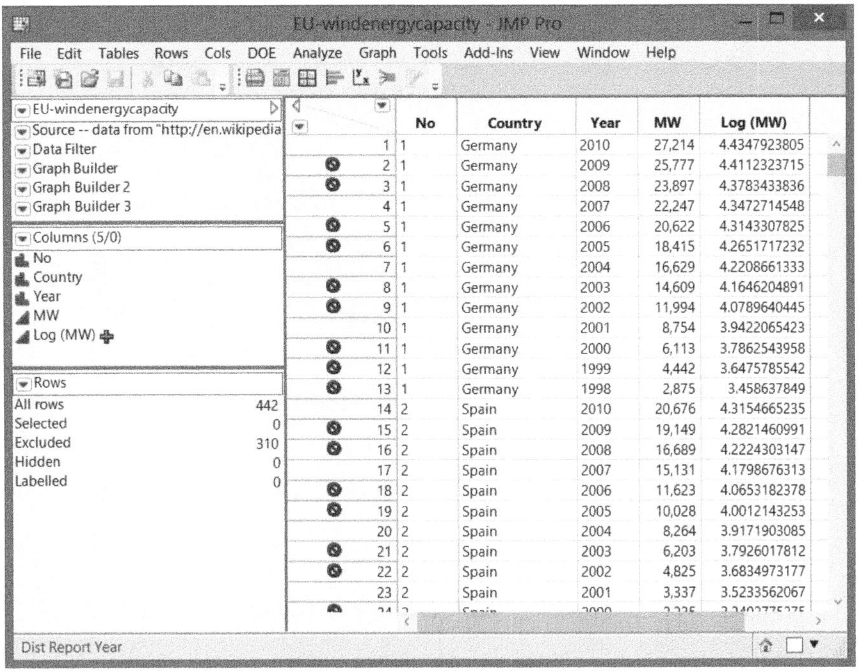

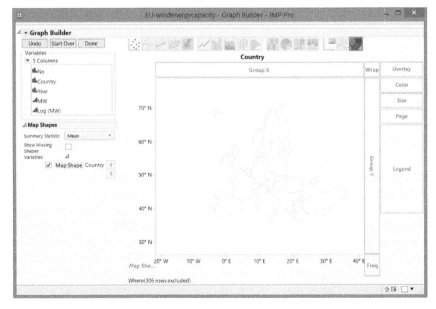

Figure 2.43   JMP data table for creating the Figures 2.41 and 2.42.

Figure 2.44   First step in the creation of Figures 2.41 and 2.42.

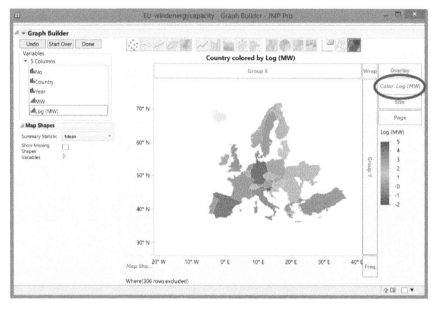

Figure 2.45   Second step in the creation of Figures 2.41 and 2.42.

Figure 2.46   Dialog window for adjusting the legend in Figures 2.41 and 2.42.

menu, the local data filter does not affect or alter the associated data table or other associated reports. After reproducing Figure 2.42, as described here, you can access the "Rows" menu and select the option "Clear Row States". This changes your report window immediately: it now contains a graph for all years from 1998 to 2010 instead of only four. In the hotspot (red triangle) menu of the "Graph Builder", you then have to select "Script", and then "Local Data Filter". This step is illustrated in Figure 2.47. In the resulting local data filter on the left side, select the column "Year" and click "Add". In the list of years that appears, you can then select the years you would like to compare, for example 2000 and 2008. The result is shown in Figure 2.48. Notice that your data table has not changed as a result of your use of the local data filter, since it does not affect the row states in your data table.

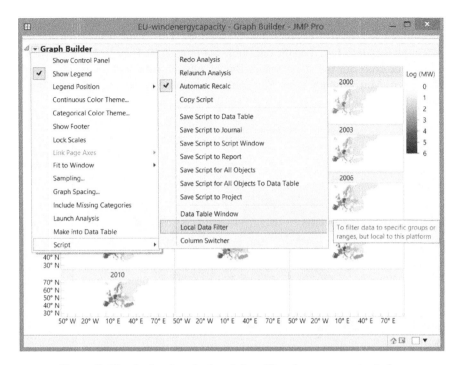

Figure 2.47   Activating the local data filter from a report window.

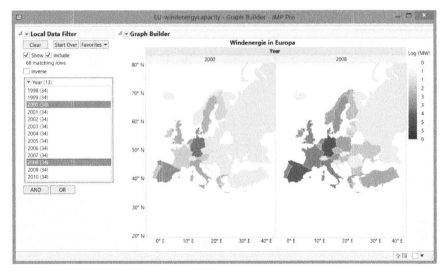

Figure 2.48   Comparing wind energy production in 2000 and 2008 with the local data filter.

Figure 2.49 shows the US states that voted predominantly for Barack Obama or for Mitt Romney in the 2012 US presidential elections. This figure was also made with the "Graph Builder", based on the data table in Figure 2.50. Here, JMP automatically takes a blue color for the states where Barack Obama won, and a red color for the states where Mitt Romney won. You can modify these colors by right-clicking on them in the legend.

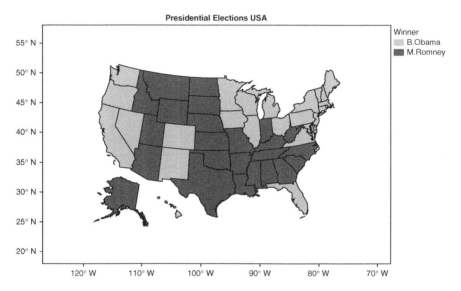

Figure 2.49   Graphical representation of the voting behavior in the presidential elections in 2012.

No	State	Winner	M.Romney	B. Obama	G. Johnson
1	1 Alabama	M.Romney	60.7%	38.4%	0.6%
2	2 Alaska	M.Romney	55.0%	41.6%	2.5%
3	3 Arizona	M.Romney	54.3%	44.0%	1.3%
4	4 Arkansas	M.Romney	60.5%	36.9%	1.5%
5	5 California	B.Obama	38.5%	59.1%	1.1%
6	6 Colorado	B.Obama	46.5%	51.2%	1.3%
7	7 Connecticut	B.Obama	40.4%	58.4%	0.8%
8	8 Delaware	B.Obama	40.0%	58.6%	0.9%
9	9 District of Columbia	B.Obama	7.1%	91.4%	0.7%
10	10 Florida	B.Obama	49.1%	50.0%	0.5%
11	11 Georgia	M.Romney	53.4%	45.4%	1.2%
12	12 Hawaii	B.Obama	27.8%	70.6%	0.9%
13	13 Idaho	M.Romney	64.5%	32.6%	1.4%
14	14 Illinois	B.Obama	41.1%	57.3%	1.1%
15	15 Indiana	M.Romney	54.3%	43.8%	1.9%
16	16 Iowa	B.Obama	46.5%	52.1%	0.8%
17	17 Kansas	M.Romney	60.0%	37.8%	1.8%
18	18 Kentucky	M.Romney	60.5%	37.8%	0.9%
19	19 Louisiana	M.Romney	57.8%	40.6%	0.9%
20	20 Maine	B.Obama	40.9%	56.0%	1.9%

Figure 2.50   Data table on the voting behavior in the US presidential elections in 2012.

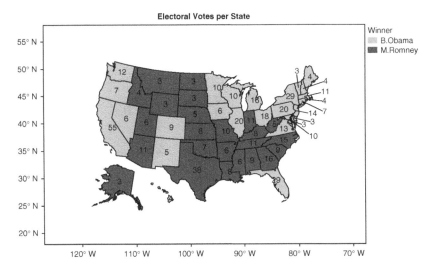

Figure 2.51   Graphical representation of the voting behavior in the US presidential elections in 2012 *showing the number of electoral votes per state.*

Figure 2.51 resembles Figure 2.49, but it also shows the number of electoral votes for each state. In order for the number of electoral votes to appear in the figure, you should use the variable "Electoral Votes" as a label. To do this, right-click on the column "Electoral Votes" first and choose "Label/Unlabel". Then, select all rows of the table, and by means of a right-click on a selected row, choose the option "Label/ Unlabel" once more. After that, each row in the data table will be marked with a symbol indicating that it is labeled.

2.8   More graphical capabilities

Nowadays, statistical software packages like JMP can not only represent univariate and bivariate data graphically, but also multivariate data. The following examples deal with the weight, price, and fuel consumption of cars. In both examples, a graphical representation of three variables is provided. Example 2.8.1 deals with two quantitative variables and a qualitative one, while Example 2.8.2 involves three quantitative variables.

Example 2.8.1   *Figure 2.52 contains a scatter plot for the weight (in kg) and the price (in dollars) of* 74 *cars. In the graphical representation, a distinction was made between US and non-US cars. For US cars, a square symbol is used, while, for non-US cars, a triangle is used. The advantage of this graphical representation is that it immediately shows*

- *that there is a positive relation between price and weight for both US and non-US cars, and*

- *that for a given price, US cars are heavier than non-US cars.*

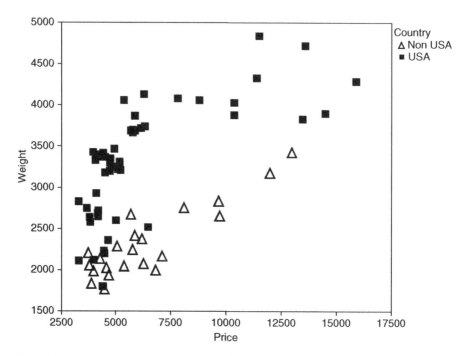

Figure 2.52   Stratified scatter plot for the weight and price (in dollars) of 74 US and non-US cars.

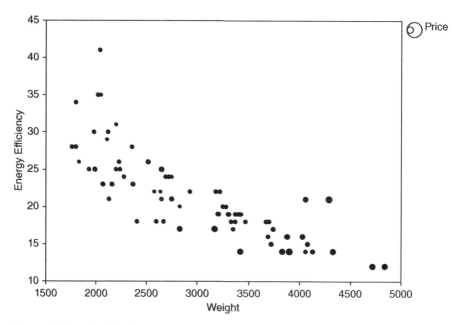

Figure 2.53   Bubble plot of weight (in kilograms), price (in dollars) and energy efficiency (in km/l fuel) of 74 cars. The area of each circle corresponds to the price of the car.

Whenever different symbols are used in a graphical representation for different categories, this is called **stratification** *or a* **stratified graphical representation***.*

To create a stratified scatter plot in JMP, you can use the "Graph Builder". Start by making a regular scatter plot and then drag the variable that indicates the origin of the cars to the "Overlay" zone.

Example 2.8.2   *Figure 2.53 contains a so-called* **bubble plot** *for the weight (in kg), energy efficiency (in km/l fuel) and the price (in dollars) of 74 cars. A bubble plot is in fact nothing more than a classic scatter plot, with the additional feature that each symbol in the scatter plot (here a circle) has a different size. In Figure 2.53, the size of each circle indicates the price of the corresponding car. The location of each symbol in the figure indicates the weight and the energy efficiency of the corresponding car. The advantage of this graphical representation is that it is immediately clear that*

- *there is a negative relation between the weight of a car and its energy efficiency,*

- *there is also a negative relation between the price and the energy efficiency of a car, and*

- *there is a positive relationship between the weight and the price of a car.*

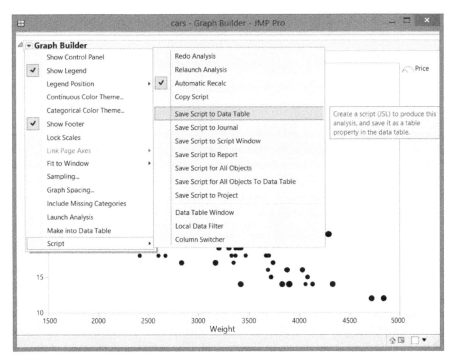

Figure 2.54   Saving a graph in a data table in JMP.

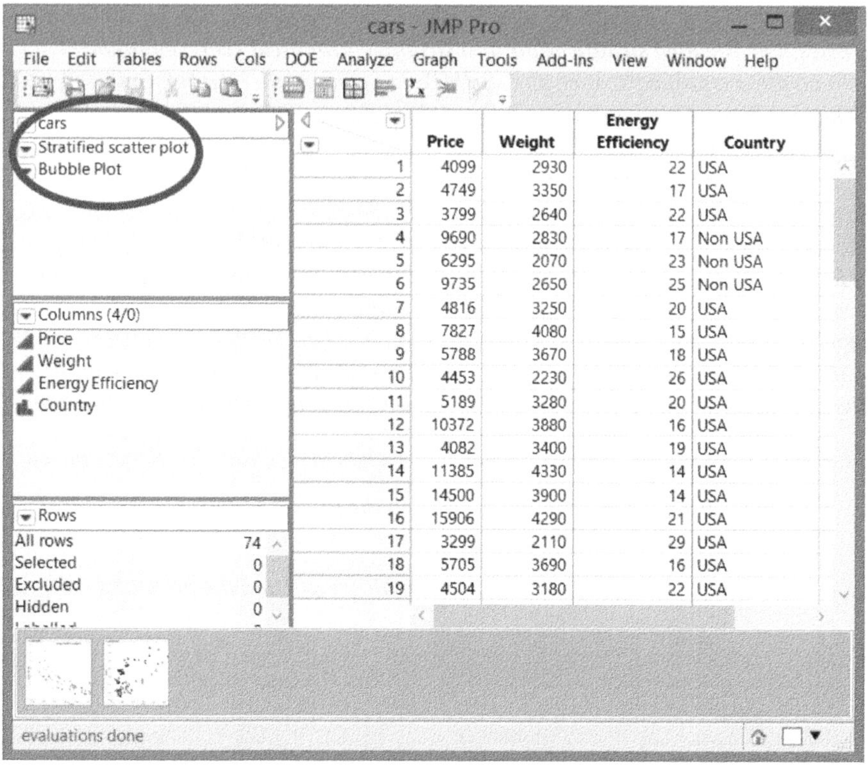

Figure 2.55   Scripts for generating graphs saved in the data table.

Indeed, the smallest circles generally appear at the top left of the figure, while the largest circles can be found at the bottom right.

There are two ways to generate bubble plots in JMP. First, you can choose the option "Bubble Plot" in the "Graph" menu. Second, you can use the "Graph Builder". When using the second approach, you have to drag one quantitative variable to the "Size" zone. In the example, this was done with the price variable.

When constructing figures in JMP, you can always edit all symbols and lines by left- or right-clicking on them. You can also change colors, as well as modify the appearance of the axes, titles, and legends. Obtaining optimal results often requires some practice. The most important is that you dare to experiment. If you are satisfied with the result, you can save the graph by clicking the hotspot (red triangle) next to the name of your graph, choosing the option named "Script", and selecting "Save Script to Data Table", as shown in Figure 2.54. The **script** is then saved at the top left of the JMP data table (see Figure 2.55), and can be run at any time – even if the rows in the data table have changed. You can change the name of your script after clicking on it.

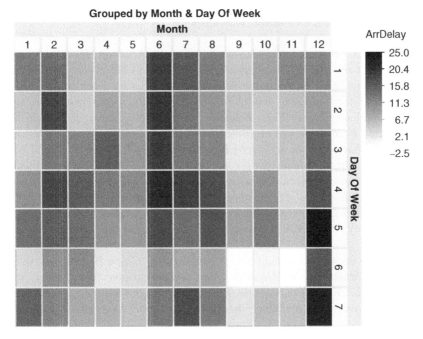

Figure 2.56   A heatmap that visualizes the times at which there were small or large delays on all flights in the USA in 2007.

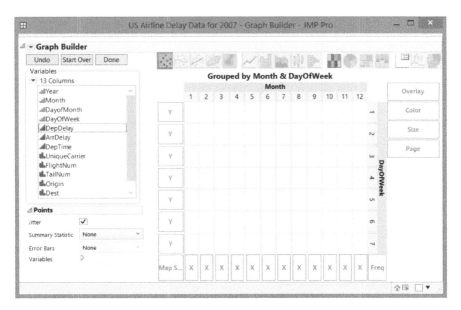

Figure 2.57   First step in the creation of the heatmap in Figure 2.56.

To reproduce the graph, you need to click on the hotspot (red triangle) next to the name of the script, and then select "Run script".

Example 2.8.3 *Another interesting display is called* **heatmap**. *Figure 2.56 shows a heatmap for the average delay at arrival of all 7,453,215 flights in the USA in 2007. Each row in the heatmap corresponds to a day of the week (with a "1" for Monday, a "2" for Tuesday …). Each column in the heatmap corresponds to a month (with a "1" for January, a "2" for February …). White colored boxes indicate times characterized by low (or even negative) delays[2]. Dark gray or black colored boxes denote times that are characterized by large delays.*

It is striking that the columns for the months 1, 2, 6, 7, 8, and 12 are predominantly colored in dark gray or black. Consequently, in summer and winter months, there are larger delays. The months of September, October, and November (columns 9, 10, and 11) score much better in terms of delay. The row corresponding to Saturday (row 6) is the least gray colored row, suggesting that there usually are no major delays on Saturdays.

In order to generate a heatmap, you should use the "Graph Builder". First, drag the variable "Month" to the "Group X" zone, and the variable "DayOfWeek" to the "Group Y" zone. You will then obtain the screen shown in Figure 2.57. The next step is to drag the variable "ArrDelay" to the "Color" zone. As a final step, you have

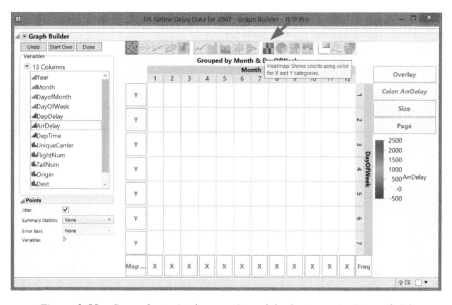

Figure 2.58   Second step in the creation of the heatmap in Figure 2.56.

---

[2] A negative delay means that the plane arrives early.

Table 2.6   Average delay (expressed in minutes) for each combination of weekday and month for the flights in the USA in 2007.

Day	Month											
	Jan	Feb	Mar	Apr	May	Jun	Jul	Aug	Sep	Oct	Nov	Dec
Monday	13	14	6	6	4	19	16	12	5	8	11	12
Tuesday	5	11	4	8	6	20	13	9	5	5	5	8
Wednesday	4	13	12	15	9	19	13	11	2	5	5	15
Thursday	11	18	15	12	11	17	18	17	5	10	3	17
Friday	14	15	13	10	9	17	13	16	8	12	5	23
Saturday	3	10	11	2	3	10	7	7	−1	0	−1	16
Sunday	15	12	7	7	6	13	17	12	2	5	5	22

to click on the "Heatmap" button at the top of the screen. In Figure 2.58, this is the eleventh button in the long row of buttons at the top of the "Graph Builder".

The heatmap provides the same information as Table 2.6, which can be constructed with the option "Tabulate" in the "Analyze" menu.

3

Descriptive statistics of sample data

Agnes lowers the latest issue of The Illustrated London News *to her lap, offended and upset. An article has just informed her that the average Englishwoman has 21,917 days to live. Why, oh why must newspapers always be so disagreeable? ... 21,917 days. Less in her case, as she's been alive for so long already. How many days are left to her?*
(from *The Crimson Petal and The White*, Michael Faber, pp. 253–254)

Sample data can be summarized based on a number of **descriptive statistics**. In this chapter we discuss the most important statistics for location, variation and skewness. Roman letters are typically used to name statistics of sample data. If the statistics are calculated for an entire population or an entire process, they are called **parameters** rather than statistics. Parameters are represented by Greek letters. A list of all Greek letters can be found in Appendix A.

The number of statistics that can be calculated depends on the nature of the data. If we want to talk about the location of nominal data, the numerical information is limited to the frequencies (the greatest frequency, in particular). For ordinal data, we can take into account the order in the data. Therefore, it can make sense to speak, for example, about the middle element of a sample. For variables measured on an interval scale or a ratio scale, the arithmetic mean plays an important role. An overview of the statistics that we address in this book can be found in Table 3.1. It is important to note that statistics that are defined for a particular measurement scale can generally be used for data on a higher measurement scale as well (see also Section 2.1.3).

The focus in this chapter is mainly on univariate descriptive statistics for quantitative variables. Statistics for bivariate quantitative data are discussed in Section 3.9.2.

Statistics with JMP: Graphs, Descriptive Statistics, and Probability, First Edition. Peter Goos and David Meintrup.
© 2015 John Wiley & Sons, Ltd. Published 2015 by John Wiley & Sons, Ltd.
Companion Website: wiley.com/go/goosandmeintrup

Table 3.1   Summary of descriptive statistics for each measurement scale.

	Nominal	*Ordinal*	*Interval/ratio*
Location	Mode	Mode Median Quartiles	Mode Median Quartiles Arithmetic mean Geometric mean
Variation	Nominal dispersion index	Range Interquartile range Ordinal dispersion index	Range Interquartile range Mean absolute deviation Variance and standard deviation Coefficient of variation
Moments			Central and non-central moments
Skewness			Pearson Fisher
Steepness			Kurtosis
Location, variation, and skewness		Box plot	Box plot
Correlation or association		Rank correlation coefficient	Covariance Correlation coefficient Rank correlation coefficient

Throughout, we denote the number of observations in a sample or the sample size by n. Whenever we discuss a single quantitative variable, we call the variable x and the n observations for that variable $x_1, x_2, \ldots, x_n$. Whenever we discuss two variables, we refer to them as x and y, and denote the n observations as $x_1, x_2, \ldots, x_n$ and $y_1, y_2, \ldots, y_n$.

3.1   Measures of central tendency or location

Location statistics are values that best describe the central tendency of data. In other words, they give an indication of how big or how small the data set is. The most commonly used statistics are the arithmetic mean, the median, and the mode. Sometimes, the geometric mean makes more sense than the arithmetic mean. Therefore, we also discuss the geometric mean.

3.1.1   Median

When data was recorded on an ordinal, interval or ratio scale, using the order of the data, we can determine the median.

Definition 3.1.1 *The **median** M_e of a set of observations is the middle element of the ordered data set:*

- *if the number of elements n is odd: the $\frac{n+1}{2}$-th element,*

- *if the number of elements n is even: the mean of the $\frac{n}{2}$-th and the $\left(\frac{n}{2}+1\right)$-th element.*

Example 3.1.1 *Suppose that a sample consists of the following 10 observations: 6, 3, 4, 7, 4, 6, 7, 6, 5, 3. So, we have that $n = 10$, $x_1 = 6$, $x_2 = 3$, ..., $x_{10} = 3$. To determine the median of this data set, we first rank the observations from small to large: 3, 3, 4, 4, 5, 6, 6, 6, 7, 7. Since there is an even number of observations in the data set, the median is the mean of the two middle elements 5 and 6. Hence, the median of the sample is $M_e = 5.5$.*

Example 3.1.2 *The median of the sample in Example 2.4.2 concerning no-shows for flights of Brussels Airlines is 2 because, after ranking the observations from small to large, both the 50th and the 51st observation equal 2. This is very easy to verify by means of the empirical cumulative distribution function in Figure 3.1. It suffices to determine which number of no-shows corresponds to a cumulative relative frequency of 0.5. This is shown graphically by the dotted lines in the figure.*

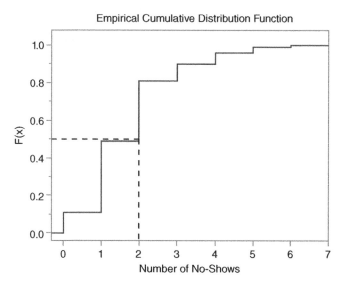

Figure 3.1   Finding the median no-show value for Example 2.4.2 using the empirical cumulative distribution function.

Some properties of the median are:

- About 50% of the observations are below/above the median.

- The median is not affected by a few extremely large or extremely small observations.

- The sum of the absolute deviations of the observations x_i from a constant c, $\sum_{i=1}^{n} |x_i - c|$, is minimal if $c = M_e$.

- The median is the mean of a truncated data set in which only the middle (the two middle) observation(s) remain(s).

3.1.2   Mode

For nominal data, the only numerical information is the frequency of the different classes or categories. Apart from determining frequencies, we cannot perform any calculations with the data. For this type of data, the most commonly used statistic is the class with the largest frequency in the sample.

Definition 3.1.2 *The **mode** M_o of a sample is the observation with the highest frequency.*

Example 3.1.3   *In Example 2.3.1, a needle diagram was used to plot the ratings of Spanish red wines. The mode is the judgment "F/G: fair to good" because this judgment is the most frequent one.*

Example 3.1.4   *In Example 2.3.2, missing parts (incomplete) is the most common defect for the defective mobile phones. Consequently, this is the mode in this example.*

Example 3.1.5   *Figure 2.12 shows that 22 was the most frequently drawn number in the Belgian lottery. Therefore, the mode is 22.*

Example 3.1.6   *The mode of the data in Example 2.4.2 concerning the number of no-shows for flights of Brussels Airlines is 1 because this value has the highest frequency (38%) in the data set.*

Definition 3.1.3 *The **mode** M_o of a set **of grouped data** is the class center of the modal class, while the modal class is the class with the highest frequency.*

Example 3.1.7   *The modal class in Figure 2.14 is the interval [0,28]. Therefore, the mode is 14. In Figure 2.20, the mode is 2.750.*

The examples made clear that the mode can not only be determined for nominal variables, but also for ordinal variables, interval variables, and variables measured on a ratio scale. However, the mode is rarely used for continuous quantitative variables because there are better statistics for this type of data. For grouped data, the mode strongly depends on which intervals are used for grouping the data (i.e., which intervals are chosen for constructing a histogram or for calculating frequencies), which does not make its use attractive.

The mode may not be unique: it sometimes happens that there are several values or classes with the same highest frequency. When two or more peaks occur in a histogram, one speaks of a bimodal or multimodal histogram (the different peaks do not need to have the same height). If a histogram has only one peak, it is called unimodal. Multimodal histograms often result from using too large a number of classes, or from a sample based on data from more than one population, or a process that can operate in more than one way.

Example 3.1.8 *Figure 3.2 shows a bimodal histogram of 143 male and female students from the University of Connecticut. The female students are indicated by means of gray dots, while the male students are indicated by means of black dots. The students have been classified according to their height. The lowest class corresponds to a height of about 5 feet (about 152 cm), while the students in the highest class are about 6 feet 5 inches (about 192.5 cm). The dots in the picture form a histogram with two main peaks. The first peak is located at the class "5:6" (5 feet 6 inches), while the second peak is located at the class "5:10" (5 feet 10 inches). Therefore, the histogram is bimodal.*

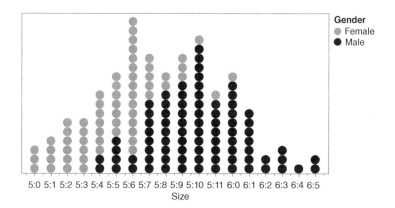

Figure 3.2   Bimodal histogram.

3.1.3   Arithmetic mean

The arithmetic mean, also called the sample mean, is undoubtedly the best known measure of central location.

Definition 3.1.4 *The **arithmetic mean** or **sample mean** $\bar{x}$ of observations $x_1, \ldots, x_n$ is*

$$\bar{x} = \frac{1}{n}(x_1 + x_2 + \cdots + x_n) = \frac{1}{n}\sum_{i=1}^{n} x_i.$$

Example 3.1.9 *The arithmetic mean of the data in Example 3.1.1 is*

$$\bar{x} = \frac{1}{10}(6 + 3 + 4 + 7 + 4 + 6 + 7 + 6 + 5 + 3) = 5.1.$$

Definition 3.1.5 *The **arithmetic mean** $\bar{x}$ for **grouped data*** *is*

$$\bar{x} = \frac{1}{n}(f_1 x_1 + f_2 x_2 + \cdots + f_k x_k) = \frac{1}{n}\sum_{i=1}^{k} f_i x_i,$$

where x_i is the center of the i-th class, f_i is the (absolute) frequency of the i-th class, n is the number of observations and k is the number of classes.

Example 3.1.10 *The arithmetic mean of the number of no-shows in Example 2.4.2 equals*

$$\bar{x} = \frac{1}{100}(11 \times 0 + 38 \times 1 + 32 \times 2 + 9 \times 3 + 6 \times 4 + 3 \times 5 + 1 \times 6) = 1.74.$$

If the arithmetic mean is calculated for a finite population with N elements, then we write

$$\mu = \frac{1}{N}\sum_{i=1}^{N} x_i.$$

In practice, the **population mean** μ is almost always unknown. Therefore, we usually have to work with arithmetic means $\bar{x}$ of samples. The connection between sample means $\bar{x}$ and the population mean μ is an issue that is discussed in the book *Statistics with JMP: Hypothesis Tests, ANOVA and Regression.*

Some properties of the arithmetic mean of a sample are:

- The sum of all observations is equal to the arithmetic mean multiplied by the sample size n: $\sum_{i=1}^{n} x_i = n\bar{x}$.
- The sum of the deviations of the observations from the mean is zero: $\sum_{i=1}^{n}(x_i - \bar{x}) = 0$.
- The sum of the squared deviations of the observations from a constant c, $\sum_{i=1}^{n}(x_i - c)^2$, is minimal if $c = \bar{x}$.
- The arithmetic mean of a sample of constant values $a, \ldots, a$ equals the constant value itself: $\bar{a} = a$.
- The arithmetic mean of a number of observations $ax_1 + b, \ldots, ax_n + b$ (where a and b are constants), obtained by linearly transforming an original set of observations, $x_1, \ldots, x_n$, can be obtained by applying the linear transformation to the original mean: $a\bar{x} + b$.

Example 3.1.11 *Table 3.2 includes 21 stages of the Tour de France 2005, a three-week professional cycling race in France. The average length of a stage is equal to 170.67 km. This is the total distance that the cyclists have to cover (3584 km) in the 2005 Tour de France, divided by 21. However, the typical American is not familiar with a statistic expressed in kilometers and prefers to express distances*

Table 3.2   The 21 stages of the Tour de France 2005.

1	Saturday 2 July	19 km	Fromentine - Noirmoutier-en-l'Ile (indiv. time trial)
2	Sunday 3 July	182 km	Challans - Les Essarts
3	Monday 4 July	208 km	La Châtaigneraie - Tours
4	Tuesday 5 July	66 km	Tours - Blois (team time trial)
5	Wednesday 6 July	179 km	Chambord - Montargis
6	Thursday 7 July	187 km	Troyes - Nancy
7	Friday 8 July	225 km	Lunéville - Karlsruhe
8	Saturday 9 July	235 km	Pforzheim - Gérardmer
9	Sunday 10 July	170 km	Gérardmer - Mulhouse
10	Tuesday 12 July	192 km	Grenoble - Courchevel
11	Wednesday 13 July	173 km	Courchevel - Briançon
12	Thursday 14 July	187 km	Briançon - Digne-les-Bains
13	Friday 15 July	162 km	Miramas - Montpellier
14	Saturday 16 July	220 km	Agde - Ax-3 Domaines
15	Sunday 17 July	205 km	Lézat-sur-Lèze - Saint-Lary Soulan (Pla d'Adet)
16	Tuesday 19 July	177 km	Mourenx - Pau
17	Wednesday 20 July	239 km	Pau - Revel
18	Thursday 21 July	189 km	Albi - Mende
19	Friday 22 July	154 km	Issoire - Le Puy-en-Velay
20	Saturday 23 July	55 km	Saint-Etienne - Saint-Etienne (indiv. time trial)
21	Sunday 24 July	160 km	Corbeil-Essonnes - Paris Champs-Élysées

in miles. To express the average length of the stages in miles, he may either first divide the 21 distances by 1.609344 (one mile is 1.609344 km) and calculate the average based on the transformed data, or he may simply divide the mean expressed in kilometers by 1.609344. Both approaches lead to a mean stage length of 106.05 miles.

Example 3.1.12   *The average temperature in Belgium is* $11.2°C$[1]. *To determine the average temperature in* $°F$[2], *we can multiply the average in* $°C$ *by 9/5 and add 32,*

---

[1] The Swedish scientist Anders Celsius (1701–1741) defined 0° as the boiling point and 100° as the freezing point of water. His successor, the Swedish astronomer Strömer, turned this around, setting 0 as the freezing point and 100 as the boiling point of water. To avoid confusion the name Celsius was kept for this scale.

[2] Gabriel Fahrenheit (1686–1736) created the first reliable thermometer. He was the first to use mercury and to close the glass tube of the thermometer at the top, so that it would not react to changes in air pressure. Negative numbers were unusual in his time. Therefore, he assigned the value of 0° to the lowest temperature that he was able to obtain with a mixture of ice, salt, and ammonium chloride. He assigned the value of 100°F to the average body temperature of a human being.

since the transformation formula is given by

$$temperature\ is\ degrees\ Fahrenheit = \frac{9}{5}\ temperature\ in\ °C + 32.$$

This yields an average temperature of 52.16° F for Belgium. In Kelvin[3], the average temperature in Belgium is 11.2 + 273.15 = 284.35 Kelvin. Indeed, the transformation formula is

$$temperature\ in\ Kelvin = temperature\ in\ °C + 273.15.$$

An advantage of the arithmetic mean in comparison with ordinal location measures such as the median or the quartiles is that all observations are used in its calculation. This advantage, however, sometimes turns into a disadvantage, namely when the data set contains extremely high or extremely low values. Such extreme values exert a major influence on the mean. These extreme values are called outliers, and may cause problems in the statistical processing of data. It is often difficult to determine the cause of outliers. Frequently, but not always, they are due to errors in recording or inputting data, faulty measurements or defective machines. If the cause can be traced, then the extreme values can possibly be corrected. If not, and if the cause is an error, it is best to drop the extreme values from the data set. If no obvious cause for an extreme value can be found, then a decision on whether or not to keep the observation in the sample is not easy. The investigator should then attempt to make a good decision, for example, based on past experience. It is also useful to compare an analysis including the extreme value(s) and an analysis excluding the extreme value(s).

Example 3.1.13 *Suppose that a sample consists of the following 10 data points: 60, 3, 4, 7, 4, 6, 7, 6, 5, and 3. The arithmetic mean of this data set is*

$$\bar{x} = \frac{1}{10}(60 + 3 + 4 + 7 + 4 + 6 + 7 + 6 + 5 + 3) = 10.5,$$

which is considerably higher than the average of 5.1 in Example 3.1.1, despite the fact that only one observation is different. The median in the current example and the median in Example 3.1.1 are identical. This illustrates that the arithmetic mean is more sensitive to extreme values than the median.

3.1.4 Geometric mean

In certain contexts, the so-called geometric mean makes more sense than the arithmetic mean. An example will clarify this.

Definition 3.1.6 *The **geometric mean** G of a set of observations $x_1, \ldots, x_n$ is*

$$G = \sqrt[n]{x_1 \times \ldots \times x_n} = (x_1 \times \ldots \times x_n)^{\frac{1}{n}} = \left(\prod_{i=1}^{n} x_i\right)^{\frac{1}{n}}.$$

---

[3] The British physicist Kelvin (1824–1907) set his temperature scale to the value 0 for the absolute zero, $-273.15°C$.

This definition implies that the geometric mean can only be calculated for strictly positive observations. The geometric mean is always smaller than or equal to the arithmetic mean. Both are equal only if all observations in a data set are equal.

Example 3.1.14 *The so-called FIRST account offered by a big European insurance company was an insurance account, offering the investor a variable return rate, namely a base interest rate of 3% plus an annual bonus. The realized annual return rates are shown in Table 3.3.*

Table 3.3   Annual return rates of the FIRST account.

Year	Return	Year	Return	Year	Return	Year	Return
1989	7.50%	1994	7.00%	1999	6.00%	2004	5.10%
1990	7.50%	1995	7.00%	2000	6.50%	2005	6.00%
1991	9.00%	1996	6.50%	2001	6.50%	2006	6.20%
1992	8.25%	1997	6.00%	2002	3.75%	2007	5.20%
1993	7.25%	1998	5.50%	2003	4.50%	2008	3.00%

An investor who deposited a sum of €10,000 in her account on January 1, 1989, ended up with a total of

$$€10{,}000 \times (1.075)(1.075)(1.09)(1.0825)\ldots(1.03) = €33{,}322.66$$

after 20 years.

To calculate the average annual return rate, we determine the geometric mean of the growth rates 1.075, 1.075, 1.09, ..., 1.03:

$$G = \sqrt[20]{(1.075)(1.075)(1.09)(1.0825)\ldots(1.03)} = 1.06203.$$

The average annual return rate therefore is $1.06203 - 1 = 0.06203$, *that is,* 6.203%. *It is not difficult to verify that*

$$(1.075)(1.075)(1.09)(1.0825)\ldots(1.03) = (1.06203)^{20}.$$

If we had invested €10,000 for 20 years at an annual return rate of 6.203%, we would have ended up with €33,322.66.

If we calculate the arithmetic mean of all return rates in Table 3.3, we obtain 6.2125%. *If we would invest €10,000 at this rate for 20 years, we would end up with* €33,382.02. *This is different from the €33,322.66, which the FIRST account actually yielded. Therefore, in this kind of application, we prefer the geometric mean over the arithmetic mean.*

3.2   Measures of relative location

A measure of relative location indicates the position of an observation in comparison to the values of the other observations.

3.2.1   Order statistics, quantiles, percentiles, deciles

Definition 3.2.1 *The i-th **order statistic** $x_{(i)}$ in a sample of n observations is the i-th observation after ranking the observations from small to large.*

The first order statistic is the minimum, while the last order statistic is the maximum. We call these two values x_{min} and x_{max}.

Example 3.2.1 *For the data in Example 3.1.1, the first order statistic is $x_{(1)} = x_{min} = 3$, the third order statistic $x_{(3)} = 4$, and the tenth order statistic $x_{(10)} = x_{max} = 7$.*

Definition 3.2.2 *The $(100 \times p)$-th **percentile** or **quantile** c_p of a sample, where $0 < p < 1$, is a real number that is greater than (about) $100 \times p\%$ of the observations, and smaller than (about) $100 \times (1 - p)\%$ of the observations.*

There are several slightly different methods to calculate percentiles. The different approaches lead to noticeable differences in small data sets, but for large data sets, there is virtually no difference between the methods of calculation. Next, we discuss the method used by JMP:

- For the calculation of percentiles, sort the n observations from small to large in positions $1, 2, \ldots, n$.

- Calculate the position of the $(100 \times p)$-th percentile as $q = p(n + 1)$.

- If q is an integer, then $x_{(q)}$ is the $(100 \times p)$-th percentile or quantile c_p of the sample.

- If q is not an integer,
  - first, determine the largest integer that is less than q, and call that integer a;
  - next, determine the difference between q and a, and call that difference f;
  - then, the $(100 \times p)$-th percentile or quantile c_p of the sample is

$$c_p = (1 - f) \cdot x_{(a)} + f \cdot x_{(a+1)},$$
$$= x_{(a)} + f \cdot (x_{(a+1)} - x_{(a)}). \tag{3.1}$$

Example 3.2.2   *Consider once more the data from Example 3.1.1. The 25th percentile is in position* $q = 0.25(10 + 1) = 2.75$*. Hence,* $a = 2$ *and* $f = 0.75$*. Consequently, using the first formula in Equation (3.1), we obtain*

$$c_{0.25} = (1 - 0.75)x_{(2)} + 0.75x_{(3)} = 0.25 \times 3 + 0.75 \times 4 = 0.75 + 3 = 3.75.$$

Of course, the second formula in Equation (3.1) provides exactly the same result:

$$c_{0.25} = x_{(2)} + 0.75(x_{(3)} - x_{(2)}) = 3 + 0.75(4 - 3) = 3 + 0.75 = 3.75.$$

The 50th percentile is in position $0.5(10 + 1) = 5.5$*. Hence,* $a = 5, f = 0.5$*, and*

$$c_{0.5} = (1 - 0.5)x_{(5)} + 0.5x_{(6)} = 0.5 \times 5 + 0.5 \times 6 = 2.5 + 3 = 5.5.$$

Finally, the 80th percentile is in position $0.8(10 + 1) = 8.8$*. Hence,* $a = 8, f = 0.8$*, and*

$$c_{0.8} = (1 - 0.8)x_{(8)} + 0.8x_{(9)} = 0.2 \times 6 + 0.8 \times 7 = 1.2 + 5.6 = 6.8.$$

If the product $(100 \times p)$ is a multiple of 10, the corresponding percentile is sometimes called the $(10 \times p)$-th **decile**. The fifth decile, that is, the 50th percentile or quantile $c_{0.5}$, is always equal to the median.

Among other things, quantiles are used to test whether a data sample might originate from a given probability density. This can be done by constructing so-called quantile diagrams (see the book *Statistics with JMP: Hypothesis Tests, ANOVA and Regression*).

3.2.2   Quartiles

Definition 3.2.3   *The first/second/third* **quartile** $Q_1/Q_2/Q_3$ *is the 25/50/75-th percentile of the sample. In other words,* $Q_1 = c_{0.25}$, $Q_2 = c_{0.5}$, *and* $Q_3 = c_{0.75}$.

It is obvious that the second quartile is the median. Together with the median and the arithmetic mean, the first and third quartile are generally shown in so-called box plots (see Section 3.7).

3.3   Measures of variation or spread

The best known statistics of variation or spread are for quantitative data. These statistics measure the variation or spread around a central value. Data with the same mean or median may still differ greatly in this respect. There exist also lesser-known measures of variation for nominal and ordinal variables.

3.3.1   Range

The easiest measure of variation of a data set is its range. To determine the range, at least an ordinal scale is needed.

Definition 3.3.1 *The **range** R of a set of observations is the difference between the value of the largest and the smallest observation:*

$$R = x_{max} - x_{min}.$$

The biggest advantage of the definition of the range is its simplicity. A major drawback is that only two observations are used in the calculation. All intermediate observations have no influence. It is clear that the range is particularly sensitive to extreme values. In industry, the range is often used in statistical process control.

3.3.2   Interquartile range

A better picture of the variation or spread of sample data is obtained by using the distance between the first and third quartile:

Definition 3.3.2 *The **interquartile range** Q is defined as the difference between the third and the first quartile:*

$$Q = Q_3 - Q_1.$$

Since half of the data is between Q_1 and Q_3, the interquartile range is a measure of spread for half of the data set. This measure of spread is insensitive to extreme values, as long as less than 25% of the data values are extremely small and less than 25% are extremely large.

3.3.3   Mean absolute deviation

A measure of variation of a sample of quantitative data around the arithmetic mean is the mean absolute deviation (MAD). Just like the arithmetic mean, the mean absolute deviation is sensitive to extreme values.

Definition 3.3.3 *The **mean absolute deviation** is the arithmetic mean of the absolute values of the deviations of the observations from the arithmetic mean:*

$$MAD = \frac{\sum_{i=1}^{n} |x_i - \bar{x}|}{n}.$$

3.3.4   Variance

For sample data measured on a ratio scale or an interval scale, the variance is used more often as a measure of spread than the mean absolute deviation.

Definition 3.3.4 *The **sample variance** s^2 of a set of observations $x_1, \dots, x_n$ is*

$$s^2 = \frac{1}{n-1} \sum_{i=1}^{n} (x_i - \bar{x})^2.$$

The sample variance is the mean of the squared deviations from the arithmetic mean $\bar{x}$, dividing by $n - 1$ instead of n. The sample variance can also be calculated with the alternative formulas

$$s^2 = \frac{1}{n-1}\left(\sum_{i=1}^{n} x_i^2 - n\bar{x}^2\right) = \frac{1}{n-1}\left\{\sum_{i=1}^{n} x_i^2 - \frac{1}{n}\left(\sum_{i=1}^{n} x_i\right)^2\right\}.$$

In order to show that these alternative formulas are valid, we need to make use of the fact that

$$\sum_{i=1}^{n} x_i = n\bar{x},$$

which follows from the definition of the arithmetic mean. Then, we can rewrite the definition of the sample variance as follows:

$$s^2 = \frac{\sum_{i=1}^{n} (x_i - \bar{x})^2}{n-1},$$

$$= \frac{1}{n-1}\sum_{i=1}^{n} (x_i - \bar{x})^2,$$

$$= \frac{1}{n-1}\sum_{i=1}^{n} (x_i^2 - 2x_i\bar{x} + \bar{x}^2),$$

$$= \frac{1}{n-1}\left\{\sum_{i=1}^{n} x_i^2 - \sum_{i=1}^{n} 2x_i\bar{x} + \sum_{i=1}^{n} \bar{x}^2\right\},$$

$$= \frac{1}{n-1}\left\{\sum_{i=1}^{n} x_i^2 - 2\bar{x}\sum_{i=1}^{n} x_i + n\bar{x}^2\right\},$$

$$= \frac{1}{n-1}\left\{\sum_{i=1}^{n} x_i^2 - 2\bar{x}(n\bar{x}) + n\bar{x}^2\right\},$$

$$= \frac{1}{n-1}\left\{\sum_{i=1}^{n} x_i^2 - n\bar{x}^2\right\},$$

$$= \frac{1}{n-1}\left\{\sum_{i=1}^{n} x_i^2 - n\left(\frac{\sum_{i=1}^{n} x_i}{n}\right)^2\right\},$$

$$= \frac{1}{n-1}\left\{\sum_{i=1}^{n} x_i^2 - n\frac{1}{n^2}\left(\sum_{i=1}^{n} x_i\right)^2\right\},$$

$$= \frac{1}{n-1}\left\{\sum_{i=1}^{n} x_i^2 - \frac{1}{n}\left(\sum_{i=1}^{n} x_i\right)^2\right\}.$$

If we compute the variance of all N elements of a population or a process with N elements, then we speak of a **population variance** or a **process variance**. For these variances, the symbol σ^2 is used. They are calculated based on the following equation:

$$\sigma^2 = \frac{\sum_{i=1}^{N}(x_i - \mu)^2}{N},$$
(3.2)

where μ represents the population mean. For large samples, the sample variance and the population variance are similar.

Since the population variance is calculated as the arithmetic mean of a set of squared deviations, it is more intuitive than the sample variance, where we divide by $n-1$ instead of n. As will be explained in the book *Statistics with JMP: Hypothesis Tests, ANOVA and Regression*, dividing by $n-1$ provides slightly better estimates for a sample variance than dividing by n. The sample variance cannot be calculated when the data set contains only one observation, because we need to divide by $n-1$. Considering the meaning of the sample variance, this makes sense, as a single observation does not contain any information about spread.

The denominator $n-1$ in the definition of the sample variance is called the number of degrees of freedom. Each degree of freedom corresponds to a unit of information. In a sample of n observations, we have n units of information. To calculate the sample variance, we need to calculate the sample mean (arithmetic mean) first. Calculating this value costs us one unit of information. Then, there are only $n-1$ units of information or degrees of freedom left to calculate the variance.

Just as the arithmetic mean, the sample variance can be calculated with the aid of frequencies if the data is grouped. The required formula is

$$s^2 = \frac{1}{n-1}\sum_{i=1}^{k} f_i(x_i - \bar{x})^2,$$

where x_i is the center of the i-th class, f_i the frequency of the i-th class, n the number of observations, and k the number of classes.

It is worth noting that a variance is always non-negative, and zero if and only if all observations in the sample have the same value, that is, if and only if the observations do not vary. A variance is always expressed in a unit that is the square of the originally measured unit. For example, if the data is measured in seconds, then the variance is measured in seconds squared.

The variance of a linear transformation $y_1 = ax_1 + b, y_2 = ax_2 + b, \ldots, y_n = ax_n + b$ of observations $x_1, x_2 \ldots, x_n$, where a and b are constants, is equal to the variance of the original data multiplied by a^2. This is easy to show. Suppose that s_y^2 is the sample variance of the newly created data set $y_1, y_2, \ldots, y_n$, and that s_x^2 is the sample variance of the original data set $x_1, x_2, \ldots, x_n$. Then,

$$s_y^2 = \frac{1}{n-1}\left\{\sum_{i=1}^{n} y_i^2 - n\bar{y}^2\right\},$$

$$= \frac{1}{n-1}\left\{\sum_{i=1}^{n} (ax_i + b)^2 - n\overline{(ax + b)}^2\right\},$$

$$= \frac{1}{n-1} \left\{ \sum_{i=1}^{n} (a^2 x_i^2 + 2abx_i + b^2) - n(a\bar{x} + b)^2 \right\},$$

$$= \frac{1}{n-1} \left\{ \sum_{i=1}^{n} (a^2 x_i^2 + 2abx_i + b^2) - n(a^2\bar{x}^2 + 2ab\bar{x} + b^2) \right\},$$

$$= \frac{1}{n-1} \left\{ a^2 \sum_{i=1}^{n} x_i^2 + 2ab \sum_{i=1}^{n} x_i + nb^2 - na^2\bar{x}^2 - 2nab\bar{x} - nb^2 \right\},$$

$$= \frac{1}{n-1} \left\{ a^2 \sum_{i=1}^{n} x_i^2 + 2ab \sum_{i=1}^{n} x_i - na^2\bar{x}^2 - 2nab\bar{x} \right\},$$

$$= \frac{1}{n-1} \left\{ a^2 \sum_{i=1}^{n} x_i^2 + 2nab\bar{x} - na^2\bar{x}^2 - 2nab\bar{x} \right\},$$

$$= \frac{1}{n-1} \left\{ a^2 \sum_{i=1}^{n} x_i^2 - na^2\bar{x}^2 \right\},$$

$$= a^2 \frac{1}{n-1} \left\{ \sum_{i=1}^{n} x_i^2 - n\bar{x}^2 \right\},$$

$$= a^2 s_x^2.$$

3.3.5 Standard deviation

The **sample standard deviation** is the (positive) square root of the sample variance:

$$s = \sqrt{s^2}.$$

Analogously, the **population standard deviation** is the (positive) square root of the population variance:

$$\sigma = \sqrt{\sigma^2}.$$

Standard deviations are expressed in the same unit as the original data.

In general, the sample standard deviation gives a better picture of the distribution of the data set than the range. However, if we only have two observations, then the sample standard deviation and the range contain the same information. Indeed,

$$s = \frac{R}{\sqrt{2}} = \frac{|x_1 - x_2|}{\sqrt{2}} = \frac{|x_2 - x_1|}{\sqrt{2}}$$

if $n = 2$. It is a useful exercise to prove this equality.

Example 3.3.1 *Kevlar is a material that is highly resistant to conditions of high pressure. For that reason, it is used in space shuttles and bicycle tires. In order to test* the strength of the material, it is exposed to a particular stress level, and the life span, the time that elapses until the material breaks, is recorded. Ten of these stress tests yielded the following life spans: 50.1, 118.0, 353.0, 29.6, 84.2, 669.7, 118.5, 166.0, 202.0, and 137.8 hours.

The arithmetic mean of these values is 192.89 hours. The range is 640.1 hours, the interquartile range $Q_3 - Q_1 = 239.75 - 75.675 = 164.075$ hours. The mean absolute deviation can be calculated as

$$MAD = \frac{|29.6 - 192.89| + |50.1 - 192.89| + \cdots + |669.7 - 192.89|}{10},$$

$$= 129.21 \text{ hours.}$$

The sample variance of the data is

$$s^2 = \frac{(29.6 - 192.89)^2 + (50.1 - 192.89)^2 + \cdots + (669.7 - 192.89)^2}{9},$$

$$= 36314.72 \text{ (hours)}^2.$$

The sample standard deviation of the life spans of kevlar is

$$s = \sqrt{36314.72} \text{ hours} = 190.56 \text{ hours.}$$

3.3.6   Coefficient of variation

Although the variance and the standard deviation play an extremely important role in statistics, they are sometimes not the best choices as measures of spread.

Example 3.3.2 *Consider the following two data sets with eight observations:*

- *Sample 1: 15, 20, 20, 30, 35, 35, 40, 45*
- *Sample 2: 1015, 1020, 1020, 1030, 1035, 1035, 1040, 1045*

The arithmetic means of the two sets are 30 and 1030, while the sample variance is equal to 114.2857 for both samples. Therefore, the sample standard deviation is also the same for both samples. It equals 10.69. Nevertheless, it is clear that – relative to the arithmetic mean – the variability in the second sample is considerably smaller than in the first sample.

Definition 3.3.5 *The **coefficient of variation** CV is defined as the ratio of the sample standard deviation s and the arithmetic mean $\bar{x}$:*

$$CV = \frac{s}{\bar{x}}.$$

Example 3.3.3 *It is easy to check that the coefficients of variation for the samples in Example 3.3.2 are 0.3563 and 0.0104.*

The coefficient of variation is unreliable when $\bar{x}$ is very small and it is sensitive to outliers. The coefficient of variation is useful for comparing spread of data with different means and indispensable when comparing the variation of data with different dimensions (i.e., data expressed in different units of measurement).

3.3.7   Dispersion indices for nominal and ordinal variables

For the spread of nominal and ordinal data, so-called nominal and ordinal dispersion indices, abbreviated as nDi and oDi, are used.

The spread for nominal data is interpreted in a different way than that for ordinal data. For nominal data, the spread is a measure of **heterogeneity**, while, for ordinal data, it is a measure of **polarization**. Also, the calculation of the nominal and ordinal dispersion indices differ. For the latter index, cumulative relative frequencies are used in the calculations, while this is not the case for the nominal dispersion index.

3.3.7.1   Nominal dispersion index

Definition 3.3.6 *The **nominal dispersion index** of a nominal variable with k different categories or classes is*

$$nDi = 1 - \frac{observed\ homogeneity}{maximal\ homogeneity},$$

$$= 1 - \frac{\sum_{i=1}^{k} \left(f_i^* - \frac{1}{k}\right)^2}{1 - \frac{1}{k}},$$

where f_i^* is the relative frequency of the i-th class.

In the definition of the nominal dispersion index, the sum of squares

$$\sum_{i=1}^{k} \left(f_i^* - \frac{1}{k}\right)^2$$

is a measure of **homogeneity**. The larger this measure, the bigger the extent to which the observations fall within a single category. The relative frequency f_i^* equals the (absolute) frequency f_i divided by the number of observations n.

The nominal dispersion index always lies between 0 and 1. If the index is equal to 0, then there is no spread at all. All observations then belong to the same category or class, and the homogeneity is maximized. If the index is equal to 1, then the observations are evenly distributed across all categories or classes, and the heterogeneity is maximized. In most practical cases, the dispersion index takes values between 0 and 1.

Example 3.3.4 *The Technical University of Munich has an international master's program and wants to examine from which European countries the students originate. To this end, it wishes to quantify the heterogeneity of the students in terms of their origin. Suppose that there are 40 students in a particular course and that 8 students come from each of the following five countries: France, Italy, Poland, Spain, and Portugal.*

Since there are 40 *students, n is of course equal to* 40. *Since there are five different categories or classes, k* = 5, *and the maximal homogeneity is*

$$1 - \frac{1}{k} = 1 - \frac{1}{5} = 0.8.$$

The frequencies $f_1, f_2, \ldots, f_5$ *for all five categories are equal to* 8. *The initial calculations for the determination of the nominal dispersion index are shown in Table 3.4.*

Table 3.4   Calculating the observed homogeneity in Example 3.3.4.

Class	Country	f_i	f_i^*	$f_i^* - (1/k)$	$(f_i^* - (1/k))^2$
1	France	8	0.20	0	0
2	Italy	8	0.20	0	0
3	Poland	8	0.20	0	0
4	Spain	8	0.20	0	0
5	Portugal	8	0.20	0	0

In this example, the observed homogeneity is

$$\sum_{i=1}^{k} \left(f_i^* - \frac{1}{k} \right)^2 = \sum_{i=1}^{5} \left(f_i^* - \frac{1}{5} \right)^2 = 0 + 0 + 0 + 0 + 0 = 0.$$

Hence, the nominal dispersion index is equal to

$$nDi = 1 - \frac{\sum_{i=1}^{5} \left(f_i^* - \frac{1}{5} \right)^2}{1 - \frac{1}{5}} = 1 - \frac{0}{0.8} = 1.$$

Example 3.3.5 *Consider again the scenario in Example 3.3.4, but now assume that all* 40 *students come from Poland. The initial calculations for the determination of the nominal dispersion index in that case are shown in Table 3.5.*

Table 3.5   Calculating the observed homogeneity in Example 3.3.5.

Class	Country	f_i	f_i^*	$f_i^* - (1/k)$	$(f_i^* - (1/k))^2$
1	France	0	0	−0.2	0.04
2	Italy	0	0	−0.2	0.04
3	Poland	40	1	0.8	0.64
4	Spain	0	0	−0.2	0.04
5	Portugal	0	0	−0.2	0.04

In this example, the observed homogeneity is

$$\sum_{i=1}^{k} \left(f_i^* - \frac{1}{k} \right)^2 = \sum_{i=1}^{5} \left(f_i^* - \frac{1}{5} \right)^2 = 0.04 + 0.04 + 0.64 + 0.04 + 0.04 = 0.8.$$

Hence, the nominal dispersion index is equal to

$$nDi = 1 - \frac{\sum_{i=1}^{5} \left(f_i^* - \frac{1}{5} \right)^2}{1 - \frac{1}{5}} = 1 - \frac{0.8}{0.8} = 0.$$

In Example 3.3.5, all students come from Poland, which leads to a nominal dispersion index of 0. If all students had come from any other single country, the nominal dispersion index would also have been 0.

Example 3.3.6 *Consider again the scenario in Example 3.3.4, but now assume that the 40 students are unevenly scattered over the five countries. The calculations for the determination of the nominal dispersion index in that case are shown in Table 3.6.*

Table 3.6   Calculating the observed homogeneity in Example 3.3.6.

Class	Country	f_i	f_i^*	$f_i^* - (1/k)$	$(f_i^* - (1/k))^2$
1	France	20	0.500	0.300	0.0900
2	Italy	2	0.050	−0.150	0.0225
3	Poland	5	0.125	−0.075	0.0056
4	Spain	3	0.075	−0.125	0.0156
5	Portugal	10	0.250	0.050	0.0025

In this example, the homogeneity is

$$\sum_{i=1}^{k} \left(f_i^* - \frac{1}{k} \right)^2 = \sum_{i=1}^{5} \left(f_i^* - \frac{1}{5} \right)^2 = 0.09 + \cdots + 0.0025 = 0.1362.$$

Therefore, the nominal dispersion index is equal to

$$nDi = 1 - \frac{\sum_{i=1}^{5} \left(f_i^* - \frac{1}{5} \right)^2}{1 - \frac{1}{5}} = 1 - \frac{0.1362}{0.8} = 0.82975.$$

3.3.7.2 Ordinal dispersion index

Definition 3.3.7 *The **ordinal dispersion index** of an ordinal variable with k different categories or classes is*

$$oDi = 1 - \frac{observed\ concentration}{maximal\ concentration},$$

$$= 1 - \frac{\sum_{i=1}^{k-1} \left(F_i^* - \frac{1}{2}\right)^2}{\frac{k-1}{4}},$$

where F_i^* is the relative cumulative frequency of the i-th class.

In the definition of the ordinal dispersion index, the sum of squares

$$\sum_{i=1}^{k-1} \left(F_i^* - \frac{1}{2}\right)^2$$

is a measure of **concentration**. The larger this measure, the larger the extent to which the observations fall within a single category. The relative cumulative frequency F_i^* is equal to the sum of the absolute frequencies $f_1, f_2, \ldots, f_i$, divided by the number of observations n. Note that in the sum of squares for the calculation of the concentration, we only add the first $k - 1$ values of $(F_i^* - \frac{1}{2})^2$.

Like the nominal dispersion index, the ordinal dispersion index always lies between 0 and 1. If the index is equal to 0, then there is no polarization at all. All observations then belong to the same category or class, and the concentration is maximal. If the index is equal to 1, the observations are evenly distributed between the two extreme categories or classes. In that case, the polarization is maximal: the observations are all at the two extremes, the two poles. In most practical cases, the ordinal dispersion index takes values between 0 and 1.

Example 3.3.7 *Suppose you want to quantify the extent to which the population is in favor of or against immigrant voting rights. You set up a survey in which you ask the question "What is your attitude towards voting rights for foreigners?". As possible answers, you offer the options: very positive (++), positive (+), neutral (+/−), negative (−), and very negative (−−). Suppose you asked 40 people to participate in the survey, 20 of whom turn out to have a very positive attitude and 20 of whom turn out to have a very negative attitude towards immigrant voting rights.*

Since there are five response categories, $k = 5$, and the maximal concentration is

$$\frac{k - 1}{4} = \frac{5 - 1}{4} = 1.$$

The frequencies f_1 and f_5 for the two extreme classes (very positive and very negative) are both equal to 20. The initial calculations for the determination of the ordinal dispersion index are shown in Table 3.7.

Table 3.7   Calculating the observed concentration in Example 3.3.7.

Class	Attitude	f_i	F_i	F_i^*	$F_i^* - 0.5$	$(F_i^* - 0.5)^2$
1	++	20	20	0.5	0	0.00
2	+	0	20	0.5	0	0.00
3	+/−	0	20	0.5	0	0.00
4	−	0	20	0.5	0	0.00
5	−−	20	40	1.0	0.5	0.25

In this example, the observed concentration is

$$\sum_{i=1}^{k-1}\left(F_i^* - \frac{1}{2}\right)^2 = \sum_{i=1}^{5-1}\left(F_i^* - \frac{1}{2}\right)^2 = 0.00 + 0.00 + 0.00 + 0.00 = 0.$$

Therefore, the ordinal dispersion index is equal to

$$oDi = 1 - \frac{\sum_{i=1}^{5-1}\left(F_i^* - \frac{1}{2}\right)^2}{\frac{5-1}{4}} = 1 - \frac{0}{1} = 1.$$

Example 3.3.8 *Consider again the scenario in Example 3.3.7, but now assume that all 40 people surveyed are neutral towards immigrant voting rights. The initial calculations for the determination of the ordinal dispersion index in that case are shown in Table 3.8.*

Table 3.8   Calculating the observed concentration in Example 3.3.8.

Class	Attitude	f_i	F_i	F_i^*	$F_i^* - 0.5$	$(F_i^* - 0.5)^2$
1	++	0	0	0	−0.5	0.25
2	+	0	0	0	−0.5	0.25
3	+/−	40	40	1.0	0.5	0.25
4	−	0	40	1.0	0.5	0.25
5	−−	0	40	1.0	0.5	0.25

In this example, the observed concentration is

$$\sum_{i=1}^{k-1} \left(F_i^* - \frac{1}{2} \right)^2 = \sum_{i=1}^{5-1} \left(F_i^* - \frac{1}{2} \right)^2 = 0.25 + 0.25 + 0.25 + 0.25 = 1.$$

Hence, the ordinal dispersion index is equal to

$$oDi = 1 - \frac{\sum_{i=1}^{5-1} \left(F_i^* - \frac{1}{2} \right)^2}{\frac{5-1}{4}} = 1 - \frac{1}{1} = 0.$$

In the preceding example, all answers are concentrated in the category "neutral (+/−)". This leads to an ordinal dispersion index of 0. If the answers had all been concentrated in a different category, this would have led to an ordinal dispersion index of 0 as well.

Example 3.3.9 *Consider again the scenario in Example 3.3.7, but now assume that all 40 people surveyed are unevenly distributed over the five answer categories. The initial calculations for the determination of the ordinal dispersion index in that case are shown in Table 3.9.*

Table 3.9   Calculating the observed concentration in Example 3.3.9.

Class	Attitude	f_i	F_i	F_i^*	$F_i^* - 0.5$	$(F_i^* - 0.5)^2$
1	++	5	5	0.125	−0.375	0.1406
2	+	10	15	0.375	−0.125	0.0156
3	+/−	2	17	0.425	−0.075	0.0056
4	−	3	20	0.500	0.000	0.0000
5	−−	20	40	1.000	0.500	0.2500

In this example, the observed concentration is

$$\sum_{i=1}^{k-1} \left(F_i^* - \frac{1}{2} \right)^2 = \sum_{i=1}^{5-1} \left(F_i^* - \frac{1}{2} \right)^2 = 0.1406 + 0.0156 + 0.0056 + 0.0000 = 0.1618,$$

so that the ordinal dispersion index is equal to

$$oDi = 1 - \frac{\sum_{i=1}^{5-1} \left(F_i^* - \frac{1}{2} \right)^2}{\frac{5-1}{4}} = 1 - \frac{0.1618}{1} = 0.8382.$$

3.4   Measures of skewness

Histograms and stem and leaf plots of sample data can be symmetric or asymmetric. A histogram that is not symmetric is called **skewed**. In a histogram that is skewed to the left (or negatively skewed), the left-hand tail is longer than the right-hand tail. In a histogram that is skewed to the right (or positively skewed), the right-hand tail is longer than the left-hand tail.

In a unimodal histogram, skewness can be determined based on the positions of the arithmetic mean, the median and the mode. In a perfectly symmetrical histogram, the three statistics are identical. In a histogram that is skewed to the left, the mean is smaller than the median, which in turn is smaller than the mode. When a histogram is skewed to the right, the mode is smaller than the median, which itself is smaller than the arithmetic mean. The reason for this is that the arithmetic mean is more sensitive to extremely large or extremely small values than the median. Based on this observation, Pearson[4] introduced a measure of skewness:

Definition 3.4.1 *Pearson's coefficient of skewness is defined as*

$$S_P = \frac{3(\bar{x} - M_e)}{s}.$$

In the definition of Pearson's coefficient of skewness, we divide by the sample standard deviation to obtain a measure that is independent of the unit of measurement. Without this division, $\bar{x} - M_e$ could easily be made artificially large or small, simply by changing the measurement unit. The factor 3 in the definition of the skewness coefficient ensures that it is always between -3 and $+3$. In a right-skewed or positively skewed distribution, $\bar{x} > M_e$, and consequently $S_P > 0$. In a left-skewed or negatively skewed distribution, the opposite is true.

Example 3.4.1 *For the data in Example 3.3.1 concerning the life span of kevlar, Pearson's coefficient of skewness is*

$$S_P = \frac{3(\bar{x} - M_e)}{s} = \frac{3(192.89 - 128.15)}{190.56} = 1.02.$$

This indicates a slightly positive skewness, which corresponds to the fact that the histogram in Figure 3.3 has a long right tail.

---

[4] Karl Pearson (1857–1936) was one of the founders of statistics. Among other things, he laid the foundations for testing hypotheses and principal component analysis. Pearson claimed that statistics is the grammar of science. Einstein called Pearson's book "The Grammar of Science", a mandatory reading for all students.

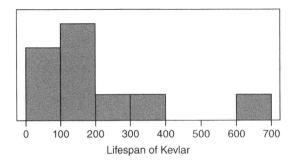

Figure 3.3   Histogram for the results of the stress tests in Examples 3.3.1 and 3.4.1.

A second measure of skewness is derived from the third central moment of the sample data. Generally, the k-th central moment of a sample of size n is defined as follows:

Definition 3.4.2 *The k-th **central sample moment**[5] of a sample is the mean of the k-th powers of the deviations from the sample mean:*

$$m_k = \frac{\sum_{i=1}^{n} (x_i - \bar{x})^k}{n}.$$

The **skewness of Fisher**[6], perhaps the most commonly used measure for skewness, is based on the third central moment m_3 and is calculated as m_3/s^3, or a function thereof. JMP calculates Fisher's skewness as

$$S_F = \frac{n^2}{(n-1)(n-2)} \frac{m_3}{s^3} = \frac{n}{(n-1)(n-2)} \sum_{i=1}^{n} \left(\frac{x_i - \bar{x}}{s}\right)^3.$$

The skewness measure is dimensionless and it is zero for a symmetric histogram ($S_F = 0$), positive for a right-skewed histogram ($S_F > 0$) and negative for a left-skewed histogram ($S_F < 0$).

Example 3.4.2 *For the data in Example 3.1.1, we have $s = 1.524$ and $\bar{x} = 5.1$. Therefore,*

$$m_3 = \frac{1}{10}\{(6 - 5.1)^3 + (3 - 5.1)^3 + \cdots + (3 - 5.1)^3\} = -0.528 \ cm^3,$$

---

[5] The k-th non-central sample moment is

$$m_k' = \frac{\sum_{i=1}^{n} x_i^k}{n}.$$

[6] Sir Ronald Aylmer Fisher (1890–1962) was an English statistician and geneticist. He is considered the founder of statistics. He laid the foundations for the analysis of variance, the maximum likelihood method, and the statistical design of experiments.

and

$$S_F = \frac{10^2}{9 \times 8} \frac{m_3}{s^3} = \frac{10^2}{9 \times 8} \frac{-0.528}{(1.524)^3} = -0.207.$$

Therefore, the data are skewed to the left.

Example 3.4.3 *For the data in Example 3.3.1 and Figure 3.3, we compute Fisher's skewness as*

$$S_F = \frac{10^2}{9 \times 8} \frac{m_3}{s^3} = \frac{10^2}{9 \times 8} \frac{10293932.3}{(190.56421)^3} = 2.066,$$

indicating that the data are skewed to the right.

3.5 Kurtosis

The extent to which a histogram has a sharp peak is quantified by a number called kurtosis. Kurtosis can be viewed as a measure of steepness.

Definition 3.5.1 *The **kurtosis** of a data sample is*

$$g = \frac{n(n+1)}{(n-1)(n-2)(n-3)} \sum_{i=1}^{n} \left(\frac{x_i - \bar{x}}{s} \right)^4 - \frac{3(n-1)^2}{(n-2)(n-3)}.$$

Like the skewness, the kurtosis is dimensionless. The kurtosis is zero for normally distributed data. A positive value for the kurtosis indicates a sharper peak than in normally distributed data, while a negative value indicates a flatter peak.

Example 3.5.1 *For the data of Example 3.3.1, the kurtosis is*

$$g = \frac{10 \times 11}{9 \times 8 \times 7} (40.70662021) - \frac{3 \times 9^2}{8 \times 7} = 4.545095681,$$

indicating a sharp peak.

3.6 Transformation and standardization of data

In previous sections, we have already seen how the mean and the variance of a data set x_i are affected by a linear transformation $y_i = ax_i + b$, where a and b are constants. The mean of linearly transformed data can be found by applying the linear transformation to the mean of the original data, that is, $\bar{y} = a\bar{x} + b$. It is not difficult to show that the median of linearly transformed data is equal to the linear transformation of the original median. The variance of the y_i values can be calculated as $s_y^2 = a^2 s_x^2$, that is, a^2 times the original variance. The standard deviation of the y_i values is equal to $s_y = |a|s_x$. The skewness coefficients for the transformed data are identical to those of the original data if a is positive, and to their opposite if a is negative. The kurtosis remains unchanged under a linear transformation of the data. It is a useful exercise to demonstrate all this.

For non-linear transformations, no simple formulas exist to calculate summary statistics. For example, if $y_i = x_i^2$, it is not true that $\bar{y} = \bar{x}^2$. If $y_i = \ln(x_i)$, then $\bar{y} \neq \ln(\bar{x})$. It is easy to check this for a small example.

By far the most commonly used linear transformation in statistics is called a **standardization**. For this kind of transformation, a and b are set to $1/s$ and $-\bar{x}/s$, respectively. Typically, the new variable obtained using this transformation is denoted by the letter z. A standardized value

$$z_i = \frac{x_i - \bar{x}}{s}$$

expresses how many standard deviations an observation x_i is away from the sample mean, since $x_i = \bar{x} + z_i s$. For any standardized variable z, the mean is zero and the variance is one.

Note that standardized variables are used in the calculation of Fisher's skewness and of the kurtosis.

3.7   Box plots

A frequently used graphical representation of univariate ordinal or quantitative data is the so-called box plot. There are a lot of different versions of box plots that can be found in the statistical literature. The central part of the data is usually represented by means of a box. The box is bounded by the first and the third quartile. Typically, the median is represented with a line in between these two quartiles. In addition, the mean is often indicated with another symbol.

In addition to these statistics, a box plot indicates extremely large and extremely small values using dots. For this purpose, rules of thumb are used. One such rule of thumb states that an observation x_i is extreme if

$$x_i < Q_1 - 1.5 \times Q$$

or

$$x_i > Q_3 + 1.5 \times Q.$$

In this expression, Q is the interquartile range, as defined in Section 3.3.2.

Some box plots also contain lines or whiskers that reach down to the smallest and up to the largest sample value that are not considered extreme values.

Example 3.7.1   *Figure 3.4 contains two box plots of the lengths of the 21 stages in the Tour de France 2005 shown in Table 3.2. The box plots in the figure indicate that there are three outliers within the 21 data points, namely three stages with a considerably shorter length than the other stages. These three data points correspond to the stages 1, 4, and 20. A closer inspection of the data in Table 3.2 reveals that the stages 1 and 20 were individual time trials, while stage 4 was a team time trial. It is common for*

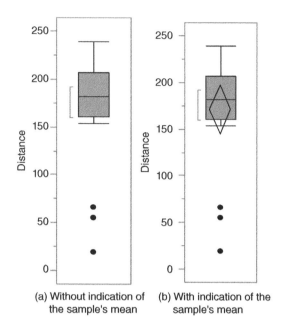

Figure 3.4   Box plots of the lengths of the 21 stages in the 2005 Tour de France (see Example 3.7.1). The three dots represent extreme values.

time trials to be substantially shorter than ordinary stages. The difference between the two box plots in Figure 3.4 is that one box plot indicates the mean, while the other one does not.

The box plots only show three extremely small observations. Thus, there are no extremely large observations in this data set. The whiskers of the box plots extend to the maximum value (239 km) on one side, and to the smallest value that is not an extremely small value (154 km) on the other. The median is 182 km, while the first and the third quartile are 161 km and 206.5 km. The interquartile range is 45.5 km. The arithmetic mean is 170.67 km.

From the first and third quartile, we learn that half the stages of the 2005 Tour de France were between 161 km and 206.5 km long. However, the interval [161, 206.5] is not the shortest possible interval that contains 50% of all stage lengths. The shortest interval is represented by the vertical bracket next to the box plots in Figure 3.4. From the figure, we can deduce that the shortest interval extends from (about) 160 km to (about) 190 km.

The box plot in Figure 3.4b indicates the arithmetic mean by means of a diamond. The center of the diamond corresponds to the mean. The end points of the diamond correspond to the lower limit and the upper limit of the 95% confidence interval for the mean. The derivation of such a confidence interval is not covered in this book, but in Statistics with JMP: Hypothesis Tests, ANOVA and Regression.

In JMP, box plots as in Figure 3.4 can be created with the option "Distribution" in the "Analyze" menu. By default, you get a graphical display with both a histogram *and a box plot. You can ensure that only the box plot is displayed via the hotspot (red triangle icon) next to the name of the variable "Distance". As shown in Figure 3.5, to this end, you have to uncheck the option "Histogram" in the "Histogram Options". You can also uncheck the option "Vertical" if you want your box plot to be displayed horizontally instead of vertically.*

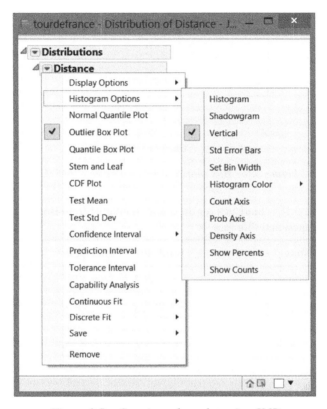

Figure 3.5   Creating a box plot using JMP.

Example 3.7.2 *Between* 28 *March and* 30 *October* 2003, *several flights of Brussels Airlines to Brussels were delayed. Figure 3.6 contains box plots for the delays (expressed in minutes) of flights from Bordeaux (BOD), Florence (FLR) and Budapest (BUD). For the first two cities, the data set contains* 215 *observations, while* 185 *observations are available for Budapest. Each of the three box plots contains several extremely large values, so the data is skewed to the right. The Budapest airport scores pretty well: the delays from this airport are notably smaller than those for the other airports. Also, the variability of the delays for flights from Budapest is small compared to that of the flights from Florence and Bordeaux. Some flights from Florence and Bordeaux had delays of more than two hours, and sometimes even more than three hours. This is shown by the dots located above the values of* 120 *and* 180 *minutes on the vertical axis in Figure 3.6.*

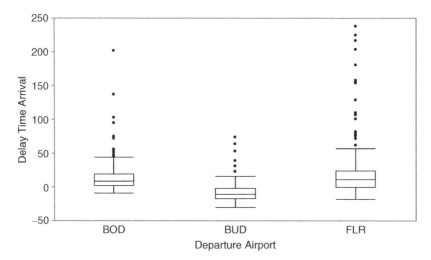

Figure 3.6   Box plots for delays in Example 3.7.2.

Table 3.10   Statistics for the delays on flights from Bordeaux, Budapest and Florence.

Statistics	Bordeaux	Budapest	Florence
Maximum	202	74	238
90% Percentile	36	10	57
Third quartile	19	−2.5	25
Median	9	−11	11
First quartile	2	−18	0
10% Percentile	−2	−23.4	−6
Minimum	−9	−32	−18
Arithmetic mean	14.67	−8.14	21.88
Standard deviation	23.20	15.54	42.11
Number of observations	215	185	215
Variance	538.45	241.39	1773.19
Skewness	3.95	2.12	3.09
Kurtosis	23.78	7.17	10.41
Interquartile range	17	15.5	25
Range	211	106	256

Table 3.10 contains a summary of the most important statistics for the delays on flights from Bordeaux, Budapest, and Florence. It is an interesting exercise to compare the information provided by the box plots in Figure 3.6 with these statistics.

The generation of the three box plots in Figure 3.6 can be done using the "Graph Builder" in the "Graph" menu of JMP. To do so, you have to drag the variable "Delay

Time Arrival" to the "Y" zone, and the variable "Departure Airport" to the "X" zone. The resulting screen is shown in Figure 3.7. Finally, you have to click the button with the box plots in the upper center of the "Graph Builder". So, there are various ways to generate box plots in JMP: via the "Distribution" platform in the "Analyze" menu, and via the "Graph Builder".

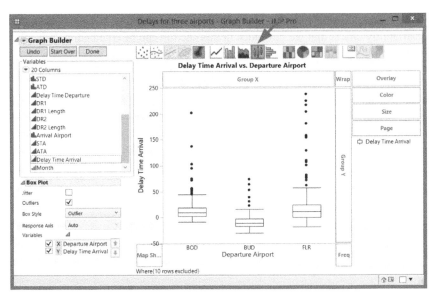

Figure 3.7   Creating the box plots in Figure 3.6 with the "Graph Builder" in JMP.

Finally, it is useful to note that, for certain data sets, the median and the first or third quartile coincide. This gives the corresponding box plots an unusual look and complicates their interpretation.

Example 3.7.3   *Figure 3.8 contains a box plot for the delays of all flights to Brussels between March 28 and October 30, 2003, with Brussels Airlines. The box plot indicates that some flights arrived more than 25 minutes early. The largest delay exceeds 350 minutes, for a flight from Edinburgh to Brussels.*

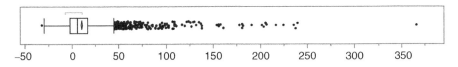

Figure 3.8   Box plot of delays of all flights to Brussels with Brussels Airlines.

An interesting feature of the box plots in JMP is that you can select the extreme values with your mouse. If you do so, the corresponding rows are highlighted in the data table. In this way, you can obtain a quick overview of the observations that led to extreme values.

3.8   Variability charts

In the introduction, we already mentioned the quality improvement program Six Sigma. If you participate in a Six Sigma training, you will undoubtedly be confronted with the following quote by the US author W. Edwards Deming: "Uncontrolled variation is the enemy of quality." In order to reduce variation, it is important to first identify the source(s) of variation in your process or product. To this end, graphics like the variability charts that we introduce in this section can be tremendously helpful.

In its simplest form, a variability chart can be considered as a systematic way to create several box plots side by side, as shown in the following example.

Example 3.8.1   *In Example 3.7.2, we looked at the delay at arrival of flights to Brussels from three different departure airports. Figure 3.9 shows a variability chart of this data set. Essentially, it contains the box plots of the delay for the three departure airports Bordeaux, Budapest, and Florence. It is very similar to Figure 3.6, but there are two differences. First, the variability chart contains all data points. Second, below the box plots, there is a second graph showing the standard deviations of the delays for each departure airport.*

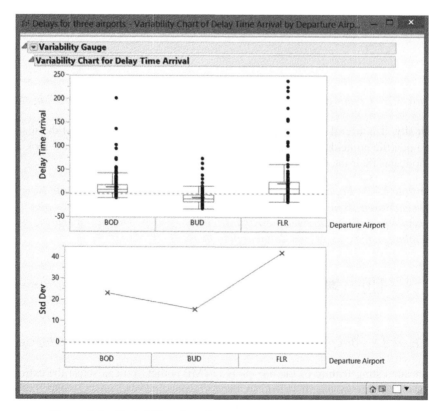

Figure 3.9   Variability chart for the delays for flights to Brussels.

If you manage the airport in Florence, you might wonder whether the poor results you are responsible for are evenly distributed over the entire period from April to October 2003, or whether there were better and worse months. To address this question, we need to split the data by a second variable, the month of departure. Splitting data by several variables and looking at the variation side by side is exactly what variability charts are built for. Figure 3.10 shows the result for our example.

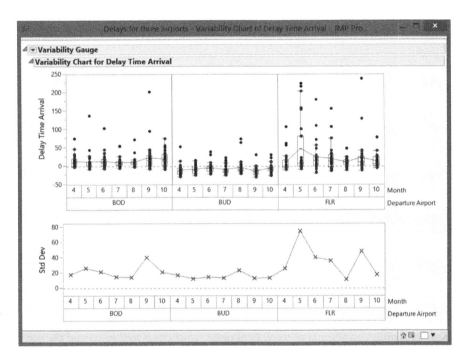

Figure 3.10   Variability chart of delays for flights to Brussels, by month and departure airport.

Now, it is easy to see that Budapest airport does not only consistently have the lowest average delay time, but the standard deviation is very stable over time as well. The situation is very different in Florence. There are severe outliers nearly every month. The average delay time is not that different from the one in Bordeaux, but the variation is much larger than that for Bordeaux. The bottom part of the variability chart tells us that Florence has good months, for example August, and very bad months, for example May. As manager of the Florence airport, one can now try to find a special cause explaining the difference in performance between August and May.

To create a variability chart in JMP, you have to select the option "Quality and Process" in the "Analyze" menu, and then pick "Variability/Attribute Gauge Chart". In the resulting dialog window, you need to drag the variable "Delay Time Arrival" to the "Y" field, and the variables "Departure Airport" and "Month" to the "X, Grouping" field. This step is shown in Figure 3.11. It is important to make sure that the order

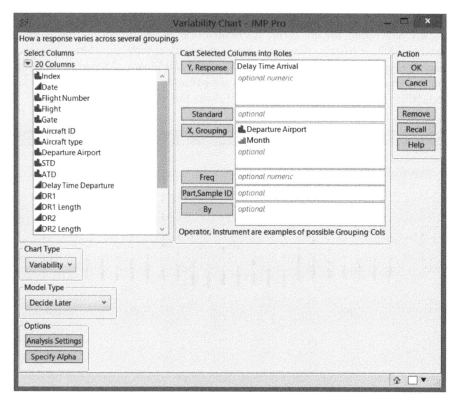

Figure 3.11   Creating a variability chart: Step 1.

of the grouping variables is correct. Otherwise, your chart will be nested the other way around. By default, the resulting variability chart contains range bars. To remove these, go to the hotspot (red triangle) menu next to the word "Variability Gauge" in the initial output, uncheck "Show Range Bars", and check "Show Box Plots". This step is shown in Figure 3.12. The result will be the variability chart shown in Figure 3.10.

Variability charts are sometimes called "multi-vari charts", indicating that they can help identify variability that comes from multiple sources. Typically, one can distinguish three types of variation, the so-called "within variation", "between variation", and the "variation over time". In our delay time example, the within variation is the variation within one airport, the between variation is the variation between the airports, and the variation over time is the spread observed over the months.

Variability charts have numerous application areas. A prominent one is measurement system analysis, which investigates part-to-part variation, operator-to-operator variation, machine-to-machine variation, lot-to-lot variation, and so on. We end this section with another typical application in quality control, the comparison of production lines.

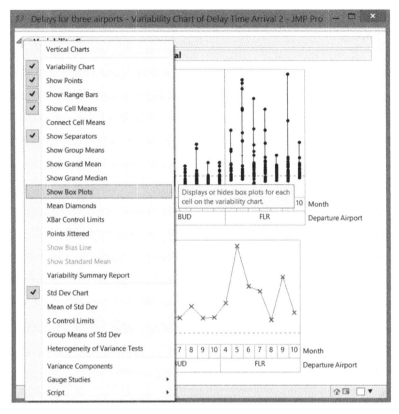

Figure 3.12   Creating a variability chart: Step 2.

Example 3.8.2 *Over the past years, solar companies based in Europe struggled to survive due to the pricing pressure coming from solar panels produced in Asia. In order to keep the production costs low, a German solar company built a small prototype factory in Germany, and then rebuilt a bigger version of the same production site twice in Malaysia. Although the Malaysian engineers were trained regularly on the German site, this strategy still involved some risks. It is easy to build a copy of an existing factory, but would it be possible to transfer the production quality from Germany to Malaysia, too?*

Figure 3.13 shows a variability chart of some yield data, comparing the three factories, named "Germany", "Malaysia A", and "Malaysia B", each containing three production lines. The chart contains the group means, which is another option in the menu for creating the variability chart in JMP. One can immediately recognize that both factories in Malaysia do not reach the same average yield as the German one. In addition, the variation in the "Germany" and "Malaysia A" sites is more or less the same and does not change much from one production line to the next. However, the "Malaysia B" site has a clear quality problem. The spread of the yield is much bigger in that factory than in the two other factories.

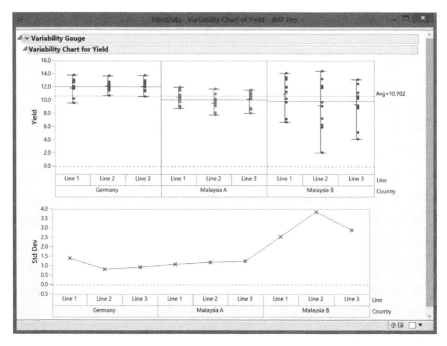

Figure 3.13   Variability chart comparing yield data of production lines in Germany and Malaysia.

3.9   Bivariate data

In this section, we look at situations where we study two variables, in general x and y, simultaneously. We have n observations of the first variable x, and n corresponding observations of the second variable y. The structure of the data is shown in Table 3.11. For example, we obtain a **bivariate data** table when we record the height (x) and weight (y) of a number of people. A data table with more than two variables, such as the data on Spanish wines in Figure 2.1, is called a **multivariate data table**.

Table 3.11   General structure of a bivariate data table for two variables x and y.

Observation	x	y
1	x_1	y_1
2	x_2	y_2
3	x_3	y_3
$\vdots$	$\vdots$	$\vdots$
n	x_n	y_n

Measuring and interpreting the relationship or association between two (or more) variables is one of the main tasks of statistics. The covariance and the correlation are frequently used to measure the strength of a linear relationship between two quantitative variables.

3.9.1 Covariance

Suppose that you have a data set with n observations of two quantitative variables x and y, as shown in Table 3.11.

Definition 3.9.1 *The **sample covariance** between the variables x and y is defined as*

$$s_{XY} = \frac{1}{n-1} \sum_{i=1}^{n} (x_i - \bar{x})(y_i - \bar{y}).$$

The covariance can be positive or negative. A term $(x_i - \bar{x})(y_i - \bar{y})$ in the definition of the sample covariance is positive if observation i has

- an x value smaller than $\bar{x}$, and a y value smaller than $\bar{y}$, or

- an x value bigger than $\bar{x}$, and a y value bigger than $\bar{y}$.

A term $(x_i - \bar{x})(y_i - \bar{y})$ in the definition of the sample covariance is negative if observation i has

- an x value smaller than $\bar{x}$, and a y value bigger than $\bar{y}$, or

- an x value bigger than $\bar{x}$, and a y value smaller than $\bar{y}$.

If the number of negative terms is dominant, this results in a negative covariance. In the other case, the covariance is positive.

An example of a scatter plot with a negative covariance is shown in Figure 3.14. In this figure, four quadrants are marked by drawing a vertical line at $\bar{x} = 5.1$, and a horizontal line at $\bar{y} = 3.8$. The observations with a positive cross-product $(x_i - \bar{x})$ $(y_i - \bar{y})$ will always lie in quadrant I or III. The observations with a negative cross-product $(x_i - \bar{x})(y_i - \bar{y})$ will always lie in quadrant II or IV. If the observations in quadrants I and III are dominant, then x and y have a positive covariance. If the observations in quadrants II and IV dominate, then the covariance between x and y is negative. It is clear that the observations in Figure 3.14 are almost all located in quadrants II and IV.

Example 3.9.1 *Suppose that a data set with 10 observations for two variables x and y is given: (6,2), (3,6), (4,5), (7,3), (4,6), (6,3), (7,2), (6,3), (5,2), and (3,6). This data is shown in tabular form in Table 3.12 and as a scatter plot in Figure 3.14. It is easy*

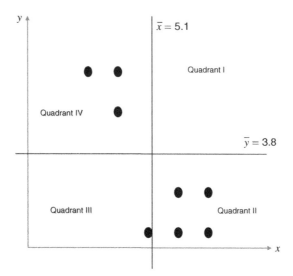

Figure 3.14   Scatter plot with negative covariance.

Table 3.12   Bivariate data
table for Example 3.9.1.

Observation	x	y
1	6	2
2	3	6
3	4	5
4	7	3
5	4	6
6	6	3
7	7	2
8	6	3
9	5	2
10	3	6

to check that $\bar{x} = 5.1$ and $\bar{y} = 3.8$. The sample covariance is

$$s_{XY} = \frac{1}{9}\{(6 - 5.1)(2 - 3.8) + (3 - 5.1)(6 - 3.8) + \cdots + (3 - 5.1)(6 - 3.8)\},$$

$$= \frac{-20.8}{9} = -2.311.$$

The calculations that we need to perform for the determination of the sample covariance are shown schematically in Table 3.13, where the deviations from the means $\bar{x} = 5.1$ and $\bar{y} = 3.8$ (in other words, $x_i - \bar{x}$ and $y_i - \bar{y}$) and the products $(x_i - \bar{x})(y_i - \bar{y})$ are determined sequentially.

Table 3.13   Calculating the sample covariance in Example 3.9.1.

Observation	x_i	y_i	$x_i - \bar{x}$	$y_i - \bar{y}$	$(x_i - \bar{x})(y_i - \bar{y})$
1	6	2	0.9	−1.8	−1.62
2	3	6	−2.1	2.2	4.62
3	4	5	−1.1	1.2	−1.32
4	7	3	1.9	−0.8	−1.52
5	4	6	−1.1	2.2	−2.42
6	6	3	0.9	−0.8	−0.72
7	7	2	1.9	−1.8	−3.42
8	6	3	0.9	−0.8	−0.72
9	5	2	−0.1	−1.8	0.18
10	3	6	−2.1	2.2	−4.62
Sum					−20.8

The concept of a covariance can also be applied to a population or a process. For a finite population of N elements, the covariance is defined as follows:

Definition 3.9.2 *The **population covariance** between the variables x and y is defined as*

$$\sigma_{XY} = \frac{1}{N} \sum_{i=1}^{n} (x_i - \mu_X)(y_i - \mu_Y),$$

where μ_X is the population mean of the variable x and μ_Y is the population mean of the variable y.

The alert reader will have noticed that the sample variance is a special case of the sample covariance, and the population variance a special case of the population covariance. A variance is simply the covariance of a variable with itself. Consequently, we have $s_X^2 = s_{XX}$ and $\sigma_X^2 = \sigma_{XX}$.

A disadvantage of the covariance as a statistic for the association between two variables is that the result depends on the unit of measurement used for each of the variables. Whenever, for example, a variable that was originally expressed in meters is re-expressed in centimeters, the covariance of that variable with any other given variable will be multiplied by 100.

Example 3.9.2 *Suppose that a data set with 10 observations for two variables x and y is given: (6,20), (3,60), (4,50), (7,30), (4,60), (6,30), (7,20), (6,30), (5,20), and (3,60). This data set is the same as the one in Example 3.9.1, except for the fact that all y values have been multiplied by a factor of 10. It is easy to verify that $\bar{x} = 5.1$ and $\bar{y} = 38$. The sample covariance is $s_{XY} = -23.11$, which is 10 times larger than the covariance of Example 3.9.1.*

Because of its sensitivity to the units of measurement, the magnitude of a covariance is hard to interpret. The correlation coefficient, which does not suffer from this problem, is a more popular measure of the association between two quantitative variables.

3.9.2   Correlation

Definition 3.9.3 *The **sample correlation coefficient**, also known as **Pearson's coefficient of correlation**, of the observations $(x_1, y_1), (x_2, y_2), \ldots, (x_n, y_n)$ is defined as*

$$r_{XY} = \frac{s_{XY}}{s_X s_Y} = \frac{\sum_{i=1}^{n}(x_i - \bar{x})(y_i - \bar{y})}{\sqrt{\sum_{i=1}^{n}(x_i - \bar{x})^2}\sqrt{\sum_{i=1}^{n}(y_i - \bar{y})^2}}.$$

The **population correlation coefficient** between two variables is

$$\rho_{XY} = \frac{\sigma_{XY}}{\sigma_X \sigma_Y}.$$

Example 3.9.3   *For the data of Example 3.9.1, we can verify that $s_X = 1.524$ and $s_Y = 1.751$, hence $r_{XY} = -2.311/(1.524 \times 1.751) = -0.866$. The two variables in Example 3.9.2 have the same sample correlation coefficient, namely -0.866.*

Correlation coefficients are bounded between -1 and 1. Variables with a correlation of $+1$ are perfectly positively correlated. Variables with a correlation of -1 are perfectly negatively correlated. In each of these cases, there is a linear relationship of the form $y = ax + b$ between the two variables under study. This can be proven easily. A correlation coefficient always has the same sign as the corresponding covariance. Variables with a correlation of 0 are called uncorrelated.

An important remark concerning correlation coefficients is that they only indicate the extent to which there is a linear relationship between two variables. A correlation of (almost) zero only indicates that there is no linear relationship between the two variables. However, there might be a quadratic, cubic, or logarithmic relationship between the variables.

Example 3.9.4   *For the sample with observations $(6,1)$, $(7,4)$, $(6.5, 2.25)$, $(4,1)$, $(8,9)$, and $(2,9)$, the correlation coefficient between the observed variables x and y is $r_{XY} = -0.0955$. Therefore, the two measured variables are almost uncorrelated. Nevertheless, there is a functional relation between the two variables, namely $y = (x - 5)^2$.*

Example 3.9.5   *The covariance and the correlation between the measured and the declared alcohol percentage of the Spanish red wines in Example 2.2.1 equal 0.203 and 0.818, respectively. The covariance and the correlation between the measured alcohol content and the price of Spanish red wines in Example 2.2.1 amount to 0.827 and 0.507, respectively. Finally, the covariance between the declared alcohol content and the price is 0.517, while the correlation between these two variables is equal to 0.391.*

All this can be verified in Figure 3.15, where a correlation matrix and a covariance matrix are shown for the variables "Alcohol measured", "Alcohol declared", and "Price". The computation of the correlation matrix and the covariance matrix can be done in JMP using the "Analyze" menu, by first choosing "Multivariate Methods",

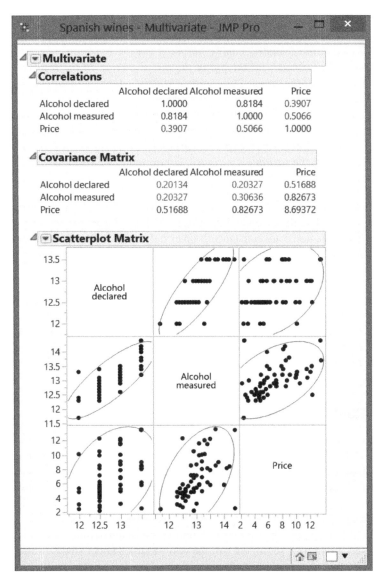

Figure 3.15   Correlation matrix, covariance matrix, and scatter plots to determine the relationship between the price, the measured, and the declared alcohol content of the Spanish red wines from Example 2.2.1.

and then selecting the option "Multivariate". The next step is to enter the names of the three variables in the field called "Y". By default, you only get to see the correlation matrix. The covariance matrix can be obtained via the hotspot (red triangle) next to the word "Multivariate" at the top of the initial output. In the resulting pop-up menu, you can select the option "Covariance Matrix". JMP automatically generates a "Scatterplot Matrix" for all selected variables.

Notice that the correlation matrix contains ones on the diagonal. This means that each variable is perfectly positively correlated with itself. Each diagonal element of the covariance matrix represents a covariance of a variable with itself. In other words, it represents the variable's variance. For example, the variance of the variable "Price" is €²8.694.

Sometimes, linear combinations of variables need to be investigated. Suppose that you want to calculate the mean and the variance of the linear combination $u = ax + by + c$, where a, b, and c are constants. It takes only a little effort to demonstrate that $\bar{u} = a\bar{x} + b\bar{y} + c$. A slightly more difficult exercise is to prove that $s_u^2 = a^2 s_x^2 + b^2 s_y^2 + 2ab s_{XY}$.

Example 3.9.6 *As part of the evaluation of exams that students take, schools and universities calculate correlations between the results of various courses. This shows over and over again that results for subjects such as mathematics, statistics, operations research, and physics are quite strongly positively correlated. Between these mathematically oriented subjects and subjects such as philosophy and sociology, the correlation is close to zero.*

3.9.3   Rank correlation

The major criticism on the correlation coefficient of Pearson introduced in Section 3.9.2 is that it only quantifies the extent to which there is a linear relationship between two (quantitative) variables. An alternative is Spearman's rank correlation coefficient[7], measuring whether there is a monotone relationship between two (quantitative or ordinal) variables. Spearman's correlation coefficient also quantifies non-linear increasing or decreasing relationships between variables.

The calculation of Spearman's rank correlation coefficient requires two sets of ranks for all observations, one set for each of the two variables under study. In some cases, the ranks of two observations with respect to a certain variable will be the same. To deal with this kind of tie, an averaged rank is used. Once the ranks have been determined, the (ordinary) Pearson correlation coefficient is computed for the ranks. The result is Spearman's rank correlation coefficient.

Example 3.9.7 *Suppose that you have the scores of 10 students for the English and mathematics exams, and that you want to quantify the association between these*

---

[7] Charles Spearman (1863–1945) was an English psychologist who became famous for his work in statistics and on human intelligence. He came up with the rank correlation coefficient and developed the first form of factor analysis. He was a colleague of Karl Pearson at the University College in London, but the two gentlemen were not on very good terms.

two scores. In other words, you wish to determine whether a high (low) score for English is associated with a high (low) score for mathematics. If you are not necessarily interested in a linear relationship, then it is better to calculate Spearman's rank correlation coefficient. Table 3.14 illustrates how the ranks are determined for the 10 pairs or scores. This is the first step in the calculation of the rank correlation coefficient. Student H is given rank 1 for English and rank 1 for mathematics because she has the best score for both courses. Student I is in rank 2 for English, but only has rank 3 for mathematics. Student C has rank 10 for both subjects because she has the worst score for both English and mathematics.

Table 3.14   Determination of ranks without ties.

Student	Score English	Score mathematics	Rank English	Rank mathematics
A	56	66	9	4
B	75	70	3	2
C	45	40	10	10
D	71	60	4	7
E	61	65	7	5
F	64	56	5	9
G	58	59	8	8
H	80	77	1	1
I	76	67	2	3
J	62	63	6	6

If we now calculate the (usual) Pearson correlation coefficient for the two columns with ranks, we find a value of 0.6606. In other words, Spearman's rank correlation coefficient is 0.6606 in this example.

If we calculate the (ordinary) Pearson correlation coefficient for the original scores for mathematics and English, we would obtain a value of 0.8038.

Note that there exists an alternative formula for Spearman's rank correlation coefficient in case there are no ties in the ranks, as in Example 3.9.7. In many cases, this formula cannot be used due to the presence of ties. Our advice is to ignore that formula, which is in many textbooks, and to always use the following procedure when calculating Spearman's rank correlation coefficient for two variables x and y:

- Sort all observations from large to small according to the variable x and assign ranks to all observations based on this ranking.

- Sort all observations from large to small according to the variable y and assign ranks to all observations based on this ranking.

- Calculate the (ordinary) Pearson correlation coefficient for the two sets of ranks.

- The result of this calculation is Spearman's rank correlation coefficient.

The following example illustrates how to deal with ties in the rankings required for the computation of the rank correlation coefficient.

Example 3.9.8 *Suppose again that you have the scores of 10 students for the English and mathematics exams, and that you want to quantify the association between these two scores. Again, you are not particularly interested in a linear relationship. Table 3.15 shows a slightly modified data table in comparison to Example 3.9.7, now with a tie between student E and student J regarding the English score. Both students come in sixth place in the ranking for the English course. The next student is student G, who occupies the eighth position. Instead of using the rank 6 twice for student E and student J, we attribute the average rank 6.5 to both students. The reasoning behind this is that both students together occupy positions 6 and 7, and the average of 6 and 7 is 6.5.*

Table 3.15   Determination of ranks with ties.

Student	Score English	Score mathematics	Rank English	Rank mathematics
A	56	66	9	4
B	75	70	3	2
C	45	40	10	10
D	71	60	4	7
E	61	65	6.5	5
F	64	56	5	9
G	58	59	8	8
H	80	77	1	1
I	76	67	2	3
J	61	63	6.5	6

If we now calculate the (usual) Pearson correlation coefficient for the two columns with ranks, we find a value of 0.6687. In other words, Spearman's rank correlation coefficient is 0.6687 in this example.

If we would calculate the (ordinary) Pearson correlation coefficient for the original scores for mathematics and English in Table 3.15, we would obtain a value of 0.8005.

The calculation of Spearman's rank correlation coefficient can also be done with JMP. In the "Analyze" menu, just as for the (ordinary) correlation coefficient of Pearson, you need to choose "Multivariate Methods" followed by "Multivariate". The next step is to enter the names of the two variables under study in the field called "Y". Once JMP has computed the (ordinary) Pearson correlation coefficient, as shown in Section 3.9.2, you can choose "Nonparametric Correlations" followed by "Spearman's ρ" via the hotspot (red triangle) next to the word "Multivariate" at the top of the output. This step is illustrated in Figure 3.16. As can be seen in Figure 3.17, JMP then generates an extra piece of output that contains Spearman's rank correlation coefficient.

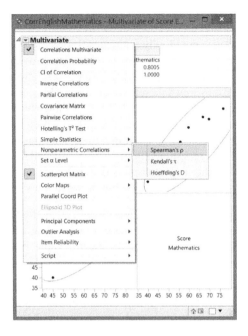

Figure 3.16   The computation of Spearman's rank correlation coefficient using JMP.

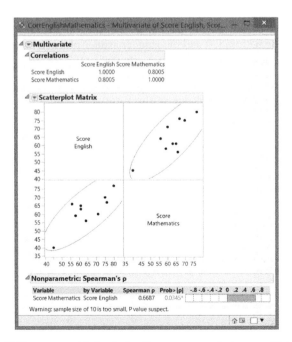

Figure 3.17   JMP output with both the Pearson correlation coefficient and Spearman's rank correlation coefficient.

3.10 Complementarity of statistics and graphics

It is highly recommended to not blindly calculate all sorts of statistics during the study of a data set. Often, a graphical representation of the data provides important insights that go unnoticed when only means, variances, and correlations are calculated. We illustrate this fact based on seven data sets with two variables x and y. The first data set has an (ordinary) correlation coefficient of 0.7, and is shown by means of a scatter plot in Figure 3.18. This scatter plot represents a typical situation in which two variables show a positive linear correlation, without the linear relationship being perfect.

Remarkably, the six data sets shown in Figure 3.19 also all have a correlation coefficient of 0.7. For none of the six scatter plots in this figure, the correlation coefficient provides a good description of the relationship between the variables x and y:

1. All but one point in Figure 3.19a are located at the bottom left and show no pattern at all. In the data, however, there is one outlier, which not only leads to an increase of the mean x-value and the mean y-value, but also to a correlation of 0.7. If this outlier is deleted, the correlation of the data would become nearly zero.

2. Except for one point, the data points in Figure 3.19b exhibit an almost perfect linear relationship. Considering only these points, the correlation coefficient would be near 1. However, the outlier in the data causes the correlation coefficient to drop to 0.7.

3. In Figure 3.19c, we recognize two clearly separated sets of data points, each having a correlation of zero. However, the fact that the larger set of points is located at the bottom left and the smaller scatter plot is located at the top right of the figure leads to a correlation coefficient of 0.7 for the complete data set.

4. The points in Figure 3.19d have a special pattern: the value of the x-variable is always larger than the value of the y-variable. This situation occurs if the

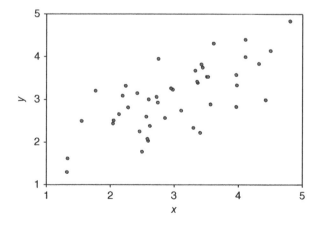

Figure 3.18   Scatter plot for two variables with correlation 0.7.

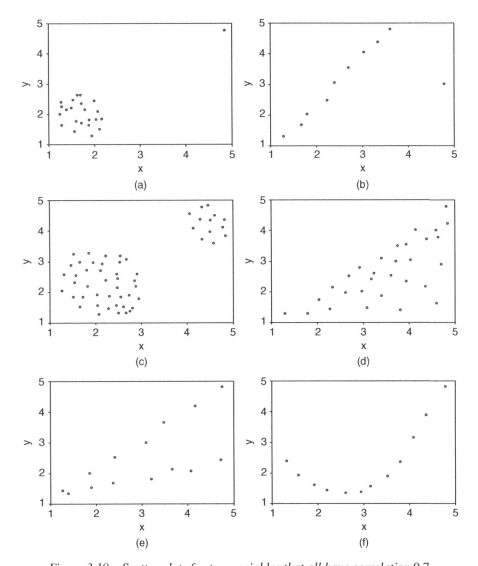

Figure 3.19   Scatter plots for two variables that all have correlation 0.7.

variable y is a component of the variable x. As an example, imagine that x represents the total energy consumption of a family, and y the heating consumption. In such a case, it does not make sense to calculate the correlation coefficient.

5. In Figure 3.19e we clearly recognize two sets of points that both lie more or less on a straight line. In such situations, the two sets of points usually correspond to two clearly identifiable groups of objects. It may be, for instance, that one set of points corresponds to a group of men, while the other set corresponds to a group of women.

6. The points in Figure 3.19f exhibit a perfect quadratic functional relation. Despite this perfect relation, the correlation is not equal to one. This is due to the fact that the correlation coefficient only measures the strength of a linear relationship.

3.11   Descriptive statistics using JMP

JMP offers the possibility to compute nearly all statistics discussed in this chapter. The easiest way to do so is to make use of the "Distribution" platform in the "Analyze" menu. For a quantitative variable, you will then automatically get a histogram, a box plot, the main percentiles (including the median, the first, and third quartile), the sample's arithmetic mean, and the standard deviation. For a qualitative variable, the output is limited: you will only see a bar chart and a table with the (absolute) frequencies for each category or class. The standard outputs for quantitative and qualitative variables are shown in Figure 3.20 and Figure 3.21, respectively.

In Figure 3.21, the bar for the most expensive wines was selected by clicking on it with the mouse. This explains the darker color of this bar. Therefore, all expensive wines are automatically highlighted in the histogram of the ratings in the same figure. This reveals that the most expensive wines either get a good (G) or a fair to good (F/G)

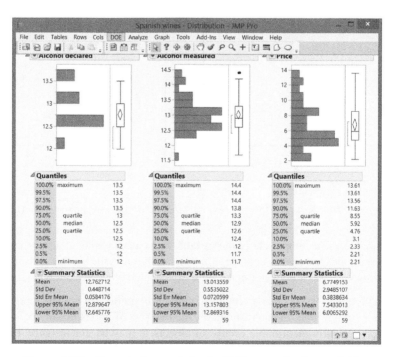

Figure 3.20   Descriptive statistics in JMP: quantitative variables.

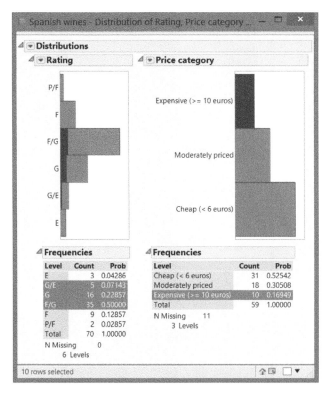

Figure 3.21   Descriptive statistics in JMP: qualitative variables.

rating. Not a single expensive wine gets an excellent (E) rating. At the same time, all observations that correspond to expensive wines are also highlighted in the data table. Using the "Shift" key one can select several bars in a histogram. Double-clicking on a bar in a histogram automatically creates a new data table with only the data corresponding to the selected bar.

By default, the standard output for quantitative variables only shows a small subset of all possible statistics that JMP can compute. A wide range of statistics can be obtained by clicking on the hotspot (red triangle) next to the term "Summary Statistics" in the output. Choosing "Customize Summary Statistics" in the resulting pop-up menu brings up the dialog window in Figure 3.22, where additional statistics can be selected.

The available statistics are the following:

- the arithmetic or sample mean: "Mean",

- the "Geometric Mean",

- the standard deviation of the sample mean: "Std Err Mean" (see the book *Statistics with JMP: Hypothesis Tests, ANOVA and Regression*),

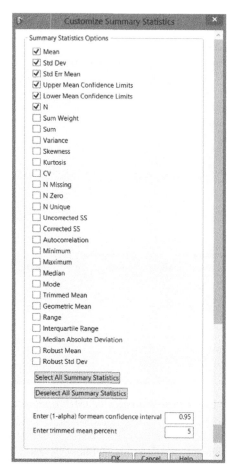

Figure 3.22   Dialog window for additional descriptive statistics in JMP.

- the lower and upper limit of a confidence interval for the arithmetic or sample mean: "Lower Confidence Limit" and "Upper Confidence Limit" (see the book *Statistics with JMP: Hypothesis Tests, ANOVA and Regression*),

- the "Median",

- the "Mode",

- the sample standard deviation: "Std Dev",

- the sample variance: "Variance",

- the "Kurtosis",

- the skewness coefficient of Fisher: "Skewness",

- the "Range",

- the "Interquartile Range",

- the coefficient of variation: "CV",

- the sum of the observations: "Sum",

- the number of observations: "N" (as well as the number of missing observations "N Missing", the number of observations equal to 0 "N Zero", and the number of unique observations "N Unique"),

- the truncated mean: "Trimmed Mean".

The truncated mean is the arithmetic mean obtained after deleting the smallest and the largest observations. At the bottom of Figure 3.22, you can see that, by default, JMP computes the truncated mean omitting the 5% smallest and 5% largest observations. The truncated mean is useful when the arithmetic mean of all data points is highly influenced by extreme values. Excluding the largest and smallest observations circumvents this problem.

As shown in Figure 3.21, JMP automatically calculates a frequency table with absolute and relative frequencies for qualitative variables. Additionally, JMP can also calculate cumulative relative frequencies, which is useful in preparing a Pareto chart or when calculating the ordinal dispersion index. To this end, right-click on the frequency table and select "Columns" followed by "Cum Prob". This is illustrated in Figure 3.23 for the rating variable. The resulting output is shown in Figure 3.24.

An alternative method to calculate descriptive statistics using JMP is to make use of the menu "Tables". In that menu, you can select the "Summary" option. This option

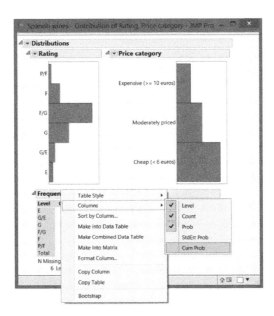

Figure 3.23   Computing relative cumulative frequencies using JMP.

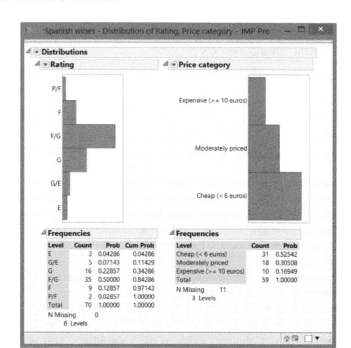

Figure 3.24   Table with cumulative relative frequencies in JMP for the rating variable.

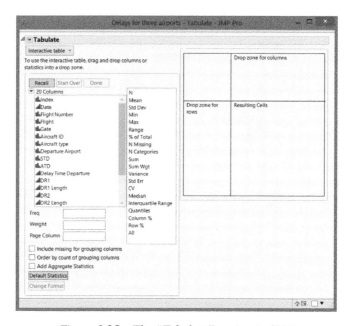

Figure 3.25   The "Tabulate" option in JMP.

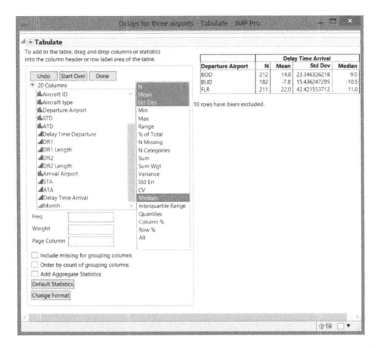

Figure 3.26   The "Tabulate" option in JMP applied to Example 3.7.2.

allows the computation of the most important statistics, including any percentile you need. Even more interesting is the option "Tabulate" in the "Analyze" menu, the use of which is illustrated in the next example.

Example 3.11.1 *If you want to get a quick overview of the delays by departure airport for the data in Example 3.7.2, you can use the "Tabulate" option in the "Analyze" menu, which produces the dialog window in Figure 3.25. Drag the variable "Delay Time Arrival" to the zone called "Drop zone for columns", select "Add Analysis Column", and then drag the variable "Departure Airport" to the zone called "Drop zone for rows". Finally, you can select the statistics "N", "Mean", "Std Dev", and "Median" simultaneously by using the "CTRL" key, and dragging these statistics to the table. You will then obtain the screen in Figure 3.26. The final step is to click on the hotspot next to the word "Tabulate" and choose the option "Make Into Data Table". This produces a new data table with the computed summary statistics.*

4

Probability

The likelihood of William's wits being nearer their end than those of his poor wife seems small, ...

(from The Crimson Petal and the White, Michel Faber, p. 588)

'Anyway, by your wonderful succession of logical inferences –'
'Well, it's more a matter of a probabilistic matrix than strict syllogistic logic –'
'By whatever screwy juju you rely upon, you decided to stake out one particular door-way. And you got lucky.'
'Lucky? Obviously, you haven't heard anything I said. It was a matter of applying Bayes' Theorem to estimate the conditional probabilities, giving due weight to the prior prob-abilities and thus avoiding the fallacies of –'

(from The Ambler Warning, Robert Ludlum, p. 302)

The founders of probability theory are people like Pascal, Fermat, Huygens, and Bernoulli. Blaise Pascal (1623–1662) was one of the greatest thinkers, theologians, philosophers, mathematicians, and physicists in Western history. Together with the French mathematician Pierre de Fermat, he laid the mathematical foundations of probability theory. His interest in probability theory was awakened during a stay in Paris, following a recommendation of his doctor to get rid of some chronic diseases. After a severe depression, he retired to a monastery in 1654 and dedicated himself to theology and philosophy. Fermat (1601–1665) is one of the most famous mathematicians in history, who also made important contributions to probability theory. His most controversial results concerned number theory and were only discovered on loose sheets of paper after his death. However, no solid proofs were found. Some of his statements were proven years later (including by Euler), others disproved.

Statistics with JMP: Graphs, Descriptive Statistics, and Probability, First Edition. Peter Goos and David Meintrup.
© 2015 John Wiley & Sons, Ltd. Published 2015 by John Wiley & Sons, Ltd.
Companion Website: wiley.com/go/goosandmeintrup

Christiaan Huygens (1629–1695) was a Dutch mathematician who became interested in probability theory after a visit to Paris, where he heard of Fermat and Pascal. In his article "Van Rekeningh in Spelen van Geluck" (On Probability in Gambling Games), he published 14 problems with solutions and five unsolved problems, that would later be addressed by Bernoulli. Jacob (aka Jacques) Bernoulli (1654–1705) was forced by his wealthy parents to study philosophy and theology in Basel. At the same time, however, he also attended lectures in mathematics and astronomy. He declined an offer to serve the Church, and accepted a professorship in mathematics at the University of Basel instead. He made important contributions to algebra, infinitesimal calculus, and the theory of series, but is best known for his work on probability theory. His book *Ars Conjectandi* was published posthumously. The work was groundbreaking because new applications of probability theory were treated, an interpretation of probability was given, and the idea that one can estimate probabilities by means of the observed relative frequencies was developed.

Probability theory deals with processes and experiments whose outcome is uncertain. Here, the words "process" and "experiment" have to be interpreted in a broad sense. Such processes or experiments take place on a daily basis. In principle, some processes are **deterministic**: if certain conditions are met, then the outcome of the process or the outcome of the experiment can be predicted with certainty. For example, pure water will boil at a pressure of 760 mm and a temperature of 100°C. Because it is not easy to accurately meet the conditions "pure water" and "pressure of 760 mm", it is, however, somewhat uncertain as to whether the water will actually boil at exactly 100°C. This illustrates that deterministic processes and experiments hardly exist in practice. There will almost always be at least a little bit of uncertainty. Therefore, almost all processes are **probabilistic** or **stochastic**: there is uncertainty about the outcome. Such processes are called **probability experiments**, **random experiments**, or **stochastic processes**. In the following, we shall use the term "random experiment" or simply "experiment". A well-known example of a (random) experiment is throwing a die. The unpredictability of the result is not due to the ignorance of the person who throws the die, but to the inherently stochastic nature of the underlying process. Another example of a (random) experiment is a student taking an exam: no matter how well the student is prepared, her result is uncertain because it depends, for example, on the set of questions or the day's form.

Probability theory studies what statements can be made about random or stochastic experiments. The aim is to understand and quantify the **likelihood** of certain outcomes. A **probability** or a **chance** is an expression of the likelihood that we attach to these outcomes. The more likely we consider a certain outcome or event, the greater the probability we assign to it.

Outcomes of random experiments can always be interpreted as a data sample from a population or a process. A die can be thrown several times, leading to a number of observations that form a sample. The exam results of a group of students can also be considered as outcomes of a number of random experiments and be treated as a sample. The sample contains the information that is necessary to better understand a process or to obtain a better description of a population.

4.1 Random experiments

A random experiment E has more than one possible outcome. The set of possible outcomes is the **sample space** Ω. Examples of experiments and the corresponding sample spaces include:

- E_1: flipping a coin: $\Omega_1 = \{\text{head/tail}\}$;

- E_2: throwing a die: $\Omega_2 = \{1, 2, 3, 4, 5, 6\}$;

- E_3: the number of times a die must be thrown before the result is a six: $\Omega_3 = \{1, 2, 3, \ldots\}$;

- E_4: throwing a red die and a blue die: $\Omega_4 = \{(1, 1), (1, 2), (1, 3), \ldots, (6, 6)\}$;

- E_5: whether an arbitrary Spanish family has a DVD player: $\Omega_5 = \{\text{yes, no}\}$;

- E_6: the time between the arrivals of two customers at a cash machine: $\Omega_6 = \{t : t > 0\}$;

- E_7: the service time of a customer at a counter in a post office: $\Omega_7 = \{t : t > 0\}$;

- E_8: the number of accidents in a company during one month: $\Omega_8 = \{0, 1, 2, \ldots\}$.

Some sample spaces, such as Ω_1, Ω_2, Ω_4, and Ω_5, are finite sets. Sample spaces Ω_3 and Ω_8 are infinite but countable. Sample spaces Ω_6 and Ω_7 are infinite and not countable. These examples show that every possible set can serve as the sample space of an experiment.

For any given experiment, multiple sample spaces can be considered. For example, in experiment E_2, one can also use the sample space $\Omega_2' = \{\text{odd, even}\}$. Which sample space is used depends on the questions one wants to answer. Quite often, the key to solving a probability problem lies in a good choice of the sample space.

An **event** G is a set of possible outcomes of an experiment. The sample space itself, Ω, and the empty set, $\varnothing$, are also considered events. An event is thus a subset of the sample space. Examples of events are:

- for E_1: obtaining a head: $G_1 = \{\text{head}\}$;

- for E_2: obtaining an even number: $G_2 = \{2, 4, 6\}$;

- for E_3: less than five throws are needed before obtaining a six: $G_3 = \{1, 2, 3, 4\}$;

- for E_4: the sum of the number of dots obtained from throwing the two dice is five: $G_4 = \{(1, 4), (2, 3), (3, 2), (4, 1)\}$;

- for E_5: a family has no DVD player: $G_5 = \{\text{no}\}$;

- for E_6: the time between two arrivals is between 2 and 5 minutes: $G_6 = \{t : 2 \leq t \leq 5\}$;

- for E_7: the service time exceeds 10 minutes: $G_7 = \{t : t > 10\}$;

- for E_8: there are no accidents in the company during one month: $G_8 = \{0\}$.

If an event contains only one outcome, it is called an **elementary event**. Examples of such elementary events are G_1, G_5, and G_8. An event G occurs if the outcome of the random experiment belongs to the set G. Event G_2 occurs, for example, when a 4 is obtained. Event G_6 occurs when a customer appears 3 minutes and 12 seconds after the previous one.

As one can guess from these examples, set theory is particularly useful in probability theory. Relations between and operations on subsets are essential for the calculation of probabilities.

Example 4.1.1 *A few years ago, the Belgian postal service introduced the so-called priority stamp. This stamp was more expensive than an ordinary stamp, but guaranteed that the letter would arrive at its destination the next (business) day. The introduction of the priority stamp was accompanied by many problems. The system of the priority stamp was regularly tested by various agencies and organizations and found to be insufficient. For a normal shipping, with a normal (non-priority) stamp the postal service promised delivery within three working days. After several years, the Belgian postal service quietly abandoned the system of the priority stamp. In Great Britain, however, the Royal Mail postal service still offers a "First Class Mail" (delivery the next working day) and a "Second Class Mail" (delivery latest at the latest after three days). In 2014, a First Class Mail stamp cost 62 pence, while a Second Class Mail stamp cost only 53 pence.*

One random experiment in this context could be to determine how much time is needed for a randomly selected letter to reach its destination and what kind of stamp the randomly selected letter has. The possible outcomes of the experiment are shown in Table 4.1. In the table, each outcome is denoted using the Greek letter ω.

Table 4.1   Sample space for the random experiment with the Royal Mail. The letter X refers to the type of stamp, while the letter Y indicates the delivery time of the letter (expressed in number of days).

		Y			
		1	2	3	> 3
X	First Class	ω_{11}	ω_{12}	ω_{13}	ω_{14}
	Second Class	ω_{21}	ω_{22}	ω_{23}	ω_{24}

For this experiment, the following events may be defined:

- $G_1 = \{\omega_{11}, \omega_{12}, \omega_{13}, \omega_{14}\}$: *the letter has a First Class stamp;*

- $G_2 = \{\omega_{11}, \omega_{12}, \omega_{21}, \omega_{22}\}$: *the letter is delivered within two days;*

- $G_3 = \{\omega_{12}, \omega_{13}, \omega_{14}\}$: *the letter has a First Class stamp and is delivered late;*

- $G_4 = \{\omega_{24}\}$: *the letter has a Second Class stamp and is delivered late;*

- $G_5 = \{\omega_{12}, \omega_{13}, \omega_{14}, \omega_{24}\}$: *the letter is delivered late.*

A set can be the **union** of a number of other sets: $G = G_1 \cup G_2 \cup \cdots = \cup_i G_i$. In Example 4.1.1, the event G_5 is the union of G_3 and G_4: $G_5 = G_3 \cup G_4$. Event G_5 occurs if the outcome of the experiment either belongs to G_3 or to G_4, or belongs to both G_3 and G_4. A union is exhaustive if it equals the entire sample space, that is, if $\cup_i G_i = \Omega$.

An event can also be the **intersection** of two or more sets: $G = G_1 \cap G_2 \cap \cdots = \cap_i G_i$. In Example 4.1.1, we have that $G_3 = G_1 \cap G_5$. If the intersection of two sets is empty, we call them "disjoint". The associated events cannot occur together: they are mutually exclusive. Examples of mutually exclusive events are G_3 and G_4.

A set can also be the **difference** between two sets. For example, in Example 4.1.1, the event G_3 is the difference between the events G_5 and G_4, which is denoted by $G_3 = G_5 \backslash G_4$. In that case, G_3 is a **subset** of G_5 (because all elements of G_3 are also elements of G_5). Similarly, G_3 is a subset of G_1. We denote this by $G_3 \subseteq G_1$.

Another important definition is the **complement** of an event. A set G^c is the complement of an event G if $G^c = \Omega \backslash G$. Each element of the sample space is either in G or in G^c. In Example 4.1.1, the complement of G_4 is equal to $G_4^c = \{\omega_{11}, \omega_{12}, \omega_{13}, \omega_{14}, \omega_{21}, \omega_{22}, \omega_{23}\}$.

Finally, we define a partition of the sample space Ω as a set of non-empty subsets of Ω, which are exhaustive and pairwise disjoint. A partition $G_1, G_2, G_3, \ldots$ satisfies the following conditions: $G_1 \cup G_2 \cup G_3 \cup \cdots = \Omega$ and $G_i \cap G_j = \emptyset$ for each $i \neq j$. Clearly, each non-empty event G and its complement G^c constitute a partition of the sample space.

4.2   Definition of probability

In this section, we define the probability that an event G occurs, in other words, that the outcome of a random experiment belongs to the set G. The **probability** quantifies the **likelihood** that the event G occurs. It is expressed by means of a function $P(G)$, which assigns a real number to the event G.

An intuitive way to introduce the concept of probability is to repeat an experiment n times and record the frequency $f_n(G)$ of the event G. The relative frequency of G is $f_n(G)/n$. For small values of n, the relative frequency is not stable, but, for large n, it stabilizes. If we let n increase towards infinity (i.e., if we repeat the experiment many times), the relative frequency of the event approaches the probability of G, namely $P(G)$. Mathematically, this is written as

$$P(G) = \lim_{n \to \infty} \frac{f_n(G)}{n}.$$

Example 4.2.1 *Suppose G is the event of obtaining an even number of dots with a die. In order to determine the probability of event G, you toss a die 10 times. If you observe an even number of dots seven times, you could argue that the probability P(G) of the event G is equal to the relative frequency of 7/10. When repeating the experiment 100 times, you may record an even number of dots 57 times. The relative* *frequency is then 57/100, and 0.57 would be your estimate of the probability P(G). After 1,000 experiments, you might have recorded an even number of dots 492 times. That suggests that the probability P(G) is 0.492. After 10,000 experiments with 5038 occurrences of the event G, the estimated probability is 0.5038. This shows how the relative frequency approaches 0.5 when the number of repetitions of the random experiment is increased. Therefore, the probability P(G) is 0.5.*

This definition is called the **empirical** or **frequency definition of probability**. This definition is imprecise because the number of times that an experiment must be repeated is not specified. Another definition that is sometimes used is the so-called **classical definition of Laplace**[1]. This definition assumes that the probability of occurrence is known for all possible elementary events. Often, it is assumed that all outcomes are equally likely. For example, in the case of throwing a die, it is typically assumed that the six possible outcomes are equally likely, and equal to 1/6. To determine the probability of obtaining an even number of dots, one can then count the number of possible outcomes with an even number of dots (three), and the total number of possible outcomes (six). This results in the following probability of an even number of dots:

$$P(G) = \frac{\text{number of even outcomes}}{\text{total number of possible outcomes}} = \frac{3}{6} = \frac{1}{2}.$$

The empirical definition and the classical definition of Laplace are inadequate if we want to make statements concerning experiments that cannot be repeated, or where we have no information about the likelihood of possible outcomes. The launch of a new product, for instance, can only happen once. Nevertheless, companies attempt to estimate the probability that the new product will be a commercial success or that it will be a failure. Obviously, it would be unrealistic to blindly apply the classical definition of Laplace and state that the probability of a commercial success is 1/2, simply because there are only two outcomes (success or failure).

The so-called **axiomatic definition of probability** does not assume that the random experiment under investigation can be repeated or that the likelihood of the elementary outcomes is known. It is quite different from the previous two definitions:

Definition 4.2.1 *A function P() is a **probability function** if it assigns a real number to each event G, and the following conditions are met:*

- *Axiom 1: $P(G) \geq 0$,*

- *Axiom 2: $P(\Omega) = 1$,*

- *Axiom 3: If $G_1, G_2, G_3, \ldots$ are pairwise disjoint events, then $P(G_1 \cup G_2 \cup G_3 \cup \ldots) = \sum_i P(G_i)$.*

---

[1] Pierre-Simon Laplace (1749–1827) began university studies in theology at Caen, but soon after moved to Paris where he published influential mathematical articles at the age of 19. He is not only a founder of probability theory, but he also made important contributions to the fields of differential equations, astronomy, and physics.

This general, abstract definition forms the basis for the derivation of calculation rules for probabilities. A disadvantage of the definition is that it does not contain any information on how initial probabilities can be determined.

Example 4.2.2 *Table 4.2 contains the relative frequencies of the delivery time of letters. The meaning of the relative frequencies is clear: 40% of all letters are First Class letters that arrive after one day, 18% of the letters are Second Class letters that arrive after one day, and so on. If we perform an experiment consisting of randomly selecting one specific letter sent via the Royal Mail, the eight relative frequencies could serve as estimates of the probabilities of the eight possible outcomes. The probability that the random letter that we draw is a First Class letter that arrives after one day would be estimated to be 0.40. This shows how relative frequencies, calculated based on sample data, can be used to estimate (unknown) probabilities.*

Table 4.2   Relative frequencies used as an approximation for probabilities in a random experiment concerning the Royal Mail.

		\multicolumn{4}{c}{Y}			
		1	2	3	> 3
X	First Class	0.400	0.060	0.035	0.005
	Second Class	0.180	0.225	0.085	0.010

Example 4.2.3 *Applying the classical probability definition of Laplace is not always easy. Suppose that we want to calculate the probability that flipping two (fair) coins yields two heads. Some will argue that this experiment has three possible outcomes: two heads, two tails, and one head and one tail. One of these outcomes is the one we are interested in, so that the desired probability is*

$$P(2\ heads) = \frac{number\ of\ good\ outcomes}{total\ number\ of\ outcomes} = \frac{1}{3}.$$

Others, however, will argue that there are four possible outcomes: two heads, two tails, a head with the first coin and a tail with the second, and a tail with the first coin and a head with the second. Then, the desired probability is

$$P(2\ heads) = \frac{number\ of\ good\ outcomes}{total\ number\ of\ outcomes} = \frac{1}{4}.$$

In order to ensure that the final reasoning is the only correct one, it is useful to label the coins or give them a color. You can also check for yourself, by flipping two coins a large number of times, that the relative frequency of the event of obtaining two heads is closer to 1/4 than to 1/3.

4.3   Calculation rules

The axioms given in the axiomatic definition of probability allow us to derive some rules for calculating probabilities by means of a few simple rules from set theory:

- $P(\varnothing) = 0$;
- If $G_1 \subseteq G_2$, then $P(G_2 \backslash G_1) = P(G_2) - P(G_1)$;
- If $G_1 \subseteq G_2$, then $P(G_1) \leq P(G_2)$;
- For any event G, the probability $P(G)$ is at least zero and at most one: $0 \leq P(G) \leq 1$;
- $P(G) + P(G^c) = 1$;
- the summation rule: $P(G_1 \cup G_2) = P(G_1) + P(G_2) - P(G_1 \cap G_2)$;
- the generalized summation rule:

$$
\begin{aligned}
P(G_1 \cup G_2 \cup G_3) = \ & P(G_1) + P(G_2) + P(G_3) \\
& - P(G_1 \cap G_2) - P(G_1 \cap G_3) - P(G_2 \cap G_3) \\
& + P(G_1 \cap G_2 \cap G_3).
\end{aligned}
$$

Example 4.3.1 *Consider an experiment in which a single card is randomly drawn from a classic 52-card deck (13 clubs, 13 spades, 13 diamonds, and 13 hearts), and the following events:*

- G_1: *the card drawn is black (clubs or spades),*
- G_2: *the card drawn is a spade,*
- G_3: *the card drawn is a seven.*

With the help of the classic definition of probability, it is easy to figure out that $P(G_1) = 26/52 = 1/2$, $P(G_2) = 13/52 = 1/4$ and $P(G_3) = 4/52 = 1/13$. We can also calculate various other probabilities:

- $G_4 = G_1 \cup G_2$: *the card drawn is black or spades (or both). $P(G_4) = 26/52 = 1/2$.*
- $G_5 = G_1 \cup G_3$: *the card drawn is black or a seven (or both). $P(G_5) = 28/52 = 7/13$.*
- $G_6 = G_2 \cup G_3$: *the card drawn is a spade or a seven (or both). $P(G_6) = 16/52 = 4/13$.*
- $G_7 = G_1 \cup G_2 \cup G_3$: *the card drawn is black or a spade or a seven. $P(G_7) = 28/52 = 7/13$.*

- $G_8 = G_1 \cap G_2$: the card drawn is black and a spade. $P(G_8) = 13/52 = 1/4$.
- $G_9 = G_1 \cap G_3$: the card drawn is black and a seven. $P(G_9) = 2/52 = 1/26$.
- $G_{10} = G_2 \cap G_3$: the card drawn is spade seven. $P(G_{10}) = 1/52$.
- $G_{11} = G_1 \cap G_2 \cap G_3$: the card drawn is spade seven. $P(G_{11}) = 1/52$.
- $G_{12} = G_1 \backslash G_2$: the card drawn is black but no spade. $P(G_{12}) = 13/52 = 1/4$.
- $G_{13} = G_2^c$: the card drawn is not a spade. $P(G_{13}) = 39/52 = 3/4$.

These examples can be used to verify the calculation rules. For example, the generalized summation rule can be verified by means of event G_7: $P(G_7) = 26/52 + 13/52 + 4/52 - 13/52 - 2/52 - 1/52 + 1/52$.

4.4   Conditional probability

It is often necessary to calculate the probability of an event G_1, given that another event G_2 already occurred. The event G_2 provides us with additional information that may have an impact on the probability of event G_1.

Insurance companies are interested in the probability of death (event G_1) of an insured person. This probability depends on the age of the insured person (event $= G_2$), their gender, smoking habits, and so on. The random drawing of the nine of spades ($= G_1$) from a deck of 52 cards has a probability of 1/52. If, however, you know that a black card has been drawn ($= G_2$), this rises the probability that it is the nine of spades to 1/26 because there are only 26 black cards. Finally, it is clear from Table 4.2 that the probability that a letter is delivered within one day ($= G_1$) given that it has a First Class stamp ($= G_2$) is different from the probability of a delivery within one day in the absence of information concerning the type of stamp.

Example 4.4.1 *Consider a small consultancy firm that had to complete 10 projects during the last year. Five of the projects have been carried out by Person X. Only 4 of the 10 projects were finished on time. This information is shown in Figure 4.1 by means of a Venn diagram. Each dot in the figure represents a completed project. The large oval represents the set of all projects, while the smaller ovals represents the set of projects that were finished on time and the set of projects carried out by Person X. The intersection of the two latter sets contains two dots, indicating that two of the projects done by Person X were finished on time.*

First, we calculate the probability that a project carried out by Person X was finished on time. Figure 4.1 shows that there are two projects that were completed by Person X and were finished on time. In total, Person X worked on five projects. The probability that a project was ready on time, given that Person X completed it, is therefore equal to 2/5. This follows from the classical definition of probability.

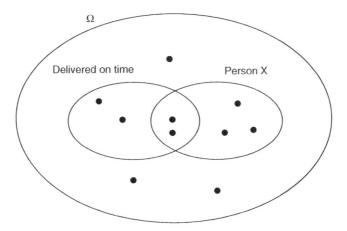

Figure 4.1   Venn diagram relevant for Example 4.4.1.

In Example 4.4.1, we have implicitly used the following definition:

Definition 4.4.1 *The **conditional probability** of an event G_1, given that event G_2 has occurred, is defined as*

$$P(G_1 \mid G_2) = \frac{P(G_1 \cap G_2)}{P(G_2)}.$$

In the example, event G_1 was an on-time completion of a project, while event G_2 was the fact that Person X worked on the project. With this notation, $P(G_1 \cap G_2) = 2/10$ and $P(G_2) = 5/10$, so that $P(G_1 \mid G_2) = (2/10)/(5/10) = 2/5$.

The conditional probability $P(G_1 \mid G_2)$ is also called an ***a posteriori* probability**, in contrast to $P(G_1)$, which is called the **unconditional** or ***a priori* probability**. It is as if the *a priori* probability is being recalculated when additional information becomes available. The result of the recalculation is the *a posteriori* probability.

Rewriting the definition of conditional probability yields the so-called **multiplication rule**:

$$P(G_1 \cap G_2) = P(G_1 \mid G_2)P(G_2).$$

This rule can be generalized to more than two events:

$$P(G_1 \cap G_2 \cap G_3) = P(G_3 \mid G_1 \cap G_2)P(G_1 \cap G_2),$$
$$= P(G_3 \mid G_1 \cap G_2)P(G_2 \mid G_1)P(G_1),$$

and

$$P(G_1 \cap G_2 \cap \cdots \cap G_k)$$
$$= P(G_k \mid G_1 \cap G_2 \cap \cdots \cap G_{k-1}) \ldots P(G_3 \mid G_1 \cap G_2)P(G_2 \mid G_1)P(G_1).$$

Example 4.4.2 *Consider again the data in Table 4.2. A user organization wants to know the probability of a delivery time of at least 2 days for First Class letters. Before calculating this conditional probability, we start with the unconditional probability of a delivery time of more than one day:*

$P(\text{"delivery} > 1 \text{ day"})$

$= P\{(\text{"}X = \text{First Class"} \cap \text{"}Y = 2\text{"}) \cup (\text{"}X = \text{First Class"} \cap \text{"}Y = 3\text{"})$

$\quad \cup (\text{"}X = \text{First Class"} \cap \text{"}Y > 3\text{"}) \cup \ldots \cup (\text{"}X = \text{Second Class"} \cap \text{"}Y > 3\text{"})\},$

$= P(\text{"}X = \text{First Class"} \cap \text{"}Y = 2\text{"}) + P(\text{"}X = \text{First Class"} \cap \text{"}Y = 3\text{"})$

$\quad + P(\text{"}X = \text{First Class"} \cap \text{"}Y > 3\text{"}) + \ldots + P(\text{"}X = \text{Second Class"} \cap \text{"}Y > 3\text{"}),$

$= 0.06 + 0.035 + 0.005 + 0.225 + 0.085 + 0.01,$

$= 0.42.$

The unconditional probability for a randomly selected letter to be a First Class letter is

$$P(\text{First Class}) = 0.4 + 0.06 + 0.035 + 0.005 = 0.5.$$

The probability that the events "delivery > 1 day" and First Class letter occur together is

$$P(\text{"delivery} > 1 \text{ day"} \cap \text{First Class letter}) = 0.06 + 0.035 + 0.005 = 0.1.$$

Therefore, the conditional probability that the delivery time is at least two days, given the knowledge that we deal with a First Class letter, is

$$P(\text{"delivery} > 1 \text{ day"} \mid \text{First Class}) = \frac{P(\text{"delivery} > 1 \text{ day"} \cap \text{First Class})}{P(\text{First Class})},$$

$$= \frac{0.1}{0.5} = 0.2.$$

The interpretation of this result is that 20% of all First Class letters are delivered late.

Example 4.4.3 *You belong to a group of 15 people and you would like to calculate the probability that at least two group members have the same birthday. For convenience, you ignore the existence of leap years and assume the probability that a person is born on a particular day to be 1/365.*

  The probability that at least two people have the same birthday is difficult to calculate. The probability of the complement G^c of this event G is, however, easier to find. The probability $P(G^c)$ is the probability that everyone has a different birthday. To calculate this probability we define the events

- G_1: *the first person is born on any day of the year;*
- G_2: *the second person is born on another day than the first person;*
- G_i: *the i-th person is born on another day than the first $i-1$ persons (i = 3, 4, \ldots, 15).*

Then, the probability that everyone has a different birthday is

$$P(G^c) = P(G_1 \cap G_2 \cap \cdots \cap G_{15}),$$
$$= P(G_1)P(G_2 \mid G_1)P(G_3 \mid G_1 \cap G_2) \ldots P(G_{15} \mid G_1 \cap G_2 \cap \cdots \cap G_{14}),$$
$$= 1 \times \frac{364}{365} \times \frac{363}{365} \times \frac{362}{365} \times \frac{361}{365} \times \cdots \times \frac{352}{365} \times \frac{351}{365} = 0.747.$$

As a result, the probability that at least two people are born on the same day is
$P(G) = 1 - P(G^c) = 1 - 0.747 = 0.253.$

It is a good exercise to determine how large the group has to be so that the probability that two individuals have the same birthday is greater than or equal to 0.5.

Example 4.4.4 *Out of a quartet of two men and two women, we want to randomly choose a committee of two people. What is the probability that the committee eventually involves two persons of the same gender?*

We have two people of the same gender in the committee if both selected members are male (event A) or female (event B). These events are mutually exclusive, so that

$$P(A \cup B) = P(A) + P(B).$$

In order to calculate the probability that both committee members are males, we define the events

- G_1: *the first person in the committee is a man;*

- G_2: *the second person in the committee is a man.*

We then have that

$$P(A) = P(G_1 \cap G_2) = P(G_1)P(G_2 \mid G_1).$$

Here, $P(G_1)$ is the probability that the first randomly drawn committee member is a man. Applying the classical probability definition yields a value of $2/4 = 1/2$ for this probability. The probability $P(G_2 \mid G_1)$ is the probability that the second committee member is also a man, and it is $1/3$. Again, this can be verified with the classical definition of probability. Combining these two results produces a probability of

$$P(A) = \frac{1}{2} \times \frac{1}{3} = \frac{1}{6}$$

for an entirely male committee. In the same way, we can calculate that the probability of a completely female committee is $P(B) = 1/6$. Hence, the probability that the committee is composed of two people of the same gender is

$$P(A \cup B) = \frac{1}{6} + \frac{1}{6} = \frac{1}{3}.$$

Example 4.4.5 *In this example, we calculate the probability that the six numbers indicated on a lottery ticket are those actually drawn on a certain Wednesday or Saturday (lottery days in many European countries) out of a drum with 42 balls numbered from 1 to 42. The six balls drawn from the drum are not replaced.*

To compute the required probability, we first define the events G_i as "the i-th number that is drawn matches a number indicated on the lottery ticket". The probability of winning the first prize, that is that the six numbers drawn are identical to those on the lottery ticket, is

$P(\text{first prize}) = P(\text{the first number drawn is one of the six on the ticket}$

$\text{AND the second number drawn is one of the six on the ticket}$

$\ldots$

$\text{AND the sixth number drawn is one of the six on the ticket}),$

$= P(G_1 \cap G_2 \cap \cdots \cap G_6),$

$= P(G_1)P(G_2 \mid G_1)P(G_3 \mid G_1 \cap G_2) \ldots P(G_6 \mid G_1 \cap G_2 \cap \ldots \cap G_5),$

$= \dfrac{6}{42} \times \dfrac{5}{41} \times \dfrac{4}{40} \times \dfrac{3}{39} \times \dfrac{2}{38} \times \dfrac{1}{37} = \dfrac{1}{5{,}245{,}786}.$

The probability that the first number drawn is indicated on the ticket, $P(G_1)$, is 6/42, because there are 42 numbers in total and 6 of them are good ones. The probability that the second number drawn is also indicated on the ticket, $P(G_2 \mid G_1)$, is 5/41, because 41 balls are in the drum when drawing the second number. Five of these are indicated on the ticket. In the same way, we can determine the remaining conditional probabilities.

Example 4.4.6 *On Wednesdays and Saturdays, in Belgium, you can not only participate in the traditional lottery game, but you can also play the Joker game. Players of this game have a form with six ordered digits ranging from 0 to 9. In every game, six numbers are drawn from a drum involving 10 balls numbered from 0 to 9. In the Joker game, each time a ball has been drawn, it is placed back into the drum. So, in contrast with the traditional lottery, the Joker game is a draw with replacement. To win the first prize, the six numbers on your form have to appear in the same order as that in which they were drawn.*

To compute the probability of winning the first prize, we first define the events G_i as "the i-th number drawn corresponds to the i-th number on our form". The probability of winning the first prize then is

$P(\text{first prize}) = P(\text{the first number drawn is the first on the form}$

$\text{AND the second number drawn is the second on the form}$

$\ldots$

$\text{AND the sixth number drawn is the sixth on the form}),$

$= P(G_1 \cap G_2 \cap \ldots \cap G_6),$

$= P(G_1)P(G_2 \mid G_1)P(G_3 \mid G_1 \cap G_2) \ldots P(G_6 \mid G_1 \cap G_2 \cap \ldots \cap G_5).$

The probability that the first number drawn is the first on the form, $P(G_1)$, is obviously 1/10. The probability that the second number drawn is the second on the form, given the fact that the first ball was already correct, $P(G_2 \mid G_1)$, is again 1/10. The reason is that, at the moment the second number is drawn, all 10 balls are back in the drum. This is due to the fact that every ball drawn is placed back into the drum before the next ball is drawn. Hence, for each following number, we also have a chance of 1/10 that it coincides with our number, and

$$P(\textit{first prize}) = \frac{1}{10} \times \frac{1}{10} \times \frac{1}{10} \times \frac{1}{10} \times \frac{1}{10} \times \frac{1}{10} = \frac{1}{10^6}.$$

An essential difference between the calculations for the lottery and those for the Joker game is that, in the latter, all conditional probabilities are the same. This is due to the fact that, in the Joker game, the events G_1, G_2, … are independent, because each ball drawn is placed back in the drum before the next drawing.

4.5 Independent and dependent events

Definition 4.5.1 *Event G_1 is **independent** of event G_2 if the occurrence of event G_2 has no impact on the probability that event G_1 occurs:*

$$P(G_1) = P(G_1 \mid G_2).$$

In case event G_1 is independent of event G_2, the information contained in G_2 does not change the probability for G_1. For independent events, the probability $P(G_1 \cap G_2)$ therefore simplifies:

$$P(G_1 \cap G_2) = P(G_1 \mid G_2)P(G_2) = P(G_1)P(G_2).$$

Moreover,

$$P(G_2 \mid G_1) = \frac{P(G_2 \cap G_1)}{P(G_1)} = \frac{P(G_2)P(G_1)}{P(G_1)} = P(G_2).$$

Therefore, if G_1 is independent of G_2, then G_2 is also independent of G_1. The events $G_1, G_2, \ldots, G_k$ are (mutually) independent if, for any subset of these k events, the following equality holds:

$$P(G_{i_1} \cap G_{i_2} \cap \cdots \cap G_{i_b}) = P(G_{i_1})P(G_{i_2}) \ldots P(G_{i_b}).$$

If a set of events is pairwise independent, this does not imply that they are independent when considered all together. The following example illustrates this.

Example 4.5.1 *You throw one red and one blue die. Define the following events:*

- *G_1: the number of dots on the red die is even, $P(G_1) = 1/2$;*
- *G_2: the number of dots on the blue die is even, $P(G_2) = 1/2$;*
- *G_3: the sum of dots on the two dice is even, $P(G_3) = 1/2$.*

Applying the classical definition of probability, it can be seen that

- $P(G_1 \cap G_2) = 9/36 = 1/4 = P(G_1)P(G_2)$;
- $P(G_1 \cap G_3) = 9/36 = 1/4 = P(G_1)P(G_3)$;
- $P(G_2 \cap G_3) = 9/36 = 1/4 = P(G_2)P(G_3)$.

Thus, the three events are pairwise independent. Nevertheless,

$$P(G_1 \cap G_2 \cap G_3) = 9/36 = 1/4 \neq P(G_1)P(G_2)P(G_3) = 1/8,$$

so that the three events, when considered together, are not independent.

Finally, it should be noted that independent events should not be confused with mutually exclusive events. For mutually exclusive events A and B, we have $P(A \cap B) = P(\varnothing) = 0$.

Example 4.5.2 *For manufacturing companies, it is essential that the quality of their products does not depend on (in other words, is independent of) the location where the production took place, the shift that assembled the products, the production line, and so on. Consider the data in Table 4.3.*

Table 4.3   Data for Example 4.5.2.

		Quality		
		excellent (E)	acceptable (A)	bad (B)
Production	P1	0.40	0.20	0.20
line	P2	0.10	0.05	0.05

The figures in the table indicate that 80% of the production is done on line P1, and only 20% on line P2. Half of the quantity produced is excellent (E). A quarter of the production has an acceptable (A) quality, and another quarter is of bad (B) quality.

The independence of the production line and the quality follows from the following calculations:

$$P(P1) = 0.80,$$

$$P(P1 \mid E) = 0.40/0.50 = 0.80,$$

$$P(P1 \mid A) = 0.20/0.25 = 0.80,$$

and

$$P(P1 \mid B) = 0.20/0.25 = 0.80,$$

and

$$P(P2) = 0.2,$$

$$P(P2 \mid E) = 0.10/0.50 = 0.20,$$

$$P(P2 \mid A) = 0.05/0.25 = 0.20,$$

and

$$P(P2 \mid B) = 0.05/0.25 = 0.20.$$

Similar calculations can be made for $P(E \mid P1)$, $P(A \mid P1)$, ...

Independence of the two properties implies that the distribution across the two production lines is the same for excellent, acceptable and bad products (here, 80–20%). Conversely, the distribution of the three qualities does not change from one production line to the other.

Example 4.5.3 *A safety system consists of two subsystems A and B, which are serially connected with each other. Subsystem A consist of two components A_1 and A_2, which operate in parallel. The safety system functions properly as long as the two subsystems A and B work. Subsystem A functions as long as one of the two components works.*

The system is presented graphically in Figure 4.2. For each component in the figure, the reliability is specified, that is the probability that the component will work the entire year. We assume for simplicity that the functioning of one component does not affect the service life of any other component. To quantify the reliability of the entire security system, we calculate the probability that the system fails in the coming year.

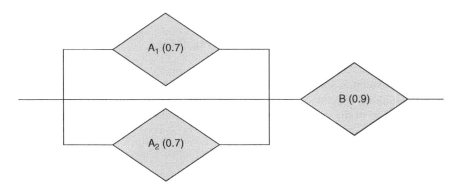

Figure 4.2   Simple safety system involving components A_1, A_2, and B, along with an indication of their reliabilities.

First, we define the following events:

- A_1: *component A_1 fails;*
- A_2: *component A_2 fails;*
- A: *component A fails;*
- B: *component B fails;*
- S: *The entire system fails.*

Since the system fails if either of the components A or B stops working, we have that

$$P(S) = P(A \cup B) = P(A) + P(B) - P(A \cap B),$$

where

$$P(A) = P(A_1 \cap A_2) = P(A_1)P(A_2) = (0.3)(0.3) = 0.09,$$

and

$$P(A \cap B) = P(A)P(B) = (0.09)(0.1) = 0.009.$$

As a result, the probability that the system fails is equal to

$$P(S) = 0.09 + 0.1 - 0.009 = 0.181.$$

4.6   Total probability and Bayes' rule

Theorem 4.6.1 *Let G_0 be any event and $G_1, G_2, \ldots, G_k$ a partition of the sample space Ω. Then, the probability of the event G_0 can be calculated as*

$$P(G_0) = \sum_{i=1}^{k} P(G_0 \mid G_i)P(G_i).$$

This proposition is called the **theorem of total probability**. The proof of the theorem builds on the concept of a partition. First, due to the fact that $G_1, G_2, \ldots, G_k$ form a partition of the sample space Ω, their union is the sample space. This allows us to write that

$$G_0 = \Omega \cap G_0,$$
$$= (G_1 \cup G_2 \cup \cdots \cup G_k) \cap G_0,$$
$$= (G_1 \cap G_0) \cup (G_2 \cap G_0) \cup \cdots \cup (G_k \cap G_0).$$

Since, for each i and j, with $i \neq j$, the sets $(G_i \cap G_0)$ and $(G_j \cap G_0)$ are disjoint (due to the fact that $G_1, G_2, \ldots, G_k$ form a partition), we have that

$$P(G_0) = P\{(G_1 \cap G_0) \cup (G_2 \cap G_0) \cup \cdots \cup (G_k \cap G_0)\},$$

$$= \sum_{i=1}^{k} P(G_i \cap G_0),$$

$$= \sum_{i=1}^{k} P(G_0 \mid G_i)P(G_i).$$

Example 4.6.1 *In order to meet the increasing demand for decaffeinated coffee, three production lines are required. Production line P1 produces half of the coffee packages. On average, 2% of those packages exhibit leakage. Production line P2 contributes 30% to the production and has 3% of poor output. Finally, production line P3 produces the remaining 20% of coffee packages. This line has the worst performance, as it delivers 6% of leaky packages. We calculate the probability that a package that is randomly selected from the warehouse has a leakage.*
  We start again by defining some events:

- *P1: the selected package comes from production line P1;*

- *P2: the selected package comes from production line P2;*

- *P3: the selected package comes from production line P3;*

- *L: the selected package has a leakage.*

Since the events P1, P2, and P3 form a partition, the theorem of total probability can be used:

$$P(L) = P(L \mid P1)P(P1) + P(L \mid P2)P(P2) + P(L \mid P3)P(P3),$$

$$= (0.02)(0.5) + (0.03)(0.3) + (0.06)(0.2),$$

$$= 0.031.$$

Consequently, 3.1% of the total production is defective.

Problems using the theorem of total probability or Bayes' rule (see next) are often represented graphically by means of so-called **probability trees**. For instance, Figure 4.3 shows the probability tree for the problem in Example 4.6.1. It is a useful exercise to draw probability trees for the next few examples in this section.

Example 4.6.2 *When the tennis player Kim Clijsters (a former no. 1 ranked tennis player) planned her comeback after numerous injuries in Spring 2005, her management organized a small exhibition tournament in her hometown. The first game scheduled for Kim Clijsters served as the opening match of the tournament. Her opponent in that game was youngster Kirsten Flipkens. In the second game, Sharapova*

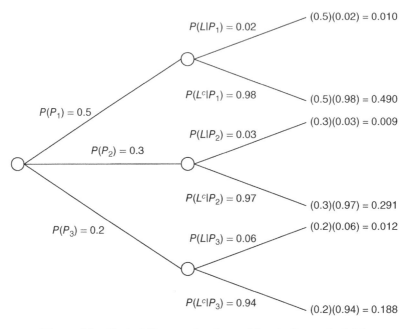

Figure 4.3   Probability tree for the problem in Example 4.6.1.

and Kuznetsova would play against each other. The winners of these matches would play the final on the next day.

Kim Clijsters' management decided to calculate the probability that she would win the tournament. This required the estimation of the probability that each of the four participating players would beat any other player in a single match. Estimates of these probabilities are shown in Table 4.4.

Table 4.4   Probabilities for each of the participating tennis players to beat any other player in a single match.

		Winner			
		Clijsters	*Flipkens*	*Sharapova*	*Kuznetsova*
	Clijsters	–	0.1	0.4	0.3
Loser	*Flipkens*	0.9	–	0.95	0.9
	Sharapova	0.6	0.05	–	0.4
	Kuznetsova	0.7	0.1	0.6	–

If we assume that the players recover quickly, so that the first match does not impact the probability of winning a possible second match, then the probability that Clijsters wins the tournament can be calculated as follows:

$$P(\text{Clijsters wins tournament}) = P(\text{Clijsters beats Flipkens AND wins her second match}),$$

$$= P(\text{Clijsters beats Flipkens})P(\text{Clijsters wins her second match}),$$

$$= 0.9\ P(\text{Clijsters wins her second match}).$$

The probability that Clijsters wins her second match needs to be calculated using the theorem of total probability since her opponent in the second match is unknown prior to the tournament:

$P(\text{Clijsters wins her second match})$

$= P(\text{Clijsters beats Sharapova} \mid \text{Sharapova reaches the final}) \times P(\text{Sharapova reaches the final})$

$+ P(\text{Clijsters beats Kuznetsova} \mid \text{Kuznetsova reaches the final}) \times P(\text{Kuznetsova reaches the final}),$

$= P(\text{Clijsters beats Sharapova} \mid \text{Sharapova beats Kuznetsova}) \times P(\text{Sharapova beats Kuznetsova})$

$+ P(\text{Clijsters beats Kuznetsova} \mid \text{Kuznetsova beats Sharapova}) \times P(\text{Kuznetsova beats Sharapova}),$

$= (0.6)(0.6) + (0.7)(0.4),$

$= 0.64.$

As a result,

$$P(\text{Clijsters wins tournament}) = (0.9)(0.64) = 0.576.$$

The probabilities for the other players to win the tournament can be calculated in a similar fashion. The success probabilities for the four players add up to 1.

Theorem 4.6.2 *Let G_0 be any event and $G_1, G_2, \ldots, G_k$ a partition of the sample space Ω. Then,*

$$P(G_j \mid G_0) = \frac{P(G_0 \mid G_j)P(G_j)}{\sum_{i=1}^{k} P(G_0 \mid G_i)P(G_i)}.$$

The proof of this theorem, which is best known as **Bayes' rule**[2], is very easy:

$$P(G_j \mid G_0) = \frac{P(G_j \cap G_0)}{P(G_0)} = \frac{P(G_0 \mid G_j)P(G_j)}{\sum_{i=1}^{k} P(G_0 \mid G_i)P(G_i)}.$$

---

[2] Thomas Bayes (1702–1761) was a British mathematician and Presbyterian priest who formulated a special case of the Bayes' rule. The interesting result was published posthumously by Bayes' friend Richard Price and was a response to a question from Abraham de Moivre, a famous French mathematician.

Example 4.6.3 *Consider again the situation described in Example 4.6.1. We now calculate the probability that a randomly chosen defective coffee package was produced on production line P1. To do this, we use Bayes' rule:*

$$P(P1 \mid L) = \frac{P(L \mid P1)P(P1)}{P(L \mid P1)P(P1) + P(L \mid P2)P(P2) + P(L \mid P3)P(P3)},$$

$$= \frac{(0.02)(0.5)}{(0.02)(0.5) + (0.03)(0.3) + (0.06)(0.2)},$$

$$= 0.3226.$$

The probability that a random defective product comes from line P2 is 0.2903. For line P3, the probability is 0.3871. These probabilities can be calculated in a similar manner. Note that the three probabilities add up to 1.

Example 4.6.4 *In order to counteract the spread of AIDS, a country's Department of Public Health might consider a compulsory AIDS test for couples who intend to marry. Suppose that the sensitivity of the available AIDS test is 98%, that is, the probability that the AIDS test is positive for someone who is infected is 98%. Moreover, assume that the specificity of the test is 95%, which means that someone who is not infected with the virus has a 95% chance of a negative test and therefore a 5% chance of a positive outcome.*

Now, suppose you spontaneously undergo an AIDS test before you become active on the "wedding market" and the test result indicates that you are infected with the AIDS virus. What is the probability that you are indeed carrier of the virus?

To calculate this probability, we use Bayes' rule:

$$P(aids \mid positive\ test)$$

$$= \frac{P(positive\ test \mid aids)P(aids)}{P(positive\ test \mid aids)P(aids) + P(positive\ test \mid no\ aids)P(no\ aids)},$$

$$= \frac{(0.98)P(aids)}{(0.98)P(aids) + (0.05)P(no\ aids)}.$$

To finalize the calculation, we still need to estimate the probability that you have AIDS, before knowing your test result. This probability, P(aids), is called the a priori probability . You can use the percentage of the total population that is infected as an estimate for the probability that you are infected. Suppose that 0.1% of the population is infected with the virus, so that P(aids) = 0.001, and, of course, P(no aids) = 0.999. Then,

$$P(aids \mid positive\ test) = \frac{(0.98)(0.001)}{(0.98)(0.001) + (0.05)(0.999)},$$

$$= 0.0192.$$

This probability is the a posteriori probability of having AIDS, after the positive test result becomes known. This probability is quite small, indicating that, even though there is an undesirable test result, you should not worry too much about your health.

Now, suppose that you are a drug user, who is not afraid of using someone else's syringe and having unsafe sex. In that case, using the a priori probability P(aids) = 0.001 in your calculations would be incorrect. Suppose that, in the environment in which you live, 0.25 is a more realistic estimate of the a priori probability of being infected, P(aids). Then,

$$P(aids \mid positive\ test) = \frac{(0.98)(0.25)}{(0.98)(0.25) + (0.05)(0.75)},$$

$$= 0.8673.$$

Clearly, in this second scenario, you have much more reason to be worried about the positive test result.

Should the Department of Public Health now require an AIDS test for everyone who wants to get married? To answer this question, we can start from a fictitious group of 100,000 tested people. Only 0.1% of them, 100 people, are carriers of the AIDS virus. In 2% of the cases, the test is falsely negative, so that we can expect that two carriers of the virus will get a wrong diagnosis. 99,900 of the 100,000 tested people do not carry the virus, but on average 5% of them, 4995 people, will get a falsely positive result. As a result, a substantial number of healthy people will unnecessarily be upset and need to undergo costly additional medical exams. These calculations indicate that making the AIDS test mandatory for all engaged couples is not advisable. The reason is that, overall, AIDS is a fairly rare disease, at least in many countries.

4.7   Simulating random experiments

In many practical situations, uncertainty and variability play an important role. Since many practical situations are more complex than most of the computed examples in books on probability theory, it is often difficult to calculate probabilities for success, defect rates or other problems analytically.

Example 4.7.1   *In the different stages of a construction project, resources such as workers, cranes, concrete mixers, and so on are needed. Once a particular phase of a project has been completed, the tools that are no longer needed can be moved to another site. When planning projects, entrepreneurs have to assess the availability of all tools properly, not to be confronted with a shortage of resources. Entrepreneurs therefore must have a clear view of the duration of each phase of a project.*

However, the durations of project phases are stochastic. As a matter of fact, the precise duration of a phase depends on all sorts of random factors such as weather, the punctuality of suppliers, and illnesses of employees. The variability in the duration complicates the task of entrepreneurs who would like to draw up a solid planning and estimate the completion dates of projects accurately. A major difficulty is to avoid shortages of resources at any point in time.

When submitting bids, entrepreneurs could try to be on the safe side by using large estimates of the duration of the various phases of the project. In that case, however, they risk that other entrepreneurs, promising a faster finishing time, will get the projects.

In such situations, the use of simulation models (for instance, to study the distribution of completion dates of projects as a function of random events such as the weather, illnesses, etc.) often makes sense. The purpose of these models is to simulate reality, and to test different alternatives using those models. A detailed treatment of this topic is beyond the scope of this book, but simulations will be used occasionally.

To simulate probabilities for simulation experiments, a computer has to generate **random numbers**. True random numbers cannot be generated by a computer. Therefore, computer-generated random numbers are usually called pseudo-random numbers. This is due to the fact that a computer can only perform user-specified tasks. However, a lot of research has been done to find good ways to generate pseudo-random numbers with a computer. Nowadays, many computer packages allow the generation of high-quality pseudo-random numbers.

In Section 6.6, we address how to generate pseudo-random numbers with JMP.

5

Additional aspects
of probability theory

Stops were rarely used alone. They tended to be piled on top of each other in combinations that were designed to take advantage of the available harmonics (...). Certain combinations in particular were used over and over again. ... The organ included an ingenious mechanism called the preset, which enables the organist to select a particular combination of stops – stops he himself had chosen – instantly.

(from *Cryptonomicon*, Neal Stephenson, pp. 8–9)

This chapter focuses on some interesting additional aspects of probability. We start with some aspects of combinatorics. Next, we turn our attention to applications of probability theory.

5.1 Combinatorics

In order to apply the classical definition of probability by Laplace, we need to know the number of elementary outcomes of a random experiment. For a simple throw of a die, this does not require much effort: there are six possible outcomes. For the Belgian lottery (see Example 4.4.5) or in the Joker Game (see Example 4.4.6), it is a lot harder to count the number of possible outcomes. In this section, we provide an overview of the main counting rules.

5.1.1 Addition rule

If a first event G_1 can occur in N_1 different ways and a second event G_2 in N_2 different ways, and the two events are mutually exclusive, then the event $G_1 \cup G_2$ can occur in $N_1 + N_2$ different ways.

Statistics with JMP: Graphs, Descriptive Statistics, and Probability, First Edition. Peter Goos and David Meintrup.
© 2015 John Wiley & Sons, Ltd. Published 2015 by John Wiley & Sons, Ltd.
Companion Website: wiley.com/go/goosandmeintrup

Example 5.1.1 *Event G_1, drawing a heart from a deck of 52 cards, can occur in 13 ways. The same applies to G_2, the event of drawing a spade. Consequently, event $G_1 \cup G_2$, drawing a heart or a spade, can occur in $13 + 13 = 26$ ways.*

The addition rule can be generalized to more than two events.

5.1.2 Multiplication principle

If an event G_1 of one experiment contains N_1 elementary outcomes, and an event G_2 of another experiment contains N_2 elementary outcomes, then the combined event $G_1 \cap G_2$ can occur in $N_1 N_2$ different ways.

Example 5.1.2 *A combined experiment consists of flipping a coin and throwing a die. As the first experiment has two possible outcomes and the second one has six possible outcomes, the number of possible outcomes of the combined experiment equals 12. The sample space of the combined experiment contains the elements (head, 1), (head, 2), ..., (head, 6), (tail, 1), ..., (tail, 6).*

Example 5.1.3 *Another combined experiment consists of throwing one red and one blue die. This experiment has 36 possible outcomes:*

				Blue			
		1	2	3	4	5	6
	1	(1,1)	(1,2)	(1,3)	(1,4)	(1,5)	(1,6)
	2	(2,1)	(2,2)	(2,3)	(2,4)	(2,5)	(2,6)
Red	3	(3,1)	(3,2)	(3,3)	(3,4)	(3,5)	(3,6)
	4	(4,1)	(4,2)	(4,3)	(4,4)	(4,5)	(4,6)
	5	(5,1)	(5,2)	(5,3)	(5,4)	(5,5)	(5,6)
	6	(6,1)	(6,2)	(6,3)	(6,4)	(6,5)	(6,6)

The result remains unchanged when we throw the same die twice, rather than one red and one blue die. In the above table, this requires replacing "Blue" by "First throw" and "Red" by "Second throw".

Example 5.1.4 *The Joker Game in Example 4.4.6 is actually composed of six random experiments, each with 10 possible outcomes. Therefore, the number of possible outcomes of the Joker Game is 10^6.*

5.1.3 Permutations

If we have a set of n different objects, we can wonder in how many different ways this set can be ordered or permuted.

Example 5.1.5 *Three tasks a, b, and c, can be performed in six different orders: abc, acb, bac, bca, cab, and cba.*

In this example, it is possible to enumerate all possibilities. If the set of objects is bigger, a complete enumeration becomes impracticable. The following reasoning is helpful in that case.

For the first position in the ordering of n objects, we have n different possibilities. Once the first position has been filled, only $n - 1$ options remain for position 2. Once the object in position 2 has also been determined, there remain $n - 2$ options for position 3. We can continue in the same way up to the final position, for which only one object remains. Applying the multiplication principle now tells us that there are $n(n - 1)(n - 2)\ldots(2)1 = n!$ (read: n factorial) possible orders or permutations of the n objects. We denote this by

$$_nP_n = n!$$

If we have n objects, but we only need to order r of them, then the number of orderings or permutations is

$$_nP_r = n(n - 1)(n - 2)\ldots(n - r + 1) = \frac{n!}{(n - r)!}.$$

This follows from a reasoning similar to the one above.

Example 5.1.6 *In order to win a prediction contest for the Olympic 100 m running final in London, you have to predict the first three athletes in the correct order. Since there are eight athletes in the final, the number of possible predictions is $_8P_3 = 8 \times 7 \times 6 = 336$.*

5.1.4 Combinations

Example 5.1.7 *Suppose that in the prediction contest of Example 5.1.6, the order was not important. In other words, your prediction was considered correct if you were able to predict who would be the first three athletes who crossed the finish line regardless of their order. In that case, the $_3P_3 = 3! = 6$ predictions (Gatlin, Blake, Bolt), (Gatlin, Bolt, Blake), (Bolt, Gatlin, Blake), (Bolt, Blake, Gatlin), (Blake, Bolt, Gatlin), and (Blake, Gatlin, Bolt) would be correct. If the order is not important, there are only $_8P_3/3! = 336/6 = 56$ different predictions.*

The task in Example 5.1.7 was no longer to determine how many rankings of three athletes from a set of eight could be made, but to determine how many sets of three can be composed. As shown in the example, the number of possible sets is equal to the number of possible permutations of three athletes out of a set of eight, divided by the number of permutations of the three selected athletes. In general, we can calculate the number of sets with r objects chosen from a set with n objects as

$$\binom{n}{r} = \frac{n!}{(n - r)!\, r!}.$$

This number is called the number of combinations of r objects out of n.

Example 5.1.8 *The indoor amateur football team The Monkeys plays against FC The Magicians. The team The Monkeys has nine team members, while its opponent FC The Magicians team has 11 members. How many different combinations of 10 players, five on each side, can be on the pitch at the start of the match? The Monkeys can start the match with $\binom{9}{5} = 126$ different teams, while FC The Magicians has $\binom{11}{5} = 462$ options. Applying the multiplication principle, we find that there are $126 \times 462 = 58212$ possible combinations of players to start the game.*

Example 5.1.9 *In Example 4.4.5, we calculated the probability that, out of the six numbers indicated on a Belgian lottery ticket, all six are good. What is the probability that, from those six numbers, only four are actually drawn, if you know that the six numbers are drawn from the set $\{1, 2, \ldots, 42\}$. The total number of possible combinations of six different numbers from 42 is $\binom{42}{6}$. How many of these combinations contain four good numbers (numbers indicated on our ticket) and two bad numbers (numbers that are not indicated)?*

Of the six numbers marked on the lottery ticket, four are actually drawn. There are $\binom{6}{4} = 15$ possible ways in which this can happen. Out of the 36 numbers not indicated on the lottery ticker, two are actually drawn. There are $\binom{36}{2} = 630$ possible combinations of two out of 36 numbers. Applying the multiplication principle, we learn that there are $15 \times 630 = 9450$ possible ways in which four good and two bad numbers can be drawn.

Therefore, the probability that we indicated four good numbers and two bad ones on the lottery ticket is

$$\frac{\text{number of combinations with four good numbers}}{\text{total number of possible combinations}} = \frac{\binom{6}{4}\binom{36}{2}}{\binom{42}{6}} = \frac{15 \times 630}{5,245,786},$$

$$= \frac{9450}{5,245,786} = 0.0018.$$

5.2   Number of possible orders

In probability theory, sequences of successes and failures are of interest. The quality inspection of manufactured goods may show that some goods are defective (failure) and some other goods are not defective (success). In a penalty shoot-out after a football match that ends in a tie, there are usually a number of penalties that are converted into a goal (success) and a number of penalties that are not converted into a goal (failure). We will learn in Chapter 8 how we can compute probabilities using the binomial distribution in such scenarios. To grasp the meaning of the mathematical expression of the binomial distribution, it is essential that we can count the number of possible sequences with a certain number of successes and failures. In Section 11.6, we study the multinomial probability distribution, a generalized version of the binomial distribution, which is useful if there are more than two possible outcomes. This is relevant when, for example, we want to calculate the probability of obtaining a 1 five times, a 2 six times, and a 3 four times after 15 throws of a die. To this end, we need to know the number of sequences involving five ones, six twos, and four threes.

5.2.1   Two different objects

Example 5.2.1 *Suppose that a football team scores two goals out of five penalty kicks, and thus has three failures. In how many orders can this happen? It may be that the team starts out with three misses and ends with two goals, but it might also be that the team starts with a miss, then scores twice, and ends with two misses.*

Counting the number of possible orders, we encounter the problem that we study five objects in total (five penalties), but that these objects are not different from each other. There are only two types of objects, goals and misses. The two goals are identical objects for us, as are the three missed penalties. If the five penalty kicks were five completely different objects, the number of possible orders would be equal to the number of permutations of five different objects, 5!.

However, we do not have five different objects, but a set of two identical objects (goals) and another set of three identical objects (misses). If we could distinguish between the two goals in the first set of identical objects, this distinction would give 2! possible orders of these two penalties. However, such a distinction is impossible. As a result, we do not have 5! possible orders of our five penalties, but only 5!/2! if we consider the first set of identical objects. If we could distinguish between the three missed penalties, we would have 3! possible orders of these three penalties. This distinction is, however, also impossible, so that we do not end up with 5!/2! possible orders, but only with

$$(5!/2!)/3! = \frac{5!}{2!\,3!} = 10,$$

if we also take into account the second set of identical objects. These 10 possible orders are shown in Table 5.1.

If we order n objects, n_1 objects of type 1 (e.g., successes) and $n_2 = n - n_1$ objects of type 2 (e.g., failures), the total number of possible orders is equal to

$$\frac{n!}{n_1!\,n_2!} = \binom{n}{n_1} = \binom{n}{n_2}.$$

5.2.2   More than two different objects

If we order n objects, n_1 objects of type 1, n_2 objects of type 2, ... , and n_k objects of type k, the total number of possible orders is equal to

$$\frac{n!}{n_1!\,n_2!\,\dots\,n_k!}.$$

Table 5.1   Ten possible orders of two converted penalty kicks (S = success) and three missed penalty kicks (F = failure) in a series of five.

Order	Penalty kick				
	1	2	3	4	5
1	S	S	F	F	F
2	S	F	S	F	F
3	S	F	F	S	F
4	S	F	F	F	S
5	F	S	S	F	F
6	F	S	F	S	F
7	F	S	F	F	S
8	F	F	S	S	F
9	F	F	S	F	S
10	F	F	F	S	S

Example 5.2.2 *The number of possible orders in which we can obtain a 1 five times, a 2 six times, and a 3 four times using 15 throws of a die is equal to*

$$\frac{15!}{5!\,6!\,4!} = 630630.$$

5.3   Applications of probability theory

5.3.1   Sequences of independent random experiments

Suppose that, due to a change in the exam regulations, a student can take the same exam many times. The probability that she passes one exam is $P(\text{success}) = p$, and the probability that she is not successful is $P(\text{failure}) = 1 - p = q$. Suppose that the student does not adopt her studying intensity as time goes by, and that the teacher assessing the exam is also not guided by the number of attempts that the student already made. In other words, we assume that the successive attempts to pass the exam are independent random experiments, and that the probabilities p and q do not change. We now calculate the probability that the student will eventually pass. We call this event H. We can determine the probability of eventually passing as follows:

$P(H) = P(\text{Students succeeds at first attempt OR}$

student succeeds for the first time at attempt 2 OR

student succeeds for the first time at attempt 3 OR

…).

The probability that a student passes at the first attempt is equal to p. The probability to fail once and succeed at the second attempt is equal to qp. The probability that the student succeeds for the first time at attempt 3 is $q^2 p$, ... In other words,

$$P(H) = p + qp + q^2 p + q^3 p + \cdots = p(1 + q + q^2 + q^3 + \dots).$$

This expression is an infinite geometric series that converges if $-1 < q < 1$. This condition is fulfilled if $p > 0$. In that case, the series converges to

$$P(H) = \frac{p}{1 - q} = \frac{p}{p} = 1.$$

This result indicates that the student will pass the exam with certainty (probability 100%), regardless of the probability p. The only requirement is a minimal study effort and minimum intelligence to ensure a strictly positive p.

In a similar way it can be argued that the Earth may not be the only planet with intelligent living beings. Suppose that we include all planets in our investigation and that the probability of finding intelligent life on any planet is p. The probability that we finally are successful in our search for life in space is, of course,

$$P(\text{"life"}) = P(\text{life is found on planet 1 OR}$$

$$\text{life is found for the first time on planet 2 OR}$$

$$\text{life is found for the first time on planet 3 OR}$$

$$\dots$$

$$\text{life is found for the first time on planet } N),$$

$$= p + qp + q^2 p + q^3 p + \cdots + q^{N-1} p,$$

$$= p(1 + q + q^2 + q^3 + \cdots + q^{N-1}),$$

where N represents the number of planets in the universe. The sum $1 + q + q^2 + q^3 + \cdots + q^{N-1}$ is a finite geometric series, which is equal to

$$\frac{1 - q^N}{1 - q}.$$

The desired probability therefore is

$$P(\text{"life"}) = \frac{p(1 - q^N)}{1 - q} = \frac{p(1 - q^N)}{p} = 1 - q^N.$$

Even if we use a very small probability p, for example $p = 1/10^6$, the probability of intelligent extraterrestrial life in an investigation of 100,000 planets is already 9.5%.

5.3.2 Euromillions

Euromillions is a European lottery in which the winning combination is determined based on two separate, independent draws. First, five numbers are drawn randomly and without replacement from a total of 50. Second, two stars are drawn from a total of 11 (also without replacement). A player who, on his simple[1] lottery ticket, checked the five numbers drawn and the two stars drawn wins the largest amount of prize money.

The total number of possible draws of five numbers from a total of 50 is

$$\binom{50}{5} = \frac{50!}{(50-5)!\,5!} = \frac{50!}{(45)!\,5!} = \frac{50 \times 49 \times 48 \times 47 \times 46}{5 \times 4 \times 3 \times 2 \times 1} = 2{,}118{,}760,$$

while the number of possible draws of two stars out of a total of 11 is equal to

$$\binom{11}{2} = \frac{11!}{(11-2)!\,2!} = \frac{11!}{9!\,2!} = \frac{11 \times 10}{2 \times 1} = 55.$$

For the total number of possible draws of five numbers out of 50 and two stars out of 11, we can use the multiplication principle. In total, there are

$$\binom{50}{5} \times \binom{11}{2} = 2{,}118{,}760 \times 55 = 116{,}531{,}800$$

possible draws.

If you are participating in the Euromillions lottery with a simple ticket, then you have a chance of 1 in 116,531,800 to win the lottery, that is to have the right five numbers and the right two stars.

In the Euromillions lottery, you also win some cash if you have not marked all five numbers and both stars correctly. For example, you also win if you have marked three out of the five numbers drawn and one of the two stars drawn. To calculate the probability that this happens is a bit more difficult than to determine the probability of winning the largest cash prize. To do so, we first have to determine the number of possible tickets with three good numbers (out of five) and one good star (out of two). To start the calculations, it is important to realize that there are five good numbers (the ones that are drawn) and 45 bad ones (those that are not drawn), and that there are two good stars (the ones that are drawn) and nine bad ones (those that are not drawn).

The number of combinations of three good numbers out of a total of five is

$$\binom{5}{3} = 10.$$

Automatically, two bad numbers are drawn from the 45 undesirable ones if only three desirable ones are drawn. In total, there are

$$\binom{45}{2} = 990$$

combinations of two bad numbers out of 45. There are

$$\binom{2}{1} = 2$$

---

[1] There are forms of the game in which a player can mark more than five numbers and more than two stars. Thus, a player can increase his chances of winning against paying a greater fee. We do not investigate this additional complication.

Table 5.2   Probabilities of success for the Euromillions lottery.

Rank	Numbers	Stars	Combinations	Probability
1	5	2	1	0.0000000086
2	5	1	18	0.0000001545
3	5	0	36	0.0000003089
4	4	2	225	0.0000019308
5	4	1	4050	0.0000347545
6	4	0	8100	0.0000695089
7	3	2	9900	0.0000849554
8	2	2	141,900	0.0012176934
9	3	1	178,200	0.0015291963
10	3	0	356,400	0.0030583926
11	1	2	744,975	0.0063928902
12	2	1	2,554,200	0.0219184806
13	2	0	5,108,400	0.0438369612

combinations of one good star from a total of two, and

$$\binom{9}{1} = 9$$

combinations of one bad star from a total of nine. Because of the multiplication principle, there are

$$\binom{5}{3} \times \binom{45}{2} \times \binom{2}{1} \times \binom{9}{1} = 10 \times 990 \times 2 \times 9 = 178,200$$

possible tickets with three good numbers and one good star. Given the total number of 116,531,800 possible draws, the probability of three good numbers and one good star is

$$\frac{178,200}{116,531,800} = 0.0015.$$

In a similar way, all the probabilities of winning various amounts of cash in Table 5.2 can be computed.

6

Univariate random variables

Therefore I will not talk to you this evening about voices, this evening I will talk mathe-matics: since no-o-obody knows if there is anything on the other side of our death or if there is nothing there, we can deduce from this complete ignorance that the chances that there is something there are exactly the same as the chances that there is nothing there. Fifty percent for cessation and fifty percent for survival. For a Jew like me, a Central European Jew from the generation of the Nazi Holocaust, such odds in favor of survival are not at all bad.

(from *A Tale of Love and Darkness*, Amos Oz, p. 515)

6.1 Random variables and distribution functions

Definition 6.1.1 *A **random variable** is a function $X()$ or X that assigns a real number to each outcome ω of a random experiment.*

If, for instance, a random experiment is carried out with possible outcomes $\{\omega_1, \omega_2, \dots, \omega_9\}$, then the function $X()$ assigns a real number to ω_1, a (not necessarily different) real number to ω_2, a (not necessarily different) real number to ω_3, and so on.

Example 6.1.1 *Testing a product, there are two possible outcomes: good and bad. We assign the value 1 to the outcome good and the value 0 to the outcome bad. We write this as $X(good) = 1$ and $X(bad) = 0$.*

To denote a random variable, we use $X()$ or simply X. The notation $X()$ indicates that a random variable is a function. In practice, however, the notation X is virtually always used. Other letters, such as Y and Z, are used for additional random variables.

Statistics with JMP: Graphs, Descriptive Statistics, and Probability, First Edition. Peter Goos and David Meintrup.
© 2015 John Wiley & Sons, Ltd. Published 2015 by John Wiley & Sons, Ltd.
Companion Website: wiley.com/go/goosandmeintrup

Typically, capital letters from the end of the alphabet are used to denote random variables, while lowercase letters represent real numbers. For example, the expression $P(X = x)$ is the probability that the random variable X takes the value x after the corresponding random experiment has been carried out. The expression $P(X < x)$ is the probability that the random variable X takes a value smaller than the real number x. When sticking to this notation, an expression like $P(x < 10)$ makes no sense.

Example 6.1.2 *A random experiment that consists of throwing a red and a blue die has* 36 *possible outcomes. These outcomes were discussed in detail in Example 5.1.3. A possible random variable X associated with that experiment is the sum of the dots on the two dice. This random variable has* 11 *possible outcomes: the integers* 2, 3, 4, ..., 12. *Another potential random variable Y is the absolute value of the difference between the number of dots on the red die and the blue die. The random variable Y can take the values* 0, 1, 2, 3, 4, *and* 5.

In some experiments, the values the random variable can take are identical to the elementary outcomes of the experiment. This is illustrated in the next example.

Example 6.1.3 *A random experiment involves drawing a random number R from the interval* [0, 1]. *The random variable X is the real number R itself. Based on this experiment, however, we may define other random variables. For example, we can define a random variable Y, which takes the value* 0 *if* $0 \le R < 0.1$, *the value* 1 *if* $0.1 \le R < 0.2$, ..., *and the value* 9 *if* $0.9 \le R < 1$.

Probabilities concerning random variables can be calculated based on the probabilities of the underlying events. If G_x is the event that $X = x$, then

$$P(X = x) = P(G_x).$$

Example 6.1.4 *Consider again the random variable X, the sum of the number of dots obtained by throwing two dice. In order to calculate the probability that the sum equals* 3, $P(X = 3)$, *we need to calculate the probability of the event G_3, which is the set of all outcomes for which X = 3. That set is $G_3 = \{(1, 2), (2, 1)\}$. Therefore,*

$$P(X = 3) = P(G_3),$$
$$= P((1, 2) \text{ is obtained or } (2, 1) \text{ is obtained}),$$
$$= P((1, 2) \text{ is obtained} + P((2, 1) \text{ is obtained}),$$
$$= \frac{1}{36} + \frac{1}{36} = \frac{1}{18}.$$

When we calculate the probability for each of the possible values of a random variable, then implicitly we have determined the probability distribution of the variable under study. For a formal definition of a probability distribution, we need to make a distinction between a discrete random variable and a continuous random variable.

6.2   Discrete random variables and probability distributions

Definition 6.2.1 *A random variable X is **discrete** if the number of possible values that the variable can take is finite or at most a countably infinite set of discrete values.*

Examples of discrete random variables are the variables introduced in Definition 6.1.1 and Example 6.1.1. The random variables X and Y in Example 6.1.2 are also discrete random variables. An illustration of a discrete random variable with a countably infinite number of values can be found in the following example.

Example 6.2.1 *A random experiment consists of repeatedly throwing a die until a six is obtained. A random variable X, defined as the number of attempts required to obtain a six, may take the values 1, 2, 3, ... There is no upper limit so that the number of possible values is countably infinite.*

Definition 6.2.2 *If a discrete random variable X can take k different values $x_1, x_2, \ldots, x_k$, then*

$$p_X(x_i) = P(X = x_i), \quad i = 1, 2, \ldots, k,$$

*is the **probability distribution** of X.*

A probability distribution for a discrete random variable with possible values $x_1, x_2, \ldots, x_k$ has the following properties:

- $p_X(x_i) \geq 0$ for each $i = 1, 2, \ldots, k$. This property follows from the fact that probabilities are always non-negative.

- $\sum_{i=1}^{k} p_X(x_i) = 1$. This property is a consequence of the fact that, when performing the experiment, the random variable X will take one of the values $x_1, x_2, \ldots, x_k$ with certainty.

As the name already suggests, the probability distribution of a random variable X involves the distribution of the probabilities across all possible values of the variable. It is clear that, for a discrete random variable with an infinite number of possible values,

$$\sum_{i=1}^{\infty} p_X(x_i) = 1.$$

Finally, we define the concept of a cumulative distribution function for a discrete random variable X:

Definition 6.2.3 *For a discrete random variable X with possible values $x_1 \leq x_2 \leq \cdots \leq x_k$, we define*

$$F_X(x) = P(X \leq x) = \sum_{x_i \leq x} p_X(x_i) \text{ for every real number x.}$$

*The function $F_X(x)$ is called the (**cumulative**) **distribution function** of X.*

Example 6.2.2 *In Example 6.1.2, the probability that the sum of the number of dots on two dice, which is the random variable X in the example, is equal to 3 was calculated. If we calculate the probabilities for all possible values of X, that is the values 2, 3, 4, ..., 12, we would obtain the probability distribution of X. This probability distribution is shown in Table 6.1, as well as the cumulative distribution function.*

It is easy to verify that the sum of all probabilities of the probability distribution $p_X(x)$ is equal to 1. It turns out that the number 7 is the most likely outcome for the random variable X. This explains why the number 7 plays a major role in the board game "Settlers of Catan". In that game, the participants have to position themselves strategically on the board, in order to collect as many resources as possible. Players earn resources if they own a village at the edge of a square with a number that matches the sum of the dots on two dice thrown. It is therefore important for the players to position their villages at the edges of squares with frequent outcomes for the sums of the dots. From the probability distribution in Table 6.1, we learn that it is a good idea to occupy squares with the numbers 6 and 8. Obviously, occupying a square with the number 7 would be better, but no such squares exist in the game. So, the numbers 6 and 8 are the best choices, which explains why the numbers 6 and 8 are colored red in the "Settlers of Catan". Squares with the numbers 2 and 12 are to be avoided because these numbers are rarely thrown. If a 7 is rolled, players must return half of the resources they collected.

The computation of the cumulative distribution function $F_X(x)$ can be best illustrated using an example. For instance,

$$F_X(5) = P(X \leq 5),$$

$$= P(X = 2) + P(X = 3) + P(X = 4) + P(X = 5),$$

$$= \frac{1}{36} + \frac{2}{36} + \frac{3}{36} + \frac{4}{36},$$

$$= \frac{10}{36}.$$

In the same way, $F_X(x) = P(X \leq x)$ can be determined for all possible values of x ranging from $-\infty$ to $+\infty$. This results in the following function:

$$F_X(x) = P(X \leq x) = \begin{cases} 0 & for -\infty < x < 2, \\ 1/36 & for \ 2 \leq x < 3, \\ 3/36 & for \ 3 \leq x < 4, \\ \vdots & \\ 35/36 & for \ 11 \leq x < 12, \\ 1 & for \ 12 \leq x < +\infty. \end{cases}$$

The probability distribution of the random variable X is shown graphically in Figure 6.1. The corresponding cumulative distribution function is shown in Figure 6.2. Cumulative distribution functions of discrete random variables are always step functions. The heights of the steps are the individual probabilities $p_X(x) = P(X = x)$. In Section 8.8.2, we describe how these graphs can be generated with JMP.

Table 6.1   Probability distribution and cumulative distribution function of the random variable X, the sum of the dots on two dice.

x	2	3	4	5	6	7	8	9	10	11	12
$p_X(x)$	1/36	2/36	3/36	4/36	5/36	6/36	5/36	4/36	3/36	2/36	1/36
$F_X(x)$	1/36	3/36	6/36	10/36	15/36	21/36	26/36	30/36	33/36	35/36	1

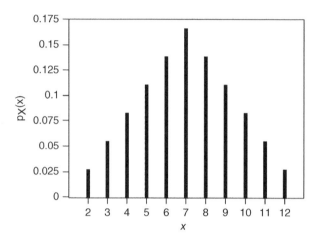

Figure 6.1   Probability distribution of the random variable X in Example 6.1.2.

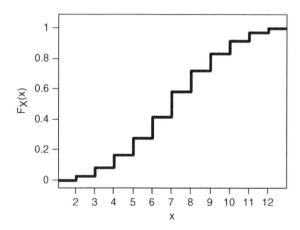

Figure 6.2   Cumulative distribution function of the random variable X in Example 6.1.2.

6.3 Continuous random variables and probability densities

Definition 6.3.1 *A random variable X is **continuous** if it can take a continuum of values.*

An example of a continuous random variable is the variable X introduced in Example 6.1.3. Another example is the net weight of a beer bottle. In order to get a probability distribution for such variables, we would have to enumerate all possible values that the random variable can take. However, this is impossible, as we are are dealing with a continuum of possible values. Therefore, we cannot assign probabilities to the individual values that a continuous random variable can take.

For this reason, for continuous random variables, there is an alternative to the probability distribution, namely a **probability density**. In order to get an intuitive idea of a probability density, it is useful to recall the histograms and polygons introduced in Section 2.4.3. Each interval in a histogram has a relative frequency, which can be regarded as the probability that the random variable under study takes a value that is within the interval. Similarly, the area under a polygon between two values a and b is approximately equal to the relative frequency of the interval $[a, b]$. Therefore, it is approximately equal to the probability that the random variable takes a value between a and b.

If now we increase the sample size and if we simultaneously reduce the size of the classes in the corresponding histogram, then the polygon converges to a continuous curve, which is the probability density. As for the polygon, the total area under this continuous curve will be equal to one. This leads to the following definition of a probability density:

Definition 6.3.2 *A non-negative function $f_X(x)$, which is defined on the real line, such that for any interval $[a, b]$,*

$$P(X \in [a, b]) = P(a \leq X \leq b) = \int_a^b f_X(x)dx,$$

*is called **probability density** or **density function** of a continuous random variable X.*

The probability density is the continuous counterpart of a probability distribution. Since $P(X \in [a, b])$ is a probability, the probability density $f_X(x)$ satisfies the following conditions:

- $f_X(x) \geq 0$, and
- $P(-\infty < X < +\infty) = \int_{-\infty}^{+\infty} f_X(x)dx = 1$.

The second condition indicates that the random variable X takes a value between $-\infty$ and $+\infty$ with certainty. In addition, the condition also implies that the area under the curve $f_X(x)$ is equal to 1, as for a polygon and a histogram that show relative frequencies.

An important insight is that the probability $P(a \leq X \leq b)$ that a continuous random variable X takes a value in the interval $[a, b]$, is equal to the area under the curve $f_X(x)$ between the lower bound a and the upper limit b of the interval. In general, we use integrals to calculate areas under curves. Here, we compute the integral of the density function $f_X(x)$.

The probability that the continuous variable X takes values less than or equal to a certain number a, $P(X \leq a)$, is equal to the probability that X takes values between $-\infty$ and a. This probability, $P(-\infty < X \leq a)$, is the integral from $-\infty$ to a of the probability density $f_X(x)$. It is the area under the curve $f_X(x)$ to the left of a.

If we calculate the probability $P(X \leq a)$ for all possible values of a, we get a function. This function is called the (cumulative) distribution function.

Definition 6.3.3 *The **(cumulative) distribution function** of a continuous random variable X is defined as*

$$F_X(x) = P(X \leq x) = \int_{-\infty}^{x} f_X(y)dy.$$

Note that in the definition of the cumulative distribution function, y is used as variable in the integral because the variable x is already used as the upper limit for the integral.

This function has a number of intuitive properties:

- $F_X(x)$ is a non-decreasing function, in other words $F_X(x_1) \leq F_X(x_2)$ if $x_1 \leq x_2$;
- $F_X(-\infty) = \lim_{x \to -\infty} F_X(x) = 0$;
- $F_X(+\infty) = \lim_{x \to +\infty} F_X(x) = 1$.

The interpretation of the latter two properties is that a cumulative distribution function increases from 0 to 1. These properties are also true for the cumulative distribution function of a discrete random variable.

Example 6.3.1 *Figures 6.3 and 6.4 clarify the relationship between the probability density and the (cumulative) distribution function. In both figures, the probability that the random variable X is less than or equal to 1 is indicated.*

Figure 6.3 shows the probability density $f_X(x)$ of a continuous random variable X. The total area under the curve is equal to 1. The shaded part of the area under the curve is the probability $P(X \leq 1)$.

Figure 6.4 shows the cumulative distribution function $F_X(x)$ that corresponds to the probability density $f_X(x)$ in Figure 6.3. This cumulative distribution function contains information about all possible probabilities of the type $P(X \leq x)$, for all possible values of x between $-\infty$ and $+\infty$. As the cumulative distribution function in Figure 6.2, the cumulative distribution function in Figure 6.4 is a non-decreasing function, which increases from 0 to 1.

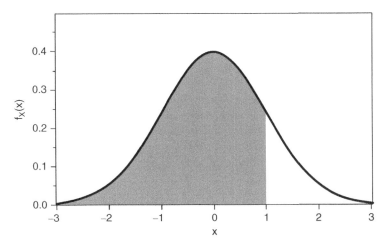

Figure 6.3   Graphical representation of a probability density. The probability $P(X \leq 1)$ *is the shaded area under the curve and is equal to* 0.8413.

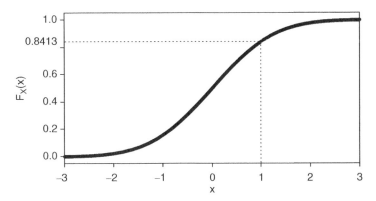

Figure 6.4   Cumulative distribution function corresponding to the probability density in Figure 6.3. The probability $P(X \leq 1)$ *is* 0.8413 *and is the value of the cumulative distribution function at* $x = 1$, $F_X(1)$.

The probability $P(X \leq 1)$ *in Figure 6.3 is the same as* $F_X(1) = 0.8413$, *the value of the function in Figure 6.4 at* $x = 1$. *Therefore, a probability can always be determined in two ways: one calculation method uses the probability density, while the other method uses the cumulative distribution function.*

The probability density in Example 6.3.1 is the standard normal probability density, which we will study in detail in Chapter 10. This probability density is nicely bell-shaped and symmetrical. However, not all probability densities are symmetric. In Example 6.3.2, we study a probability density that is not symmetrical.

Example 6.3.2 *Suppose that the probability density of a random variable X is given by*

$$f_X(x) = \begin{cases} \frac{1}{2}(3-x), & 1 \le x \le 3, \\ 0, & otherwise. \end{cases}$$

This density is depicted graphically in Figure 6.5. The (cumulative) distribution function corresponding to this density can be calculated by integrating the probability density $f_X(x)$. The result of the integral depends on the value of x. If $x < 1$, then

$$\int_{-\infty}^{x} f_Y(y)dy = \int_{-\infty}^{x} 0\,dy = 0.$$

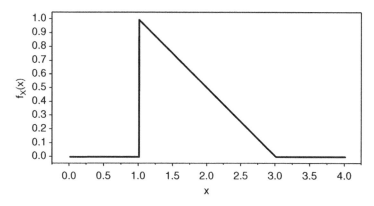

Figure 6.5   Probability density in Example 6.3.2.

If x lies between 1 and 3, then

$$\int_{-\infty}^{x} f_Y(y)dy = \int_{-\infty}^{x} 0\,dy + \int_{1}^{x} \frac{1}{2}(3-y)dy = 0 + \left[\frac{-1}{4}y^2 + \frac{3}{2}y\right]_{1}^{x} = \frac{-1}{4}x^2 + \frac{3}{2}x - \frac{5}{4}.$$

Finally, if $x > 3$, then

$$\int_{-\infty}^{x} f_Y(y)dy = \int_{-\infty}^{x} 0\,dy + \int_{1}^{3} \frac{1}{2}(3-y)dy + \int_{3}^{x} 0\,dy = 0 + 1 + 0 = 1.$$

Hence, the cumulative distribution function is equal to

$$F_X(x) = \begin{cases} 0, & x < 1, \\ -\frac{1}{4}x^2 + \frac{3}{2}x - \frac{5}{4}, & 1 \le x \le 3, \\ 1, & x > 3. \end{cases}$$

This cumulative distribution function is shown in Figure 6.6.

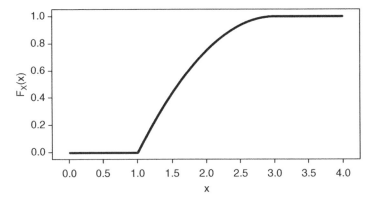

Figure 6.6   Cumulative distribution function corresponding to the probability density in Figure 6.5.

We now compute the probability that the random variable X takes values in intervals of width 0.1 *around the values* $x = 1.25$ *and* $x = 2.65$. *It follows from Definition 6.3.2 that these probabilities can be calculated as*

$$P(1.2 \leq X \leq 1.3) = \int_{1.2}^{1.3} \frac{1}{2}(3 - x)dx = 0.0875$$

and

$$P(2.6 \leq X \leq 2.7) = \int_{2.6}^{2.7} \frac{1}{2}(3 - x)dx = 0.0175.$$

The probability that X takes values in the interval [1.2,1.3] *is five times as large as the probability for the interval* [2.6,2.7]. *In other words, the probability to get a value in the vicinity of* 1.25 *is five times bigger than the probability to get a value near* 2.65.
Note that the probability $P(1.2 \leq X \leq 1.3)$ *may also be computed using the cumulative probability distribution:*

$$P(1.2 \leq X \leq 1.3) = P(X \leq 1.3) - P(X \leq 1.2) = F_X(1.3) - F_X(1.2),$$

$$= 0.2775 - 0.19 = 0.0875.$$

This is true in general:

$$P(a \leq X \leq b) = P(X \leq b) - P(X \leq a) = F_X(b) - F_X(a).$$

We conclude this section with some remarks on continuous random variables and their densities:

- If a random variable X only takes values in an interval $[a, b]$, then $f_X(x)$ equals zero outside this interval. In this case, we have that $\int_a^b f_X(x)dx = 1$. The cumulative distribution function $F_X(x)$ is zero for all x values smaller than a

and 1 for all x values larger than b. Between a and b, the function $F_X(x)$ is non-decreasing.

- For a continuous random variable X it is always true that

$$P(X = c) = P(c \leq X \leq c) = \int_c^c f_X(x)dx = 0.$$

This result makes sense because c is only one of an uncountably infinite number of possible values of the continuous random variable.

- From the previous remark, it follows that, for any a and b, with $a < b$,

$$P(a \leq X \leq b) = P(a < X \leq b) = P(a \leq X < b) = P(a < X < b).$$

The interpretation of this result is that, when calculating probabilities for continuous random variables, it does not matter whether we use strict inequalities or not, that is, whether we use $\leq$ or $<$, or $\geq$ or $>$.

- In contrast with the probability distribution $p_X(x)$ of a discrete random variable, a probability density $f_X(x)$ can take values greater than 1.

Example 6.3.3   *The probability density*

$$f_X(x) = \begin{cases} 2, & 0 \leq x \leq \frac{1}{2}, \\ 0, & otherwise, \end{cases}$$

is a valid probability density function that takes values greater than 1.

Example 6.3.4   *The probability density*

$$f_X(x) = \begin{cases} \frac{2}{3}x^{-1/3}, & 0 < x \leq 1, \\ 0, & otherwise, \end{cases}$$

is a valid probability density that goes to infinity when x approaches zero.

- A probability $P(a \leq X \leq b)$ for a continuous random variable X can always be calculated in various ways. As mentioned earlier, an integral can be used for this purpose. An alternative is to proceed as follows:

$$P(a \leq X \leq b) = P(X \leq b) - P(X \leq a) = F_X(b) - F_X(a),$$

or

$$P(a \leq X \leq b) = P(X \geq a) - P(X \geq b).$$

This is illustrated graphically in Figures 6.7 and 6.8 for $a = -1$ and $b = 1$, using the probability density of Example 6.3.1.

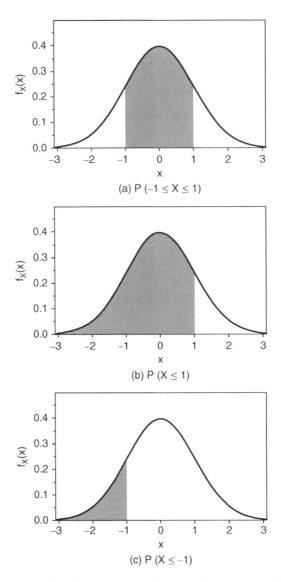

$$\text{(a) } P\ (-1 \le X \le 1)$$

$$\text{(b) } P\ (X \le 1)$$

$$\text{(c) } P\ (X \le -1)$$

Figure 6.7 Graphical illustration of the fact that $P(-1 \le X \le 1) = P(X \le 1) - P(X \le -1)$.

- The probability density $f_X(x)$ can be determined from the cumulative distribution function $F_X(x)$ by taking the derivative with respect to x:

$$f_X(x) = \frac{d}{dx} F_X(x).$$

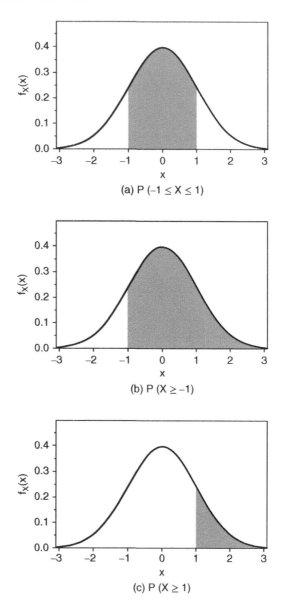

(a) P $(-1 \leq X \leq 1)$

(b) P $(X \geq -1)$

(c) P $(X \geq 1)$

Figure 6.8 Graphical illustration of the fact that $P(-1 \leq X \leq 1) = P(X \geq -1) - P(X \geq 1)$.

- A probability density that is frequently used is the so-called **uniform probability density**:

$$f_X(x) = \begin{cases} 1, & 0 \leq x < 1, \\ 0, & \text{otherwise.} \end{cases}$$

The corresponding distribution function is given by

$$F_X(x) = \begin{cases} 0, & x < 0, \\ x, & 0 \le x < 1, \\ 1, & x \ge 1. \end{cases}$$

The uniform density describes a process where numbers are randomly drawn from the interval $[0, 1]$.

6.4   Functions of random variables

Sometimes, we are not interested in a random variable X itself but in a function $Y = g(X)$ of it. This new variable Y is also a random variable with a probability distribution or density, depending on whether the original variable X is discrete or continuous. A natural question to ask is what the probability distribution or probability density of the new random variable Y is.

6.4.1   Functions of one discrete random variable

For a discrete random variable X, it is very easy to determine the probability distribution of $Y = g(X)$. We illustrate this with an example.

Example 6.4.1   *The probability distribution of X is given by the following table:*

x	-3	-2	-1	0	1	2	3
$p_X(x)$	0.08	0.14	0.19	0.28	0.17	0.08	0.06

Then, the probability distribution of the function $Y_1 = 2X + 1$ *is*

y_1	-5	-3	-1	1	3	5	7
$p_{Y_1}(y_1)$	0.08	0.14	0.19	0.28	0.17	0.08	0.06

The probability distribution of a second function, $Y_2 = X^2$, *is given by*

y_2	0	1	4	9
$p_{Y_2}(y_2)$	0.28	0.36	0.22	0.14

To construct this probability distribution, we first need to find out which values Y_2 can take. In this example, the only values Y_2 can take are 0, 1, 4, and 9. The probability that Y_2 takes the value 9, for instance, is calculated as follows:

$$P(Y_2 = 9) = P(X = -3 \text{ or } X = 3),$$
$$= P(X = -3) + P(X = 3) = 0.08 + 0.06 = 0.14.$$

6.4.2   Functions of one continuous random variable

A first approach to find the probability density of a function $Y = g(X)$ of a continuous random variable X is based on the cumulative distribution function. The approach involves three steps:

1. Determine the values that Y can take, using $Y = g(X)$ and the possible values of X.

2. Compute the cumulative distribution function $F_Y(y) = P(Y \le y)$ of Y.

3. Take the derivative of $F_Y(y)$ with respect to y. This yields the probability density function $f_Y(y)$ of $Y = g(X)$.

Example 6.4.2 *We start with the random variable X given in Example 6.3.2, and we consider the functions $V = 2X - 1$ and $Y = X^2$. The probability density of X is only strictly positive in the interval $[1, 3]$. We start with the probability density of V:*

1. *Since the probability density of X is only non-zero between 1 and 3, X can only take values between 1 and 3. As a result, $V = 2X - 1$ can only take values in the interval $[1,5]$, and the probability density of V is only non-zero between 1 and 5. As a result, the cumulative probability distribution of V is zero for v values smaller than 1 and one for v values larger than 5.*

2. *The cumulative distribution function is given by*

$$F_V(v) = \begin{cases} 0, & v < 1, \\ P(V \le v) = P(2X - 1 \le v) = P\left(X \le \frac{v+1}{2}\right), \\ \quad = F_X\left(\frac{v+1}{2}\right) = -\frac{1}{4}\left(\frac{v+1}{2}\right)^2 + \frac{3}{2}\left(\frac{v+1}{2}\right) - \frac{5}{4}, \\ \quad = \frac{1}{16}(-v^2 + 10v - 9), & 1 \le v \le 5, \\ 1, & v > 5. \end{cases}$$

3. *Taking the derivative results in*

$$f_V(v) = \begin{cases} \frac{1}{8}(5 - v), & 1 \le v \le 5, \\ 0, & \text{otherwise.} \end{cases}$$

Next, we compute the probability density of Y:

1. *Since the probability density of X is only non-zero between 1 and 3, X can only take values between 1 and 3. As a result, $Y = X^2$ can only take values in the interval [1,9], and the probability density of Y is only non-zero between 1 and 9. As a result, the cumulative probability distribution of Y is zero for y values smaller than 1 and one for y values larger than 9.*

2. *The cumulative distribution function is given by*

$$F_Y(y) = \begin{cases} 0, & y < 1, \\ P(Y \leq y) = P(X^2 \leq y) = P\left(X \leq +\sqrt{y}\right), & \\ \quad = F_X\left(\sqrt{y}\right) = -\frac{1}{4}\left(\sqrt{y}\right)^2 + \frac{3}{2}\left(\sqrt{y}\right) - \frac{5}{4}, & \\ \quad = \frac{-1}{4}y + \frac{3}{2}\sqrt{y} - \frac{5}{4}, & 1 \leq y \leq 9, \\ 1, & y > 9. \end{cases}$$

3. *Taking the derivative results in*

$$f_Y(y) = \begin{cases} -\frac{1}{4} + \frac{3}{4\sqrt{y}}, & 1 \leq y \leq 9, \\ 0, & otherwise. \end{cases}$$

A second approach for determining the probability density of a function of a continuous random variable is only usable under certain conditions and is based on a theorem from differential calculus. Let X be a continuous random variable with probability density $f_X(x)$ on the interval $[a, b]$ and $g()$ be a differentiable, strictly increasing or strictly decreasing function on the same interval. Then, the probability density function of $Y = g(X)$ is given by

$$f_Y(y) = f_X(x)\left|\frac{dx}{dy}\right| = f_X\{g^{-1}(y)\}\left|\frac{dg^{-1}(y)}{dy}\right|, \tag{6.1}$$

where $g^{-1}()$ is the inverse function of $g()$.

Example 6.4.3 *We start again with the random variable X discussed in Example 6.3.2. Since both $V = 2X - 1$ and $Y = X^2$ are strictly increasing functions on the interval [1,3], we can use the second approach for the calculation of their probability density.*

- *To determine the probability density of V, we first write $x = g^{-1}(v) = \frac{1}{2}(v + 1)$. Consequently, $\frac{dx}{dv} = \frac{1}{2}$, and*

$$f_V(v) = f_X\left(\frac{1}{2}(v + 1)\right)\left|\frac{dx}{dy}\right| = \frac{1}{2}(3 - \frac{1}{2}(v + 1))\left|\frac{1}{2}\right|,$$

$$= \frac{1}{8}(5 - v), \quad for\ 1 \leq v \leq 5.$$

- To determine the probability density of Y, we first write $x = g^{-1}(y) = +\sqrt{y}$. Consequently, $\frac{dx}{dy} = \frac{1}{2\sqrt{y}}$, and

$$f_Y(y) = f_X\left(\sqrt{y}\right)\left|\frac{dx}{dy}\right| = \frac{1}{2}\left(3 - \sqrt{y}\right)\left|\frac{1}{2\sqrt{y}}\right|,$$

$$= -\frac{1}{4} + \frac{3}{4\sqrt{y}}, \quad \text{for } 1 \leq y \leq 9.$$

6.5   Families of probability distributions and probability densities

Definition 6.5.1 *A (parametric)* **family of probability distributions** *or* **probability densities** *is a set of densities or distributions indexed by one or more parameters. For each value of the parameter(s) in a given domain, the resulting function is a probability distribution or a probability density.*

Example 6.5.1 *Consider the function*

$$f_X(x) = \begin{cases} 0, & x \leq 0, \\ \lambda e^{-\lambda x}, & x > 0. \end{cases}$$

This function is a probability density for all values of $\lambda > 0$. Therefore, $f_X(x)$ defines a family of probability densities. In this case, the parameter λ is the only parameter of the family of probability densities.

Descriptive statistics, such as the expected value and variance, typically depend on the parameter(s). In this example, the expected value of X is equal to $E(X) = 1/\lambda$, and the variance is $\sigma_X^2 = 1/\lambda^2$ (for more details on expected values and variances, see Chapter 7). To indicate this dependence, one sometimes writes $f_X(x; \lambda)$ instead of simply $f_X(x)$.

In many applications, the researcher has an idea of the family to which the probability density of a population or a process belongs. For example, a random variable X that represents the number of customers per week typically follows a Poisson distribution. The time between the appearance of two successive customers typically follows an exponential density. Both the Poisson distribution and the exponential density are families that we will study in detail in the next few chapters. However, to determine the exact values of the parameters of the distributions or densities, we need information from a sample of data. Based on a sample of data, the researcher can try to get an idea about the parameter values, or – as we say in the statistical jargon – to estimate the parameters. This problem is discussed in the book *Statistics with JMP: Hypothesis Tests, ANOVA and Regression*.

6.6   Simulation of random variables

Simulation techniques are often used to imitate a process with a given distribution. This was already explained in Section 4.7. In this section, we explain how the cumulative distribution function can be used to generate pseudo-random numbers from a probability distribution or a probability density.

Example 6.6.1 *Suppose that we want to simulate the random selection of* 10 *First Class letters of the Royal Mail in Great Britain with a computer. Of all the First Class letters,* 80% *are already delivered after one day,* 12% *arrive after two days,* 7% *after three days, and* 1% *after four days. The probability distribution is given in Table 6.2, together with the associated cumulative distribution function.*

The random selection of a letter is shown graphically in Figure 6.9. The step function in the figure is the cumulative distribution function $F_X(x)$. The starting point of the selection of a letter is a pseudo-random number between 0 *and* 1, *namely* 0.908546. *Virtually every software package has a function to generate such pseudo-random numbers. Next, the number* 0.908546 *is converted into a pseudo-random number of*

Table 6.2   Probability distribution and cumulative distribution function of the number of days required to deliver a First Class letter via the British Royal Mail.

x	1	2	3	4
$f_X(x)$	0.800	0.120	0.070	0.010
$F_X(x)$	0.800	0.920	0.990	1.000

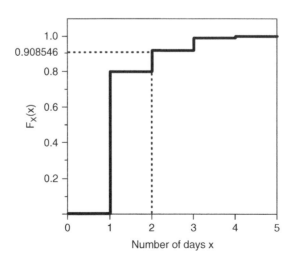

Figure 6.9   Illustration of the generation of a pseudo-random observation of a discrete random variable with the cumulative distribution function from Table 6.2.

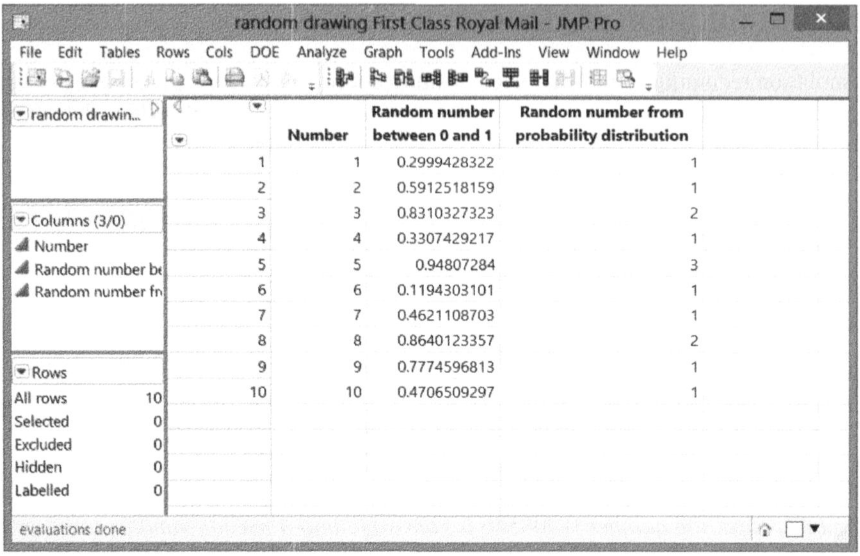

Figure 6.10   The drawing of pseudo-random numbers in JMP.

the probability distribution $f_X(x)$. This conversion, shown by means of the dotted line in Figure 6.9, yields the value 2. So, the randomly selected letter requires two days to be delivered.

JMP allows pseudo-random numbers from the probability distribution in Table 6.2 to be generated in a fairly simple manner. First, one needs a column with pseudo-random numbers between 0 and 1. Next, a formula must be programmed to perform the conversion to pseudo-random numbers from the desired distribution. Figure 6.10 shows the resulting JMP file, while Figure 6.11 shows the formula needed to generate the pseudo-random numbers. In Figure 6.10, it can be seen that seven of the 10 pseudo-random numbers from the probability distribution are equal to 1. Two of the pseudo-random numbers are equal to 2, and one pseudo-random number is equal to 3.

Creating a new data table in JMP is done via the menu "File". In that menu, you first have to select the option "New", followed by "Data Table". Creating a formula in JMP can be done by right-clicking on a column header in a data table, and then choosing "Formula". Creating a new column can be done by right-clicking on the header to the right of the last column, and then choosing the option "New Column". Alternatively, you can choose the option "New Column" in the "Cols" menu.

In a similar way, data can be simulated for a continuous random variable with a given cumulative probability distribution.

Example 6.6.2   Consider the cumulative probability distribution $F_X(x)$ derived in Example 6.3.2 and represented graphically in Figure 6.6. The drawing of a pseudo-random observation from the corresponding probability density function is

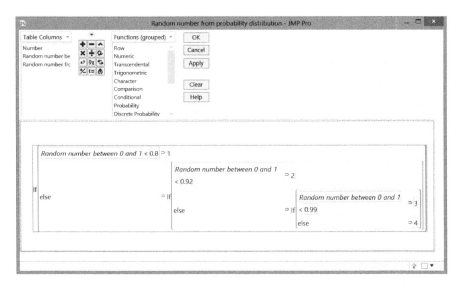

Figure 6.11   Formula to generate pseudo-random numbers using the cumulative distribution function in Table 6.2.

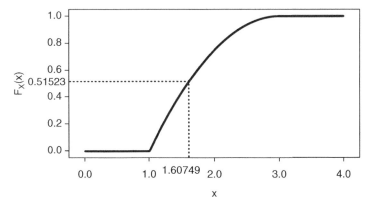

Figure 6.12   Illustration of the generation of a pseudo-random observation of a continuous random variable with the cumulative distribution function from Figure 6.6.

illustrated graphically in Figure 6.12. First, a pseudo-random number R is drawn from the interval [0,1]. Suppose that this number is equal to 0.51523. Then, by means of the inverse distribution function $x = F_X^{-1}(R)$, this number is converted into a pseudo-random observation of a process with density $f_X(x)$. Here, the inverse distribution function is

$$x = F_X^{-1}(R) = 3 - 2\sqrt{1 - R}.$$

This function, which we obtained by solving the equation

$$\frac{-1}{4}x^2 + \frac{3}{2}x - 5/4 = R$$

for x, yields the result

$$x = F_X^{-1}(0.51523) = 3 - 2\sqrt{1 - 0.51523} = 1.60749$$

for the pseudo-random number 0.51523. Hence, the pseudo-random number from the probability density in Example 6.3.2 and Figure 6.5 is 1.60749. The formula required to calculate this outcome in JMP is "$3 - 2 * Sqrt(1 - 0.51523)$"[1].

---

[1] The easiest way to enter this formula in JMP is to use the "Log" screen. You can access this screen via the "View" menu, by selecting the option "Log". An alternative method is to press "CTRL + SHIFT + L" on your keyboard. In the "Log" screen, you can type in the formula you want to evaluate, for example, "$3 - 2 * Sqrt(1 - 0.51523)$". Finally, you need to select the formula you entered and click on the white button with the red runner in the menu bar. Alternatively, you can press "CTRL + R" after selecting the typed formula.

7

Statistics of populations and processes

Ira Titcomb, the beekeeper, kidded with Homer. Ira was sixty-five, but he had another number marked on the trailer he used to carry his hives: the number of times he'd been stung by his bees.

"Only two hundred and forty-one times," Ira said. "I been keepin' bees since I was nineteen," he said, "so that amounts to only five point two stings a year. Pretty good, huh?"

(from *The Cider House Rules*, John Irving, p. 456)

In Chapter 3, we introduced a series of statistics for sample data. For some special cases, we already mentioned the corresponding statistics of populations and processes. This chapter provides more details on this topic.

7.1 Expected value of a random variable

Definition 7.1.1 *The **expected value** μ_X or $E(X)$ **of a discrete random variable** X with possible values $x_1, x_2, \ldots, x_k$ and probability distribution $p_X(x_i), i = 1, 2, \ldots, k$, is*

$$\mu_X = E(X) = \sum_{i=1}^{k} x_i p_X(x_i).$$

Definition 7.1.2 *The **expected value** μ_X or $E(X)$ **of a continuous random variable** X with probability density $f_X(x)$ is*

$$\mu_X = E(X) = \int_{-\infty}^{+\infty} x f_X(x) dx.$$

Statistics with JMP: Graphs, Descriptive Statistics, and Probability, First Edition. Peter Goos and David Meintrup.
© 2015 John Wiley & Sons, Ltd. Published 2015 by John Wiley & Sons, Ltd.
Companion Website: wiley.com/go/goosandmeintrup

If the sum (for discrete random variables) or the integral (for continuous random variables) is not finite, then one says that the expected value does not exist. The expected value is sometimes called the **mean value**. As a matter of fact, the definition of the arithmetic or sample mean for grouped data (see Definition 3.1.5) is similar in structure to the expression for the expected value of a discrete random variable in Definition 7.1.1.

Example 7.1.1   *The probability distribution for the First Class letters in Example 6.6.1 has expected value*

$$E(X) = (0.8)1 + (0.12)2 + (0.07)3 + (0.01)4 = 1.29 \; days.$$

This example illustrates that the expected value of a random variable does not necessarily have to be a value that can be taken by the variable. The random variable in this example can only take the values 1, 2, 3, or 4, while the mean value is 1.29.

Example 7.1.2   *It is easy to verify that the expected value of the probability distribution in Example 6.2.2 (the sum of the dots on two dice) is equal to 7.*

Example 7.1.3   *The expected value of the random variable X in Example 6.3.2 equals*

$$E(X) = \int_{-\infty}^{+\infty} x f_X(x) dx,$$

$$= \int_{-\infty}^{1} 0 x dx + \int_{1}^{3} x \frac{1}{2}(3-x) dx + \int_{3}^{+\infty} 0 x dx,$$

$$= 0 + \int_{1}^{3} \frac{1}{2}(3x - x^2) dx + 0,$$

$$= \left[\frac{3}{4}x^2 - \frac{x^3}{6} \right]_{1}^{3},$$

$$= \frac{27}{4} - \frac{27}{6} - \frac{3}{4} + \frac{1}{6},$$

$$= \frac{5}{3}.$$

Example 7.1.4   *You are sitting in a bar with a fellow student, who offers you two different bets. In the first bet, you can make €1 if you win, or you have to pay €1 if you lose. In the second bet, you can make €100 if you win, but you also need to pay €100 if you lose. What bet do you choose if the probability of winning is 50% in both bets?*

   Let us denote the gain in the two bets by X_1 and X_2. The expected profit in the first bet is, of course,

$$E(X_1) = -1(0.5) + 1(0.5) = 0,$$

while the expected profit in the second bet is equal to

$$E(X_2) = -100(0.5) + 100(0.5) = 0.$$

Therefore, the expected profit is the same in both bets. However, it is unlikely that you do not have a preference, because the two possible values of X_2 are much further apart than the two possible outcomes of X_1.

Example 7.1.5 *The scores of some multiple-choice tests involve a correction for guessing. The idea is that the expected score of someone who does not know anything about the subject and therefore provides random answers to all questions should be zero. Without a correction for guessing, the good students would have a relative disadvantage. We calculate the correction for guessing on a multiple choice question with four possible answers. We assume that the first option is the correct answer.*

A student who randomly answers the questions uses the following distribution:

Answer	1	2	3	4
Probability	0.25	0.25	0.25	0.25

Assume that the student earns one point for a correct answer and loses c points for a wrong answer. Then, the expected score on one question is

$$E(points) = 1(0.25) - c(0.25) - c(0.25) - c(0.25) = 0.25 - 0.75c.$$

As we want this to be zero, we set the correction for guessing to $c = 1/3$.

Suppose that a student who studied all night before the exam can eliminate one wrong answer, for example option 4, so that she only needs to select one of three possible answers. The probability distribution for this student will be

Answer	1	2	3	4
Probability	1/3	1/3	1/3	0

For that student, the expected score, taking into account a correction for guessing of $c = 1/3$, is

$$E(points) = 1(1/3) - c(1/3) - c(1/3) - c(0) = \frac{1}{3} - \frac{2}{3}c = \frac{1}{3} - \frac{2}{9} = \frac{1}{9}.$$

The positive expected value suggests that it is rational to guess as soon as one can exclude one of the answers with certainty.

7.2 Expected value of a function of a random variable

The expected value of a function $Y = g(X)$ of a discrete random variable X can be calculated as

$$\mu_Y = E(Y) = E\{g(X)\} = \sum_y y p_Y(y),$$

where $p_Y(y)$ represents the probability distribution of Y. The summation runs over all possible values of Y. We explained in Section 6.4.1 how the probability distribution $p_Y(y)$ for a function $Y = g(X)$ can be computed starting from $p_X(x)$.

Example 7.2.1 *The expected value of the function $Y_2 = X^2$ in Example 6.4.1 is equal to*

$$E(Y_2) = 0(0.28) + 1(0.36) + 4(0.22) + 9(0.14) = 2.5.$$

An alternative, more direct way to calculate $E\{g(X)\}$, is as follows:

$$\mu_Y = E(Y) = E\{g(X)\} = \sum_x g(x)p_X(x).$$

Example 7.2.2 *The expected value of $Y_2 = X^2$ in Example 6.4.1 can also be calculated as*

$$E(Y_2) = (-3)^2(0.08) + (-2)^2(0.14) + (-1)^2(0.19)$$
$$+ 0^2(0.28) + 1^2(0.17) + 2^2(0.08) + 3^2(0.06) = 2.5.$$

The expected value of a function $Y = g(X)$ of a continuous random variable X can be calculated as

$$\mu_Y = E(Y) = E\{g(X)\} = \int_{-\infty}^{+\infty} y f_Y(y) dy,$$

where $f_Y(y)$ represents the probability density of Y. We explained in Section 6.4.2 how to compute the probability density $f_Y(y)$ of the function $Y = g(X)$ starting from $f_X(x)$. Alternatively,

$$\mu_Y = E(Y) = E\{g(X)\} = \int_{-\infty}^{+\infty} g(x) f_X(x) dx.$$

Example 7.2.3 *The expected value of the random variable $Y = X^2$ in Example 6.4.2 is*

$$E(Y) = \int_1^9 y f_Y(y) dy = \int_1^9 y \left[\frac{-1}{4} + \frac{3}{4\sqrt{y}} \right] dy = \int_1^9 \left[\frac{-y}{4} + \frac{3\sqrt{y}}{4} \right] dy = 3,$$

or, alternatively,

$$E(Y) = \int_1^3 x^2 f_X(x) dx = \int_1^3 x^2 \left[\frac{1}{2}(3-x) \right] dx = \int_1^3 \left[\frac{3x^2}{2} - \frac{x^3}{2} \right] dx = 3.$$

7.3   Special cases

Theorem 7.3.1 *If Y is a linear function of a random variable X, that is, $Y = aX + b$ (where a and b are constants), then*

$$\mu_Y = E(Y) = E(aX + b) = a\mu_X + b.$$

Proof: For a continuous random variable, we have that

$$\mu_Y = \int\limits_{-\infty}^{+\infty} (ax+b)f_X(x)dx = a \int\limits_{-\infty}^{+\infty} xf_X(x)dx + b \int\limits_{-\infty}^{+\infty} f_X(x)dx = a\mu_X + b.$$

The last step follows from the definition of the expected value, and from the fact that the area under the curve $f_X(x)$ is equal to 1. ∎

This theorem can be used to calculate the expected values of the functions Y_1 and V in Examples 6.4.1 and 6.4.2.

A similar feature of the arithmetic mean of a sample was described and illustrated in Section 3.1.3. Two consequences of Theorem 7.3.1 are:

- The expected value of a constant is the constant itself. A simple proof is obtained by setting $a = 0$ in the previous proof.

- $E(X - \mu_X) = 0$.

These consequences are similar to some properties of the arithmetic or sample mean given on page 59.

Theorem 7.3.2 *For a random variable $Y = \sum\limits_{i=1}^{k} a_i g_i(X)$, with real constants a_1, $a_2, \ldots, a_k$ and functions $g_1(X), g_2(X), \ldots, g_k(X)$ of a random variable X, we have that*

$$\mu_Y = E(Y) = E\left\{ \sum_{i=1}^{k} a_i g_i(X) \right\} = \sum_{i=1}^{k} a_i E\{g_i(X)\}.$$

The proof of this theorem is an easy exercise.

Example 7.3.1 *For a random variable X with uniform density,*

$$f_X(x) = \begin{cases} 1, & 0 \le x < 1, \\ 0, & otherwise, \end{cases}$$

we can calculate the expected value of $3X^2 - 2X + 4$:

$$E(3X^2 - 2X + 4) = 3E(X^2) - 2E(X) + 4 = 3\int_0^1 x^2 dx - 2\int_0^1 x dx + 4 = 4.$$

7.4   Variance and standard deviation of a random variable

Definition 7.4.1 *The **variance** σ_X^2 or var(X) of a random variable X is defined as*

$$\sigma_X^2 = \text{var}(X) = E\{(X - \mu_X)^2\}.$$

The variance is the expected value of the function $g(X) = (X - \mu_X)^2$ and can also be defined as

$$\sum_{i=1}^{k} (x_i - \mu_X)^2 p_X(x_i)$$

for a discrete random variable X, and as

$$\int_{-\infty}^{+\infty} (x - \mu_X)^2 f_X(x) dx$$

for a continuous random variable. It is useful to compare the expression for a discrete random variable with Equation (3.2) for a population variance.

Definition 7.4.2 *The **standard deviation** σ_X of a random variable X is the (positive) square root of the variance:*

$$\sigma_X = \sqrt{\sigma_X^2}.$$

As with sample data, the population or process variance and the population or process standard deviation are used as measures of spread. The larger the variance or the standard deviation of a random variable X, the bigger the spread around the expected value.

Example 7.4.1 *The variance of the random variable X in Examples 6.6.1 and 7.1.1 is*

$$\sigma_X^2 = (0.8)(1 - 1.29)^2 + (0.12)(2 - 1.29)^2 + (0.07)(3 - 1.29)^2 + (0.01)(4 - 1.29)^2,$$
$$= 0.4059 \ days^2.$$

Therefore, the standard deviation is $\sigma_X = \sqrt{0.4059} = 0.6371 \ days$.

Example 7.4.2 *The variance of the random variable in Example 6.2.2 (the sum of the numbers of dots on two dice) is 5.833.*

Example 7.4.3 *The variances of the gains in the bets in Example 7.1.4 are*

$$\sigma_{X_1}^2 = (-1 - 0)^2(0.5) + (1 - 0)^2(0.5) = 1$$

for the first bet, and

$$\sigma_{X_2}^2 = (-100 - 0)^2(0.5) + (100 - 0)^2(0.5) = 10,000$$

for the second bet. The variances clearly indicate that the spread around the expected value in the second bet is much bigger than in the first. This example illustrates that the concept of variance can be associated with the concept of risk. Variances are often used in the calculation of risks, for example, of equity portfolios.

Example 7.4.4 *The continuous random variable X in Example 6.3.2 (see also Example 7.1.3) has variance*

$$\sigma_X^2 = \int_1^3 (x - \frac{5}{3})^2 [\frac{1}{2}(3 - x)]dx = \frac{2}{9} = 0.2222,$$

and standard deviation $\sigma_X = \sqrt{0.2222} = 0.4714$.

The variance of a random variable can sometimes be calculated more easily than by literally applying the definition. This is because

$$\sigma_X^2 = E\{(X - \mu_X)^2\},$$
$$= E(X^2 - 2X\mu_X + \mu_X^2),$$
$$= E(X^2) - 2\mu_X E(X) + \mu_X^2,$$
$$= E(X^2) - 2\mu_X \mu_X + \mu_X^2,$$
$$= E(X^2) - \mu_X^2.$$

Example 7.4.5 *We recalculate the variance of the continuous random variable X in Example 6.3.2 (see also Examples 7.1.3 and 7.4.4):*

$$\sigma_X^2 = \int_1^3 x^2 \left[\frac{1}{2}(3 - x)\right] dx - \mu_X^2,$$
$$= 3 - \left(\frac{5}{3}\right)^2,$$
$$= \frac{2}{9}.$$

Theorem 7.4.1 *If Y is a linear function of a random variable X, that is, $Y = aX + b$, then*
$$\sigma_Y^2 = \text{var}(Y) = \text{var}(aX + b) = a^2 \sigma_X^2.$$

Proof: First, we know from Theorem 7.3.1 that $\mu_Y = a\mu_X + b$. Hence,

$$\sigma_Y^2 = E\{(Y - \mu_Y)^2\},$$
$$= E\{(aX + b - \mu_Y)^2\},$$
$$= E\{(aX + b - a\mu_X - b)^2\},$$
$$= E\{(aX - a\mu_X)^2\},$$
$$= E\{a^2(X - \mu_X)^2\},$$
$$= a^2 E\{(X - \mu_X)^2\},$$
$$= a^2 \sigma_X^2.$$

∎

As a consequence, $\sigma_Y = |a|\sigma_X$.

Note that the constant b has no impact on the variance and standard deviation of $Y = aX + b$. To understand this, it is useful to consider $V = X + b$ for a moment. Here, the values of the random variable V are simply shifted values of the original random variable X. Therefore, the values of V lie at the same distance from their expected value, so that their spread does not change. A similar result exists for the variance of a sample (see page 67).

Two consequences of Theorem 7.4.1 are:

- The variance and standard deviation of a constant are zero.

- A **standardized random variable**

$$Z = \frac{X - \mu_X}{\sigma_X}$$

has an expected value of zero $(E(Z) = 0)$ and a standard deviation of 1 $(\sigma_Z^2 = \sigma_Z = 1)$. This applies to any standardized random variable, regardless of the probability distribution or probability density of the original random variable X. Standardized random variables will be used extensively later in this book.

Example 7.4.6   *For the random variable X in Example 6.6.1, we calculated in Examples 7.1.1 and 7.4.1 that $\mu_X = 1.29$ days and that $\sigma_X = 0.6371$ days. First, we compute the values of the standardized random variable $Z = (X - \mu_X)/\sigma_X$:*

x	1	2	3	4
z	-0.4552	1.1144	2.6840	4.2536
$p_X(x) = p_Z(z)$	0.80	0.12	0.07	0.01

It is easy to verify that, up to some rounding errors, $E(Z) = 0$ and $\sigma_Z^2 = 1$.

7.5   Other statistics

As for sample data, some other statistics can also be defined for random variables. First, we discuss statistics of central location, then statistics of relative location.

Definition 7.5.1   *The **mode** of a random variable X is the value x for which the probability distribution or probability density takes a maximum value.*

The mode does not exist if there is no maximum. Sometimes, a random variable has several modes.

Example 7.5.1   *The mode of X in Example 6.6.1 is 1 because the probability density $f_X(x)$ reaches its maximum at this value.*

Definition 7.5.2 *The **median** of a random variable X is the value $\gamma_{0.5}$ for which*

$$F_X(\gamma_{0.5}) = P(X \leq \gamma_{0.5}) \geq \frac{1}{2} \quad and \quad P(X \geq \gamma_{0.5}) \geq \frac{1}{2},$$

if X is discrete, or

$$F_X(\gamma_{0.5}) = \int_{-\infty}^{\gamma_{0.5}} f_X(x)dx = \int_{\gamma_{0.5}}^{+\infty} f_X(x)dx = \frac{1}{2},$$

if X is continuous.

Example 7.5.2 *The median of the continuous random variable X in Example 6.3.2 can be calculated as follows:*

$$\frac{1}{2} = \int_{-\infty}^{\gamma_{0.5}} f_X(x)dx,$$

$$= \int_{1}^{\gamma_{0.5}} f_X(x)dx,$$

$$= \int_{1}^{\gamma_{0.5}} \frac{1}{2}(3 - x)dx,$$

$$= \frac{1}{2} \left[3x - \frac{x^2}{2} \right]_{1}^{\gamma_{0.5}},$$

$$= \frac{1}{2} \left(3\gamma_{0.5} - \frac{\gamma_{0.5}^2}{2} - 3 + \frac{1}{2} \right),$$

$$= \frac{1}{2} \left(3\gamma_{0.5} - \frac{\gamma_{0.5}^2}{2} - \frac{5}{2} \right).$$

Solving this equation gives two roots, namely 1.586 and 4.414. Only one of them makes sense as a median, namely 1.586, because the random variable X can only take values between 1 and 3, so that the median should also be located between 1 and 3. Therefore, we conclude that the median, $\gamma_{0.5}$, is equal to 1.586.

As the arithmetic mean for sample data, the expected value is usually used as a measure of central location. In some cases, however, the median and the mode give a better view of the central location of a random variable. If, for example, the random variable X represents the amount of a money transfer to a bank account opened to help the victims of the Tsunami in December 2012, then the expected value will be quite large. This does not necessarily mean that an ordinary person transfers a large amount. The expected value of this random variable may be strongly influenced by a few large amounts that wealthy individuals and businesses transfer to the account. In that case, the median and the mode would give a better picture of what an ordinary person contributes.

Definition 7.5.3 *The* $100 \times p$-*th* **quantile** *value or the* $100 \times p$-*th* **percentile** γ_p *of a continuous random variable X satisfies*

$$p = \int_{-\infty}^{\gamma_p} f_X(x)dx = F_X(\gamma_p).$$

Definition 7.5.4 *The first, second and third* **quartile** *are* $\gamma_{0.25}$, $\gamma_{0.5}$, *and* $\gamma_{0.75}$, *respectively.*

The second quartile is the same as the median.

Example 7.5.3 *The first and third quartile of the random variable X in Example 6.3.2 are 1.268 and 2, respectively.*

Definition 7.5.5 **Pearson's population skewness coefficient** SP^{pop} *of a random variable X is equal to*

$$SP^{pop} = \frac{3(\mu_X - \gamma_{0.5})}{\sigma_X}.$$

This coefficient lies between -3 and $+3$. It is negative for a left-skewed, 0 for a symmetric, and positive for a right-skewed probability distribution or density.

Definition 7.5.6 *The* **skewness coefficient** *of Fisher of a random variable X is*

$$skewness\ coefficient = \frac{E\{(X - \mu_X)^3\}}{\sigma_X^3}.$$

This coefficient is also negative for a left-skewed, 0 for a symmetric, and positive for a right-skewed probability distribution or density.

Example 7.5.4 *The probability density in Figure 6.5 (Example 6.3.2) is obviously skewed to the right. Pearson's skewness coefficient* SP^{pop} *has a positive value for that density, namely* 0.513. *The skewness coefficient of Fisher from Definition 7.5.6 equals* 0.5657 *and is also positive.*

Definition 7.5.7 *The r-th* **non-central moment** *of a random variable X is*

$$\mu'_r = E(X^r), \qquad r = 0, 1, 2, \dots.$$

Obviously, the zeroth and the first non-central moment are equal to $\mu'_0 = 1$ and $\mu'_1 = \mu_X$, respectively.

Definition 7.5.8 *The r-th* **central moment** *of a random variable X is*

$$\mu_r = E[(X - \mu_X)^r], \qquad r = 0, 1, 2, \dots.$$

It is a useful exercise to check that the zeroth, first, and second central moment are equal to $\mu_0 = 1$, $\mu_1 = 0$, and $\mu_2 = \sigma_X^2$, respectively.

7.6   Moment generating functions

Finding the moments of a random variable is not always obvious. In some cases, the use of the so-called moment generating function enables us to determine the non-central moments relatively easily. The approach involving moment generating functions works both for discrete and for continuous random variables. The moment generating function is the expected value of a special function, namely e^{tX}:

Definition 7.6.1 *The **moment generating function** of a random variable X is*

$$m_X(t) = E(e^{tX}),$$

provided this expected value exists.

Theorem 7.6.1 *If the moment generating function $m_X(t)$ exists in an interval around $t = 0$, then the r-th non-central moment can be calculated as*

$$\mu'_r = E(X^r) = \left. \frac{d^r m_X(t)}{dt^r} \right|_{t=0}.$$

   This theorem means that, if we are looking for the r-th non-central moment of a random variable, we can take the r-th derivative of its moment generating function, and set the variable t in this derivative to 0.

Proof: To prove Theorem 7.6.1 for a continuous random variable X, we use the series expansion of the exponential function:

$$m_X(t) = \int_{-\infty}^{+\infty} e^{tx} f_X(x)\, dx = \int_{-\infty}^{+\infty} \left(\sum_{n=0}^{\infty} \frac{t^n}{n!} x^n \right) f_X(x)\, dx.$$

Interchanging the order of the integration and the summation[1], we can rewrite this as

$$m_X(t) = \sum_{n=0}^{\infty} \left(\int_{-\infty}^{+\infty} x^n f_X(x)\, dx \right) \frac{t^n}{n!} = \sum_{n=0}^{\infty} E(X^n) \frac{t^n}{n!}.$$

The first derivative of this expression with respect to t is

$$\frac{dm_X(t)}{dt} = \sum_{n=1}^{\infty} E(X^n) \frac{t^{n-1}}{(n-1)!}.$$

Setting $t = 0$ in this expression makes all terms in this sum zero, except for the first one:

$$\left. \frac{dm_X(t)}{dt} \right|_{t=0} = E(X^1) \frac{0^0}{0!} = E(X),$$

---

[1] To justify this step, one needs an argument from measure theory that is beyond the scope of this book.

since 0^0 and $0!$ are both one. The second derivative with respect to t is

$$\frac{d^2 m_X(t)}{dt^2} = \sum_{n=2}^{\infty} E(X^n) \frac{t^{n-2}}{(n-2)!}.$$

Setting $t = 0$ in this expression yields $E(X^2)$. The r-th derivative equals

$$\frac{d^r m_X(t)}{dt^r} = \sum_{n=r}^{\infty} E(X^n) \frac{t^{n-r}}{(n-r)!}.$$

Setting $t = 0$ in this expression yields $E(X^r)$, which completes the proof for a continuous random variable. The proof for a discrete random variable is completely analogous, and requires replacing the integral by a sum. ∎

We illustrate Theorem 7.6.1 using two examples. The first example shows how the moment generating function can be used for a discrete random variable, while the second example deals with the case of a continuous random variable.

Example 7.6.1 *A discrete random variable X has the following distribution:*

x_i	1	2	3	4
$p_X(x_i)$	0.15	0.25	0.35	0.25

The moment generating function of this random variable is

$$m_X(t) = E(e^{tX}),$$

$$= \sum_{i=1}^{4} e^{tx_i} p_X(x_i),$$

$$= 0.15 e^{tx_1} + 0.25 e^{tx_2} + 0.35 e^{tx_3} + 0.25 e^{tx_4},$$

$$= 0.15 e^{t} + 0.25 e^{2t} + 0.35 e^{3t} + 0.25 e^{4t}.$$

To find the first non-central moment of X, we determine the first derivative of the moment generating function with respect to t. This results in

$$\frac{dm_X(t)}{dt} = 0.15 e^{t} + 0.5 e^{2t} + 1.05 e^{3t} + 1 e^{4t}.$$

If we evaluate this derivative at $t = 0$, we obtain the first non-central moment, the expected value

$$\mu_1' = \mu_X = 0.15 + 0.5 + 1.05 + 1 = 2.7.$$

For the second non-central moment, we first need the second derivative of the moment generating function:

$$\frac{d^2 m_X(t)}{dt^2} = 0.15 e^{t} + 1 e^{2t} + 3.15 e^{3t} + 4 e^{4t}.$$

If we set t to 0 in this second derivative, we find the second non-central moment:

$$\mu_2' = E(X^2) = 0.15 + 1 + 3.15 + 4 = 8.3.$$

The variance of X can now easily be determined based on the first and second non-central moment:

$$\sigma_X^2 = E(X^2) - \mu_X^2 = 8.3 - (2.7)^2 = 1.01.$$

This example shows how Theorem 7.6.1 can be applied. However, in this example the use of the moment generating function to calculate the expected value and the variance is not the simplest or the quickest method. This is different in the following example. In this example, the method of the moment generating function is the fastest and easiest way to find the expected value and the variance.

Example 7.6.2 *A random variable X has the following probability density:*

$$f_X(x) = \begin{cases} \lambda e^{-\lambda x}, & x \geq 0, \\ 0, & \text{otherwise.} \end{cases}$$

The moment generating function of the probability density is

$$m_X(t) = E(e^{tX})$$

$$= \int_{-\infty}^{+\infty} e^{tx} f_X(x) \, dx,$$

$$= \int_0^{+\infty} e^{tx} \lambda e^{-\lambda x} \, dx,$$

$$= \lambda \int_0^{+\infty} e^{(t-\lambda)x} \, dx,$$

$$= \lambda \frac{1}{t-\lambda} \int_0^{+\infty} e^{(t-\lambda)x} \, d(t-\lambda)x,$$

$$= \frac{\lambda}{t-\lambda} [e^{(t-\lambda)x}]_0^{+\infty},$$

$$= \frac{\lambda}{t-\lambda} (e^{-\infty} - e^0),$$

$$= \frac{-\lambda}{t-\lambda},$$

$$= \frac{\lambda}{\lambda-t},$$

if $t < \lambda$.

To find the first non-central moment of X, we determine the first derivative of the moment generating function with respect to t. This results in

$$\frac{dm_X(t)}{dt} = \frac{\lambda}{(\lambda - t)^2}.$$

Therefore,

$$\mu_1' = \mu_X = E(X) = \frac{\lambda}{\lambda^2} = \frac{1}{\lambda}.$$

For the second non-central moment, we first need the second derivative of the moment generating function:

$$\frac{d^2 m_X(t)}{dt^2} = \frac{2\lambda}{(\lambda - t)^3}.$$

Setting t = 0 in this expression yields

$$\mu_2' = E(X^2) = \frac{2\lambda}{\lambda^3} = \frac{2}{\lambda^2},$$

so that

$$\sigma_X^2 = E(X^2) - \mu_X^2 = \frac{2}{\lambda^2} - \left(\frac{1}{\lambda}\right)^2 = \frac{1}{\lambda^2}.$$

8

Important discrete probability distributions

Of course, he was right to realize it was easy to win money on one of those machines, easier than earning it by working for a boss. But that was balanced by the fact that its much easier to lose money on that same machine, and probability was something our Heavy had never heard of.

(from *The Misfortunates,* by Dimitri Verhulst, p. 62)

In the previous chapters, we introduced random variables, probability distributions, and probability densities. In these chapters, we paid little attention to practical applications, and the examples were sometimes artificial. In this chapter, we focus on a number of probability distributions with important practical applications. Some important probability densities follow in the next chapter.

8.1 The uniform distribution

A fundamental distribution is the discrete uniform probability distribution. In this distribution, all possible outcomes have the same probability. If there are k possible outcomes, then each of these outcomes has probability $1/k$.

Example 8.1.1 *A die has six possible outcomes. In case the die is fair, each outcome is equally likely and has probability of 1/6. If X represents the number of dots obtained when throwing a die, then X is uniformly distributed.*

The mathematical expression for the discrete uniform distribution is

$$p_X(x; k) = P(X = x) = \frac{1}{k}, \qquad x = 1, \ldots, k.$$

Statistics with JMP: Graphs, Descriptive Statistics, and Probability, First Edition. Peter Goos and David Meintrup.
© 2015 John Wiley & Sons, Ltd. Published 2015 by John Wiley & Sons, Ltd.
Companion Website: wiley.com/go/goosandmeintrup

The corresponding cumulative distribution function is

$$F_X(x;k) = P(X \leq x) = \sum_{i=1}^{x} \frac{1}{k} = \frac{x}{k}, \qquad x = 1, \ldots, k.$$

The number of outcomes k is the parameter of the discrete uniform probability distribution.

Example 8.1.2 *The probability distribution of the number of dots obtained by throwing a fair die is*

$$p_X(x;6) = P(X = x) = \frac{1}{6}, \qquad x = 1, \ldots, 6,$$

and the cumulative distribution function is

$$F_X(x;6) = \frac{x}{6}, \qquad x = 1, \ldots, 6.$$

The graphical representations of these functions are shown in Figure 8.1.

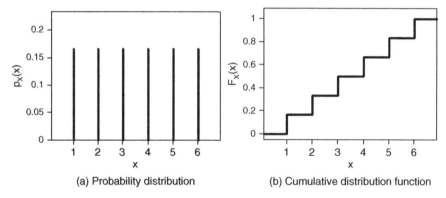

(a) Probability distribution   (b) Cumulative distribution function

Figure 8.1   Discrete uniform probability distribution for $k = 6$.

The expected value of a discrete uniformly distributed random variable X is

$$\mu_X = E(X) = \sum_{x=1}^{k} x p_X(x),$$

$$= \sum_{x=1}^{k} x \frac{1}{k},$$

$$= \frac{1}{k} \sum_{x=1}^{k} x,$$

$$= \frac{1}{k} \frac{k(k+1)}{2},$$

$$= \frac{k+1}{2}.$$

In this derivation, we made use of the fact that the sum of all integer numbers up to and including k is equal to $k(k + 1)/2$:

$$\sum_{x=1}^{k} x = \frac{k(k + 1)}{2}.$$

The variance of a discrete uniformly distributed random variable X is

$$\sigma_X^2 = \text{var}(X) = E(X^2) - (E(X))^2,$$

$$= \frac{1}{k} \sum_{x=1}^{k} x^2 - \left(\frac{k + 1}{2}\right)^2.$$

By induction, it can be shown that

$$\frac{1}{k} \sum_{x=1}^{k} x^2 = \frac{1}{k} \frac{k(k + 1)(2k + 1)}{6},$$

so that

$$\sigma_X^2 = \frac{1}{k} \frac{k(k + 1)(2k + 1)}{6} - \frac{(k + 1)^2}{4},$$

$$= \frac{2(k + 1)(2k + 1) - 3(k^2 + 2k + 1)}{12},$$

$$= \frac{k^2 - 1}{12}.$$

8.2   The Bernoulli distribution

A random variable X is Bernoulli[1] distributed if it takes the values 0 and 1 with probabilities $1 - \pi$ and π, respectively. The probability distribution is given by

$$p_X(x; \pi) = \pi^x (1 - \pi)^{1-x}, \quad x = 0, 1.$$

Since the parameter π represents a probability, it is required that $0 \leq \pi \leq 1$. It is a good exercise to show that the expected value of a Bernoulli distributed random variable is equal to

$$\mu_X = E(X) = \pi,$$

and that its variance is equal to

$$\sigma_X^2 = \text{var}(X) = \pi(1 - \pi).$$

---

[1] The distribution was named after Jacob Bernoulli whose life has already been described briefly on page 107.

Each random variable that can only take two values is Bernoulli distributed: flipping a coin (head/tail), the gender of a newborn (male/female), passing an exam (yes/no), the quality of a product (good/defective), and so on.

Example 8.2.1 *Assembled products often undergo inspection to verify that there are no defects. If an assembly line delivers $100\pi\%$ defective products, then the probability that a randomly drawn product is defective, is equal to π. When assigning the value 1 to a defective product and the value 0 to a good product, we obtain a Bernoulli distributed random variable.*

In Bernoulli processes, the terms "success" (abbreviated S) and "failure" (F) are typically associated with the outcomes 1 and 0. Therefore, the probability π, which is associated with the outcome 1, is also called the success rate. The probability $1 - \pi$ is called the failure rate. A process in which one observation is generated from a Bernoulli distribution is called a Bernoulli experiment.

Figure 8.2 shows three different Bernoulli distributions, namely those with success rates $\pi = 0.25$, $\pi = 0.5$, and $\pi = 0.75$, and the corresponding cumulative distribution functions.

8.3 The binomial distribution

8.3.1 Probability distribution

A binomial process consists of n consecutive Bernoulli experiments with the same success rate π that are performed independently of each other. The number of "successes" in a binomial process is a random variable. In the worst case, none of the n Bernoulli experiments yields a success. In the best case, each of the n Bernoulli experiments leads to success. Therefore, the number of successes can take the values $0, 1, 2, \ldots, n$. The probability distribution of the number of successes, which we call X, is the binomial distribution

$$p_X(x; n, \pi) = \frac{n!}{x!(n-x)!}\pi^x(1-\pi)^{n-x}, \quad x = 0, 1, 2, \ldots, n. \tag{8.1}$$

This expression represents the probability that n consecutive Bernoulli experiments yield exactly x successes. Of course, there are automatically $n - x$ failures if x successes are recorded. Consequently, the formula for the binomial distribution also gives the probability for $n - x$ failures. A logical consequence of this is that the number of failures is also binomially distributed[2].

---

[2] If we represent the number of failures in n Bernoulli experiments with success probability π using the random variable Y, then the probability distribution of Y equals

$$p_Y(y; n, 1 - \pi) = \frac{n!}{y!(n-y)!}(1-\pi)^y\pi^{n-y}, \quad y = 0, 1, 2, \ldots, n,$$

which is a binomial distribution with parameters n and $1 - \pi$.

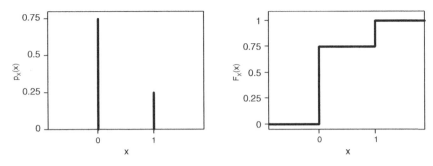

(a) Probability distribution with $\pi = 0.25$.   (b) Cumulative distribution function with $\pi = 0.25$.

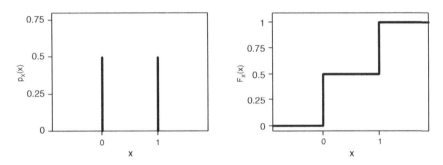

(c) Probability distribution with $\pi = 0.5$.   (d) Cumulative distribution function with $\pi = 0.5$.

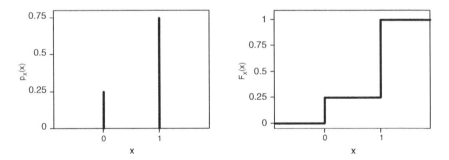

(e) Probability distribution with $\pi = 0.75$.   (f) Cumulative distribution function with $\pi = 0.75$.

Figure 8.2   Bernoulli probability distributions and corresponding cumulative distribution functions.

The values n and π are the parameters of the binomial distribution. If a random variable X is binomially distributed with parameters n and π, this is sometimes abbreviated to

$$X \sim \text{bin}(n, \pi).$$

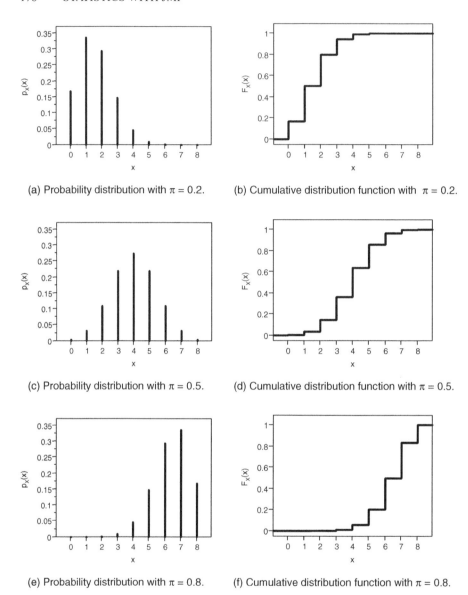

(a) Probability distribution with $\pi = 0.2$.

(b) Cumulative distribution function with $\pi = 0.2$.

(c) Probability distribution with $\pi = 0.5$.

(d) Cumulative distribution function with $\pi = 0.5$.

(e) Probability distribution with $\pi = 0.8$.

(f) Cumulative distribution function with $\pi = 0.8$.

Figure 8.3   Binomial distributions and cumulative distribution functions for $n = 8$.

The parameter n has to be a positive integer, while the success probability π needs to be between 0 and 1, just as in a Bernoulli distribution. If $n = 1$, the binomial distribution is simply a Bernoulli distribution.

Figure 8.3 shows three binomial distributions and the corresponding cumulative distribution functions. The corresponding probabilities are listed in Table 8.1.

Table 8.1   Binomial probability distributions for $n = 8$ and $\pi = 0.2, 0.5$ and 0.8.

x	$\pi = 0.2$		$\pi = 0.5$		$\pi = 0.8$	
	$p_X(x)$	$F_X(x)$	$p_X(x)$	$F_X(x)$	$p_X(x)$	$F_X(x)$
0	0.1677722	0.1677722	0.0039063	0.0039063	0.0000026	0.0000026
1	0.3355443	0.5033165	0.0312500	0.0351563	0.0000819	0.0000845
2	0.2936013	0.7969178	0.1093750	0.1445313	0.0011469	0.0012314
3	0.1468006	0.9437184	0.2187500	0.3632813	0.0091750	0.0104064
4	0.0458752	0.9895936	0.2734375	0.6367188	0.0458752	0.0562816
5	0.0091750	0.9987686	0.2187500	0.8554688	0.1468006	0.2030822
6	0.0011469	0.9999155	0.1093750	0.9648438	0.2936013	0.4966835
7	0.0000819	0.9999974	0.0312500	0.9960938	0.3355443	0.8322278
8	0.0000026	1.0000000	0.0039063	1.0000000	0.1677722	1.0000000

The probability distributions show two interesting properties, which are valid for any value of the parameter n:

1. First, it is clear from Figure 8.3c and Table 8.1 that the binomial distribution is symmetrical if $\pi = 0.5$. The probability of one success, for example, is identical to the probability of seven successes. The same is true for the probability of three and five successes. In general, one can say that $P(X = x) = P(X = n - x)$ for a binomially distributed random variable X with success rate $\pi = 0.5$. An alternative way of writing this equality is $p_X(x) = p_X(n - x)$. This implies that $P(X \leq x) = P(X \geq n - x)$ if $\pi = 0.5$.

2. In addition, the probability distributions in Figure 8.3a and Figure 8.3e mirror each other. In Table 8.1, for example, the probability of one success with $\pi = 0.2$ is identical to the probability of seven successes with $\pi = 0.8$. The reason for this is that the probability of success for the probability distribution in Figure 8.3a, namely 0.2, is the complement of the success probability of the probability distribution in Figure 8.3e, namely 0.8. Therefore, we can interpret Figure 8.3a as the probability distribution of the number of successes in $n = 8$ Bernoulli experiments with success rate $\pi = 0.2$, and Figure 8.3e as the probability distribution of the number of failures in the same eight Bernoulli experiments. Conversely, Figure 8.3e can also be interpreted as the probability distribution of the number of successes in $n = 8$ Bernoulli experiments with success rate $\pi = 0.8$, and Figure 8.3a as the probability distribution of the number of failures in the same eight Bernoulli experiments.

To understand the origin of the mathematical expression of the binomial probability distribution, it is useful to first calculate the probability of a particular series of outcomes, for example, S, S, F, S, F, ... , F, S. Suppose that, in this series, outcome S appears x times and outcome F appears $n - x$ times. The multiplication rule (see Section 5.1.2) can be used to compute the probability of this sequence:

$$\pi\pi(1-\pi)\pi(1-\pi)\ldots(1-\pi)\pi = \pi^x(1-\pi)^{n-x}.$$

It is not difficult to see that the same probability is obtained for any sequence involving x occurrences of the outcome S and $n-x$ occurrences of the outcome F. From combinatorics (see Section 5.2.1), we know that there are exactly

$$\binom{n}{x} = \frac{n!}{x!(n-x)!}$$

such sequences. Since all different sequences with x occurrences of outcome S and $n-x$ occurrences of outcome F are mutually exclusive events, the probability of x successes (and, consequently, $n-x$ of failures) equals

$P(x \text{ successes}) = P(\text{first possible sequence with } x \text{ successes and } n-x \text{ failures}$

$\qquad$ OR second possible sequence with x successes and $n-x$ failures

$\qquad \ldots$

$\qquad$ OR last possible sequence with x successes and $n-x$ failures$)$,

$\qquad = P(\text{first possible sequence with } x \text{ successes and } n-x \text{ failures})$

$\qquad + P(\text{second possible sequence with } x \text{ successes and } n-x \text{ failures})$

$\qquad + \ldots$

$\qquad + P(\text{last possible sequence with } x \text{ successes and } n-x \text{ failures})$,

$\qquad = \pi^x(1-\pi)^{n-x} + \pi^x(1-\pi)^{n-x} + \cdots + \pi^x(1-\pi)^{n-x},$

$\qquad = \dfrac{n!}{x!(n-x)!}\pi^x(1-\pi)^{n-x}.$

Example 8.3.1 *Historical data of a production process indicate a defect rate of $\pi = 0.10$. Assume that the defects appear independently of each other and calculate*

1. the probability that exactly 2 out of 20 produced units are defective, and

2. the probability that at least 3 out of 20 produced units are defective.

Let X be the number of defects in the 20 products under investigation. The random variable X is binomially distributed with parameters $\pi = 0.10$ and $n = 20$. The first probability we look for is

$$P(X = 2) = p_X(2; 20, 0.1) = \frac{20!}{2!(20-2)!}0.1^2(1-0.1)^{20-2} = \frac{20!}{2!18!}0.1^2(0.9)^{18}$$

$$= \frac{20 \times 19}{2}0.1^2(0.9)^{18} = 0.2852.$$

This value can easily be computed with a calculator. Alternatively, one can use JMP. For the probability $P(X = 2)$, we need the formula "Binomial Probability(0.1,

20, 2)". *Generally, the formula for a probability of the type* $P(X = x)$ *for a binomially distributed random variable with parameters n and π is "Binomial Probability(π, n, x)". More details on this formula can be found in Section 8.8.*

The second probability is

$$P(X \geq 3) = 1 - P(X \leq 2) = 1 - P(X = 0) - P(X = 1) - P(X = 2),$$

$$= 1 - 0.1216 - 0.2702 - 0.2852 = 0.3231.$$

The probability $P(X \leq 2)$ can also be calculated directly in JMP using the function "Binomial Distribution(0.1, 20, 2)".

There are large tables to determine probabilities based on the binomial distribution. Some of them are given in Appendix B. These tables contain probabilities of the type $P(X \geq x)$. As computers and appropriate software are now widespread, these tables become increasingly less useful. Nevertheless, it is a good exercise to reconstruct the probabilities calculated in Example 8.3.1 with the help of the tables. Another useful exercise is to reconstruct one of the tables in Appendix B by yourself, using the binomial distribution in Equation (8.1).

A limitation of the tables in Appendix B is that they only contain probabilities for $\pi \leq 0.5$. To determine probabilities for larger values of π, we have to make use of the fact that the number of failures, Y, in n consecutive Bernoulli experiments is binomially distributed with parameter $1 - \pi$ if the number of successes, X, is binomially distributed with parameter π. This is illustrated in the following example:

Example 8.3.2 *A football player takes 10 consecutive penalty kicks. For each of the penalties, his probability to score a goal is 0.75. We determine the probability that the player scores at least 8 times. We denote the number of goals by X and the number of missed penalties by Y. The random variables X and Y are both binomially distributed with n = 10 as the first parameter. The random variable X has $\pi = 0.75$ as its second parameter, while Y has $\pi = 0.25 = 1 - 0.75$ as its second parameter. Note that X + Y = 10 in this example. Indeed, if, for example, four penalty kicks are scored, inevitably six penalties are missed.*

First, the probability of at least 8 goals is

$$P(X \geq 8) = P(X = 8) + P(X = 9) + P(X = 10),$$

$$= 0.2816 + 0.1877 + 0.0563,$$

$$= 0.5256.$$

This can be verified using a calculator, or using JMP. If, however, in the absence of a calculator or computer, we want to use the tables in Appendix B, we have to rewrite the probabilities in terms of binomially distributed random variables with a success rate smaller than or equal to 0.5. This can be done as follows:

$$P(X \geq 8) = P(Y \leq 2) = 1 - P(Y > 2) = 1 - P(Y \geq 3).$$

Indeed, the probability to make at least eight goals is equal to the probability that at most two penalty kicks are missed. The probability $P(Y \geq 3)$ can be easily retrieved in *the table in Appendix B, for* $n = 10$ *and* $\pi = 0.25$. *This probability is equal to* 0.4744, *so that*

$$P(X \geq 8) = 1 - P(Y \geq 3) = 1 - 0.4744 = 0.5256.$$

Example 8.3.3 *The manager of a hotel has 200 rooms available for rent. The past few years, the manager noted that 5% of all people who reserved a room did not show up. Therefore, he begins to systematically accept more bookings than he has rooms. However, he does not want to overbook too much so that he does not have to disappoint too many potential customers. More specifically, the manager does not want to run out of rooms on more than 1% of the evenings. How many overbookings can the manager afford?*

First, we need to translate the risk that the manager accepts into a probability. Suppose that we are using the random variable X to describe the number of guests that show up at the hotel on a certain day with a valid room reservation. Then, the probability that the manager has to disappoint one or more guests with a valid reservation is $P(X > 200)$ *or* $P(X \geq 201)$. *The manager now needs to determine the maximum number of bookings for which this probability does not exceed* 1%:

$$P(X \geq 201) \leq 0.01.$$

The probability $P(X \geq 201)$ *can be calculated using the binomial distribution. The number of bookings is the parameter n of the binomial distribution, while* $\pi = 0.95$ *is the probability that a guest who booked a room actually shows up. What we need to do to solve the hotel manager's problem, is to compute the probability*

$$P(X \geq 201) = 1 - P(X \leq 200) = 1 - F_X(200)$$

for different values of n. The biggest n for which this probability is smaller than or equal to 1% *is the maximal number of bookings that the manager can accept. As shown in Figure 8.4, it is particularly easy to perform these calculations in JMP.*

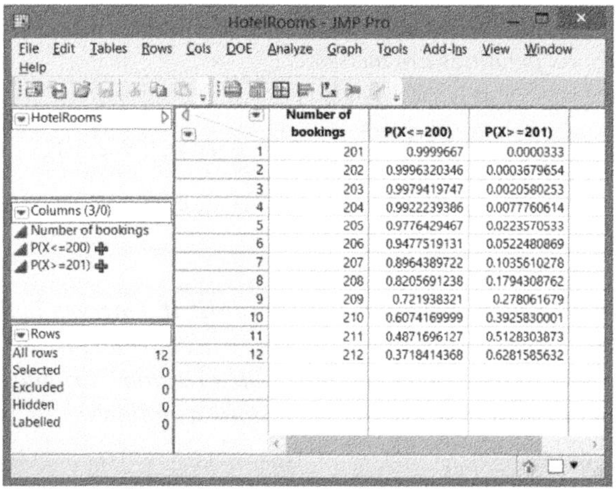

	Number of bookings	P(X<=200)	P(X>=201)
1	201	0.9999667	0.0000333
2	202	0.9996320346	0.0003679654
3	203	0.9979419747	0.0020580253
4	204	0.9922239386	0.0077760614
5	205	0.9776429467	0.0223570533
6	206	0.9477519131	0.0522480869
7	207	0.8964389722	0.1035610278
8	208	0.8205691238	0.1794308762
9	209	0.721938321	0.278061679
10	210	0.6074169999	0.3925830001
11	211	0.4871696127	0.5128303873
12	212	0.3718414368	0.6281585632

Figure 8.4 Solution of the problem of the hotel manager in Example 8.3.3 using the binomial distribution.

*It can be seen in the figure that the largest number of bookings that meets the condi-
tion is 204. The formula needed for the calculation of the column "$P(X \le 200)$" in
Figure 8.4 in JMP is "Binomial Distribution(0.95, Number of bookings, 200)".*

8.3.2   Expected value and variance

The expected value of a binomially distributed random variable X is equal to

$$\mu_X = E(X) = n\pi,$$

while the variance is given by

$$\sigma_X^2 = \text{var}(X) = n\pi(1 - \pi).$$

This is easy to prove using the fact that a binomially distributed random variable
with parameters n and π is the sum of n independent Bernoulli distributed random
variables $X_1, X_2, \ldots, X_n$. The expected value of X can be computed as[3]

$$
\begin{aligned}
E(X) &= E(X_1 + X_2 + \cdots + X_n), \\
&= E(X_1) + E(X_2) + \cdots + E(X_n), \\
&= \pi + \pi + \cdots + \pi, \\
&= n\pi.
\end{aligned}
$$

The variance of X can be calculated as[4]

$$
\begin{aligned}
\text{var}(X) &= \text{var}(X_1 + X_2 + \cdots + X_n), \\
&= \text{var}(X_1) + \text{var}(X_2) + \cdots + \text{var}(X_n), \\
&= \pi(1 - \pi) + \pi(1 - \pi) + \cdots + \pi(1 - \pi), \\
&= n\pi(1 - \pi).
\end{aligned}
$$

Example 8.3.4   *A football player takes 10 consecutive penalty kicks. For each of the
penalties, his probability to score a goal is 0.75. The expected value of the number
of goals in the 10 penalties is $10(0.75) = 7.5$. The variance is $10(0.75)(1 - 0.75) =$
1.875.*

   A random variable that is closely linked to a binomially distributed random variable
X is the fraction of successes $\hat{P} = X/n$. This fraction is a linear function of X, so that
$E(\hat{P}) = E(X)/n = \pi$ and $\text{var}(\hat{P}) = \text{var}(X)/n^2 = \pi(1 - \pi)/n$ (see Theorems 7.3.1 and
7.4.1 for the expected value and variance of a linear function of a random variable).

---

[3] In Chapter 12, we will learn that the expected value of the sum of a number of random variables is
equal to the sum of the individual expected values.

[4] In Chapter 13, we will learn that the variance of the sum of a number of independent random variables
is equal to the sum of the individual variances.

Finally, it is interesting to note that the binomial distribution can be approximated by the normal probability density. This is discussed in detail in Section 14.4 and is important for some hypothesis tests (see the book *Statistics with JMP: Hypothesis Tests, ANOVA and Regression*).

8.4   The hypergeometric distribution

In the context of the binomial distribution, it is assumed that a series of Bernoulli experiments are executed independently of each other. The success rate, π, always remains the same. This is not the case for the hypergeometric distribution, where the success probability is different for each Bernoulli experiment. We illustrate this in the following example.

Example 8.4.1 *Suppose that an urn contains 4 red and 5 blue balls, and we randomly draw three balls from the urn. Drawing a ball from the urn is of course a Bernoulli experiment because there are only two possible outcomes: we either obtain a red ball or a blue one. Suppose that, for us, drawing a red ball is a success, while drawing a blue ball corresponds to a failure. The probability of success in the first draw is 4/9, since there are 4 red balls and a total of 9. Suppose now that we do not put the first ball drawn back into the urn before drawing a second ball. The probability of success in the second drawing has two possible values. If the first ball drawn is red, then the probability of success in the second drawing is 3/8 (there are still 3 red balls in the urn, and a total of 8). If the first ball drawn is blue, then the success probability is 4/8 (there are still 4 red balls in the urn, and a total of 8). In the third drawing, the success rates are 2/7, 3/7, or 4/7, depending on the number of red balls that we obtain in the first two drawings. This shows that the probability of success changes with each drawing, and that the successive Bernoulli experiments are not independent. The probability of success, in any given Bernoulli experiment, depends on the results of the previous Bernoulli experiments.*

We consider a population of N elements: A elements of type I (successes) and $B = N - A$ elements of type II (failures). From this population, a sample of n elements is randomly drawn without replacement. Of course, the number of drawings, n, cannot be bigger than the number of elements in the population, N. The random variable X, the number of successes in the sample, then follows a so-called hypergeometric distribution:

$$p_X(x; N, A, n) = \frac{\binom{A}{x}\binom{N-A}{n-x}}{\binom{N}{n}}, \quad \max(0, n - (N - A)) \leq x \leq \min(n, A).$$

This distribution has three independent parameters: N, n, and A. The fourth parameter, B, is not independent: B can always be calculated from N and A.

A hypergeometrically distributed random variable X has expected value

$$\mu_X = E(X) = n\frac{A}{N}$$

and variance

$$\sigma_X^2 = n\left(\frac{A}{N}\right)\left(\frac{B}{N}\right)\left(\frac{N-n}{N-1}\right).$$

If we rewrite the fraction of successes A/N as π and the fraction of failures B/N as $1 - \pi$, then we get the expected value

$$\mu_X = E(X) = n\pi$$

and the variance

$$\sigma_X^2 = n\pi(1 - \pi)\left(\frac{N-n}{N-1}\right).$$

The expected values of the hypergeometric distribution and of the binomial distribution are the same, but the variances of the two distributions differ (with the variance of the hypergeometric distribution being the smallest). If N is large compared to n, the variances of the two distributions are almost the same, and the binomial distribution can be used as an approximation of the hypergeometric distribution.

Hypergeometric probabilities can be calculated in JMP. For the probability $P(X = x)$, we need the function "Hypergeometric Probability(N, A, n, x)". For the probability $P(X \leq x)$, we can use the command "Hypergeometric Distribution(N, A, n, x)".

Example 8.4.2 *In the classic Belgian lottery, 6 numbers were randomly drawn from a total of 42. On the eve of the lottery drawing, you had 6 of the 42 numbers marked on your lottery ticket. This meant that just before the draw, the population of 42 balls in the drum consisted of 6 good balls and 36 bad ones. If the 6 numbers drawn matched the numbers marked on your ticket, you won a large sum of money. The hypergeometric distribution with $N = 42$, $A = 6$, $B = 36$ and $n = 6$ can be used to calculate the probability that you had all 6 numbers right, or 5 out of the 6, or 4 out of the 6, and so on:*

$$P(X = 6) = \frac{\binom{6}{6}\binom{36}{0}}{\binom{42}{6}} = \frac{(1)(1)}{\frac{42!}{36!6!}} = \frac{1}{5245786},$$

$$P(X = 5) = \frac{\binom{6}{5}\binom{36}{1}}{\binom{42}{6}} = \frac{(6)(36)}{\frac{42!}{36!6!}} = \frac{216}{5245786},$$

$$P(X = 4) = \frac{\binom{6}{4}\binom{36}{2}}{\binom{42}{6}} = \frac{(15)(630)}{\frac{42!}{36!6!}} = \frac{9450}{5245786}.$$

These results match those in Examples 4.4.5 and 5.1.9. The reasoning followed in Example 5.1.9 is the basis for understanding the logic behind the mathematical expression of the hypergeometric probability distribution.

Table 8.2   Probabilities for two hypergeometric distributions with $N = 9$, $A = 4$ and $B = 5$ (see Figure 8.5).

x	$n = 3$		$n = 6$	
	$p_X(x)$	$F_X(x)$	$p_X(x)$	$F_X(x)$
0	0.119047619	0.119047619	–	–
1	0.476190476	0.595238095	0.047619048	0.047619048
2	0.357142857	0.952380952	0.357142857	0.404761905
3	0.047619048	1	0.476190476	0.880952381
4	–	–	0.119047619	1
5	–	–	–	–
6	–	–	–	–

Example 8.4.3   *The expected value of the number of red balls that we obtain from $n = 3$ drawings in Example 8.4.1, where $N = 9$, $A = 4$ and $B = 9 - 4 = 5$, is $3(4/9) = 4/3$. The variance is $3(4/9)(5/9)\{(9 - 3)/(9 - 1)\} = 5/9$. The probability distribution of the number of red balls is shown in Figure 8.5a. The probability distribution clearly shows that there are four possible outcomes if we perform three drawings, namely 0, 1, 2, or 3 red balls. Each of these outcomes has a strictly positive probability. Table 8.2 shows the probabilities of the four possible outcomes in detail.*

Example 8.4.4   *The expected value of the number of red balls that we obtain from $n = 6$ draws instead of $n = 3$ draws in Example 8.4.1 is $6(4/9) = 8/3$. The variance is $6(4/9)(5/9)\{(9 - 6)/(9 - 1)\} = 5/9$. The probability distribution of the number of red balls is shown in Figure 8.5b. The probability distribution shows that there still are four possible outcomes if we perform six drawings, namely 1, 2, 3, or 4 red balls. Of course, it is impossible to draw more than four red balls, since the urn contained only four such balls at the start of the experiment. It is not possible not to draw at least one red ball because the urn contains only five non-red balls and we perform six draws. Table 8.2 provides the details of the probabilities for the four possible outcomes.*

Example 8.4.5   *In a chemical plant, 5000 bottles were filled with a certain substance. There is a suspicion that the content of 300 of the bottles is contaminated. How many bottles should be randomly drawn from the lot of 5000 and be inspected to have a chance of at least 99% that the contamination is detected? The contamination is considered as being detected as soon as one contaminated bottle is found. Suppose that the random variable X represents the number of contaminated bottles discovered.*

In this problem, one wants to know the minimum number of bottles n to inspect for which $P(X \geq 1) \geq 0.99$. In this expression, X is a hypergeometric random variable with $N = 5000$ and $A = 300$. The condition to be satisfied can be rewritten as $1 - P(X = 0) \geq 0.99$ or $P(X = 0) \leq 0.01$.

The attentive reader will quickly realize that this problem is similar to the one in Example 8.3.3. With the formula "Hypergeometric Probability(5000, 300, Number

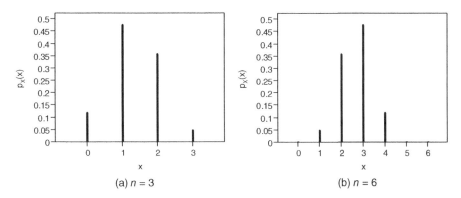

Figure 8.5 Hypergeometric distributions for $N = 9$, $A = 4$, and $B = 5$ (see Examples 8.4.1, 8.4.3, and 8.4.4).

	Number of bottles	P(X=0)	P(X>=1)
62	62	0.0210549058	0.9789450942
63	63	0.0197757499	0.9802242501
64	64	0.0185740637	0.9814259363
65	65	0.01744517	0.98255483
66	66	0.0163846734	0.9836153266
67	67	0.0153884427	0.9846115573
68	68	0.0144525958	0.9855474042
69	69	0.0135734842	0.9864265158
70	70	0.012747679	0.987252321
71	71	0.0119719582	0.9880280418
72	72	0.0112432936	0.9887567064
73	73	0.0105588399	0.9894411601
74	74	0.0099159229	0.9900840771
75	75	0.0093120299	0.9906879701
76	76	0.0087447997	0.9912552003
77	77	0.0082120133	0.9917879867
78	78	0.0077115859	0.9922884141
79	79	0.0072415584	0.9927584416
80	80	0.0068000897	0.9931999103
81	81	0.00638545	0.99361455

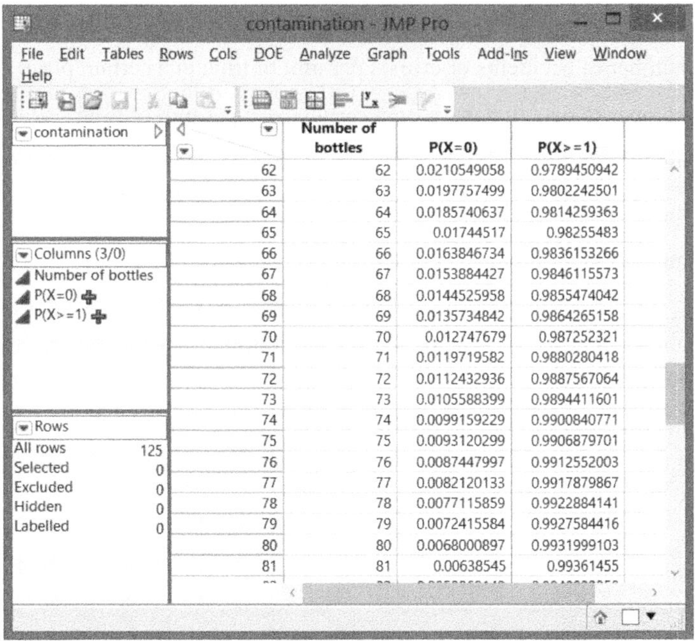

Figure 8.6 Solution to the problem of the chemical production plant in Example 8.4.5 based on the hypergeometric probability distribution.

of bottles, 0)" in JMP, it is not difficult to find out that $P(X = 0)$ drops below 0.01 as soon as $n \geq 74$. Hence, a sample of 74 bottles is the minimum required in order to have a sufficiently high probability to discover the contamination. The data table, which contains hypergeometric probabilities for several numbers of bottles to inspect, is shown in Figure 8.6. If the company wants a 99.9% probability to discover the contamination, a sample of 111 bottles is needed. This naturally leads to higher costs.

Since the sample in this example is made without replacement (indeed, repeating the inspection of a bottle makes no sense), the hypergeometric rather than the binomial distribution was used. Since N is a lot bigger than n, the binomial distribution can be used as an approximation. With JMP, it is a small exercise to find out that, using the binomial distribution, the minimum sample size is 75 for $P(X \geq 1)$ to be larger than or equal to 0.99, and 112 for $P(X \geq 1)$ to be larger than or equal to 0.999. Since we were able to find the correct probability by means of the hypergeometric distribution, the approach using the binomial distribution is only of theoretical interest.

8.5   The Poisson distribution

A particularly interesting probability distribution is the so-called Poisson distribution[5]. The reason for this is that a whole range of random variables in practice are (approximately) Poisson distributed. Typical examples are

- the number of defects in a material per unit length or per unit area,
- the number of accidents or crimes per unit of time in a certain place,
- the number of cars that drive on a certain road per unit of time,
- the number of customers that enter in a bank per unit of time,
- the number of bacteria in a liquid per unit volume, and
- the number of earthquakes per unit of time in a specific area.

The reason why these random variables are often Poisson distributed is that the processes underlying them typically satisfy the following mathematical conditions[6] for a Poisson process quite well:

1. Events do not occur in clusters: the probability of at least two events in a very short time interval is negligibly small in comparison with the probability of one or no event.

2. The probability of an event in a short time interval is constant over time. This condition implies a constant occurrence rate over time, for example five events per minute.

3. The occurrences of events in two non-overlapping time intervals are independent.

---

[5] Siméon Denis Poisson (1781–1840) was a leading French mathematician who made important contributions in mechanics and physics. In Paris, he was a student of Laplace and Lagrange. He published the Poisson distribution for the first time in "Recherches sur la probabilité des jugements en matière criminelle et matière civile" in 1837. He also introduced the "Law of large numbers", a theorem that is related to the central limit theorem (see Chapter 14).

[6] In the description of a Poisson process, the term time interval is used, but this can be replaced by length, volume, area …

The probability distribution of a Poisson distributed random variable X is

$$p_X(x; \lambda) = \frac{e^{-\lambda} \lambda^x}{x!}, \quad x = 0, 1, 2, \ldots, \tag{8.2}$$

where λ is a strictly positive parameter. No upper limit is imposed on X. Examples of the Poisson distribution for values of λ equal to 1, 2, 3, and 4.5 are given in Figures 8.7–8.10. The corresponding probabilities are listed in Table 8.3. Note that the largest value of x on the horizontal axis in the figures and in the table is 10. It is possible that a Poisson distributed random variable takes values greater than 10, but this is not likely for λ equal to 1, 2, 3, or 4.5. Therefore, we did not include larger values of x in the figures or in the table. A random variable X that is Poisson distributed with parameter λ is sometimes denoted by

$$X \sim \text{Poisson}(\lambda).$$

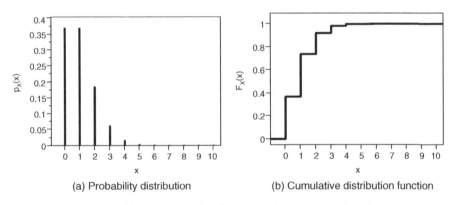

(a) Probability distribution   (b) Cumulative distribution function

Figure 8.7   Poisson distribution with parameter $\lambda = 1$.

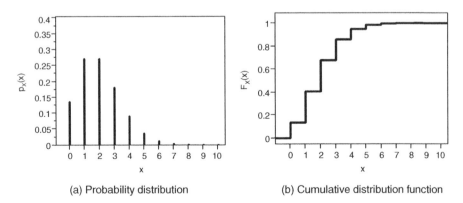

(a) Probability distribution   (b) Cumulative distribution function

Figure 8.8   Poisson distribution with parameter $\lambda = 2$.

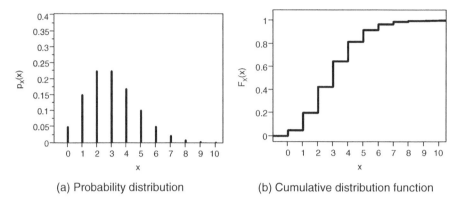

(a) Probability distribution   (b) Cumulative distribution function

Figure 8.9   Poisson distribution with parameter λ = 3.

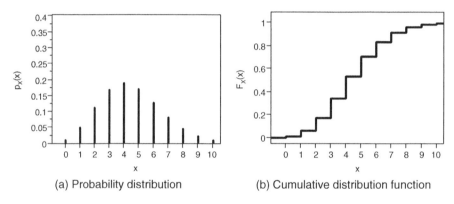

(a) Probability distribution   (b) Cumulative distribution function

Figure 8.10   Poisson distribution with parameter λ = 4.5.

Table 8.3   Probabilities for the Poisson distributions in Figures 8.7–8.10.

x	$\lambda = 1$		$\lambda = 2$		$\lambda = 3$		$\lambda = 4.5$	
	$p_X(x)$	$F_X(x)$	$p_X(x)$	$F_X(x)$	$p_X(x)$	$F_X(x)$	$p_X(x)$	$F_X(x)$
0	0.3678794	0.3678794	0.1353353	0.1353353	0.0497871	0.0497871	0.0111090	0.0111090
1	0.3678794	0.7357589	0.2706706	0.4060058	0.1493612	0.1991483	0.0499905	0.0610995
2	0.1839397	0.9196986	0.2706706	0.6766764	0.2240418	0.4231901	0.1124786	0.1735781
3	0.0613132	0.9810118	0.1804470	0.8571235	0.2240418	0.6472319	0.1687179	0.3422960
4	0.0153283	0.9963402	0.0902235	0.9473470	0.1680314	0.8152632	0.1898076	0.5321036
5	0.0030657	0.9994058	0.0360894	0.9834364	0.1008188	0.9160821	0.1708269	0.7029304
6	0.0005109	0.9999168	0.0120298	0.9954662	0.0504094	0.9664915	0.1281201	0.8310506
7	0.0000730	0.9999898	0.0034371	0.9989033	0.0216040	0.9880955	0.0823629	0.9134135
8	0.0000091	0.9999989	0.0008593	0.9997626	0.0081015	0.9961970	0.0463292	0.9597427
9	0.0000010	0.9999999	0.0001909	0.9999535	0.0027005	0.9988975	0.0231646	0.9829073
10	0.0000001	1.0000000	0.0000382	0.9999917	0.0008102	0.9997077	0.0104241	0.9933313

The expected value of a Poisson distributed random variable is equal to

$$
\begin{aligned}
E(X) &= \sum_{x=0}^{+\infty} x \frac{\lambda^x e^{-\lambda}}{x!}, \\
&= \sum_{x=1}^{+\infty} x \frac{\lambda^x e^{-\lambda}}{x!}, \\
&= \sum_{x=1}^{+\infty} \frac{\lambda^x e^{-\lambda}}{(x-1)!}, \\
&= e^{-\lambda} \lambda \sum_{x=1}^{+\infty} \frac{\lambda^{x-1}}{(x-1)!}, \\
&= e^{-\lambda} \lambda \sum_{z=0}^{+\infty} \frac{\lambda^z}{z!} \text{ with } z = x - 1, \\
&= e^{-\lambda} \lambda \, e^{\lambda}, \\
&= \lambda.
\end{aligned}
$$

The one but last step in the derivation uses the series expansion of the exponential function e^{λ}. It can also be demonstrated that

$$
\sigma_X^2 = \text{var}(X) = \lambda
$$

for a Poisson distributed random variable X.

An important property of the Poisson distribution is the following: if the number of events per time unit is Poisson distributed with parameter λ, then the number of events in t time units is also Poisson distributed, but with parameter λt. Of course, the same applies to lengths: if the number of events per unit of length is Poisson distributed with parameter λ, the number of events over d units of length is Poisson distributed with parameter λd. This property is used in Example 8.5.1.

Probabilities for Poisson distributed random variables can be calculated in JMP. For a probability $P(X = x)$, we have the function "Poisson Probability(λ, x)", while the function "Poisson Distribution(λ, x)" is needed for a probability $P(X \le x)$. Appendix C contains some tables with probabilities of the type $P(X \ge x)$ for Poisson distributed random variables. As with the binomial distribution, these tables can be reconstructed by hand, using the probability distribution in Equation (8.2).

Example 8.5.1 *The number of calls for a taxi service is Poisson distributed with an average of 30 calls per hour or 0.5 calls per minute. The probability of not receiving any calls during a period of three minutes is equal to*

$$
p_X(0; \lambda = 1.5) = \frac{(1.5)^0 e^{-1.5}}{0!} = e^{-1.5} = 0.223.
$$

Note that λ was set to 1.5 here because a calling rate of 0.5 per minute corresponds to a rate of 1.5 per 3 minutes. The JMP function needed to compute this probability is "Poisson Probability(1.5, 0)". The probability of more than 5 calls in an interval of 5 minutes ($\lambda = 2.5$) is given by

$$P(X \geq 6) = \sum_{x=6}^{+\infty} \frac{(2.5)^x e^{-2.5}}{x!} = 1 - \sum_{x=0}^{5} \frac{(2.5)^x e^{-2.5}}{x!} = 0.042.$$

The latter can also be calculated with the function "1−Poisson Distribution(2.5, 5)" in JMP.

Example 8.5.2 *A street hawker, whose foreign diploma in statistics is not recognized by the German authorities, sells flowers every evening in the restaurants and bars of Berlin. Because the flowers are worthless the next day, the vendor wants the stock he buys per day to be as small as possible. However, to maximize his profit, he wants to avoid the flowers selling out too often. Suppose that he is selling an average of 20 flowers per night and that he wants to be able to meet with the demand for flowers on at least 90% of evenings. The man explores his statistical knowledge and argues that his stock v must satisfy*

$$P(X \leq v) \geq 0.90 \quad or \quad P(X > v) \leq 0.10,$$

where the random variable X represents the number of flowers demanded during one night. That random variable is Poisson distributed with parameter $\lambda = 20$. He uses a similar approach as the hotel manager in Figure 8.4 and finds out that a stock of 26 units is large enough. The solution approach for the flower vendor is shown in Figure 8.11.

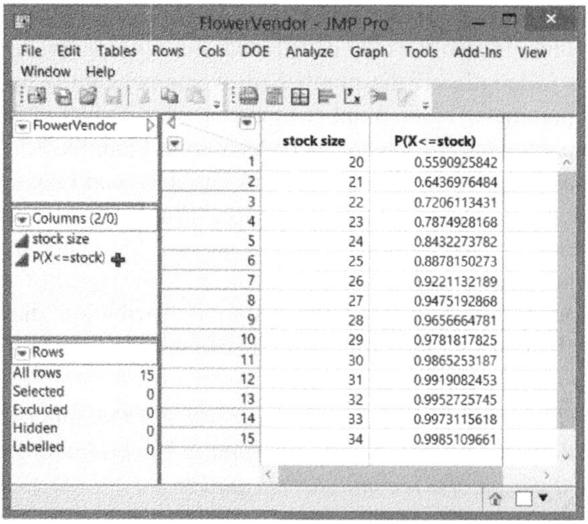

	stock size	P(X<=stock)
1	20	0.5590925842
2	21	0.6436976484
3	22	0.7206113431
4	23	0.7874928168
5	24	0.8432273782
6	25	0.8878150273
7	26	0.9221132189
8	27	0.9475192868
9	28	0.9656664781
10	29	0.9781817825
11	30	0.9865253187
12	31	0.9919082453
13	32	0.9952725745
14	33	0.9973115618
15	34	0.9985109661

Figure 8.11 *Solution to the problem of the flower vendor in Example 8.5.2.*

Example 8.5.3 *At the Football World Cup in France in* 1998, *in* 48 *games involving* $48 \times 2 = 96$ *playing teams,* 126 *goals were scored. The frequencies of the numbers of goals scored by all playing teams are shown in Table 8.4. On average, the number of goals per match per team was* $126/96 = 1.3125$. *This average is simply the total number of goals,* 126, *divided by the number of occasions on which a team was playing,* 96.

Table 8.4   Frequency of the numbers of goals scored by all teams playing in the Football World Cup in 1998 compared with a Poisson distribution with parameter $\lambda = 1.3125$.

x	0	1	2	3	4	5	6
Frequency	26	34	24	8	1	2	1
$p_X(x; \lambda)$	0.2691	0.3533	0.2318	0.1014	0.0333	0.0087	0.0019
$96 \times p_X(x; \lambda)$	25.84	33.91	22.26	9.74	3.19	0.84	0.18

Table 8.4 also shows the probability distribution $p_X(x, \lambda)$ *of a Poisson distributed random variable with parameter* $\lambda = 1.3125$. *The probabilities* $p_X(x, \lambda)$ *indicate how likely it is that a team scores x goals in one match at the World Cup. In the table, these probabilities are multiplied by 96 to get an idea of the number of goals that a team would score in 96 games. If we compare the numbers that we obtain in this way with the initial data, we see that the Poisson distribution with parameter* $\lambda = 1.3125$ *is actually a good summary of the data.*

It should be mentioned that the Poisson distribution is a limiting case of the binomial distribution. If n and π are the parameters of the binomial distribution, n is large and $n\pi = \lambda$, then the Poisson distribution and the binomial distribution create (almost) the same probabilities. This is illustrated in Table 8.5.

Finally, the Poisson distribution, as well as the binomial distribution, can be approximated by a normal distribution with mean λ and variance λ. This

Table 8.5   Comparison between probabilities of the Poisson distribution with parameter $\lambda = 1$ and several binomial distributions with $n\pi = 1$.

	Binomial distribution							Poisson
n	5	10	20	50	100	200	500	$\lambda = 1$
π	0.2	0.1	0.05	0.02	0.01	0.005	0.002	
$P(X = 0)$	0.3277	0.3487	0.3585	0.3642	0.3660	0.3670	0.3675	0.3679
$P(X = 1)$	0.4096	0.3874	0.3774	0.3716	0.3697	0.3688	0.3682	0.3679
$P(X = 2)$	0.2048	0.1937	0.1887	0.1858	0.1849	0.1844	0.1841	0.1839
$P(X = 3)$	0.0512	0.0574	0.0596	0.0607	0.0610	0.0612	0.0613	0.0613
$P(X = 4)$	0.0064	0.0112	0.0133	0.0145	0.0149	0.0151	0.0153	0.0153

approximation works well for large values of λ. With increasing λ, the graphical representation of the Poisson distribution approaches a perfect bell shape. This can be seen in Figures 8.7–8.10. The normal distribution will be discussed in detail in Chapter 10.

If the parameter λ of the Poisson distribution is an integer, the distribution has a remarkable feature: the probability $P(X = \lambda - 1)$ is then equal to the probability $P(X = \lambda)$. This can be verified in Figures 8.7– 8.9 and in Table 8.3. For example, for the Poisson distribution with $\lambda = 2$, we have that $P(X = 1) = P(X = 2)$. Such an equality of probabilities does not occur for non-integer values of λ. We can prove this property of the Poisson distribution as follows:

Consider the two consecutive probabilities

$$P(X = x) = \frac{e^{-\lambda}\lambda^x}{x!}$$

and

$$P(X = x + 1) = \frac{e^{-\lambda}\lambda^{x+1}}{(x + 1)!}.$$

These two probabilities are the same if

$$\frac{e^{-\lambda}\lambda^x}{x!} = \frac{e^{-\lambda}\lambda^{x+1}}{(x + 1)!},$$

which can be simplified to

$$\frac{\lambda^x}{x!} = \frac{\lambda^{x+1}}{(x + 1)!}.$$

This equation can be rewritten as

$$\frac{\lambda^x}{x!} = \frac{\lambda \cdot \lambda^x}{(x + 1) \cdot x!},$$

which can be simplified to

$$1 = \frac{\lambda}{x + 1}$$

and

$$x = \lambda - 1.$$

Consequently, the probabilities for $x = \lambda - 1$ and $x = \lambda$ are equal. This is only possible if λ is an integer, as a Poisson distributed random variable X can only take non-negative integer values.

8.6   The geometric distribution

If a random variable X counts the number of attempts that we need until the first success in a series of independent Bernoulli experiments, then X is geometrically distributed:

$$p_X(x; \pi) = (1 - \pi)^{x-1}\pi, \quad x = 1, 2, 3, \ldots,$$

where π represents the probability of success. Obviously, π again lies between 0 and 1. The expected value and variance of X are equal to

$$\mu_X = E(X) = \frac{1}{\pi}$$

and

$$\sigma_X^2 = \text{var}(X) = \frac{1 - \pi}{\pi^2}.$$

The probability distribution and the cumulative distribution function for $\pi = 0.2$ are shown in Figure 8.12. For comparison, the probability distribution and the cumulative distribution function for $\pi = 0.1$ are displayed in Figure 8.13. The smaller success rate implies that the probabilities for small values of the random variable are significantly smaller in Figure 8.13 than in Figure 8.12. The opposite is true for larger values.

It is a useful exercise to calculate some probabilities for a geometrically distributed random variable by hand and draw the associated probability distribution and cumulative distribution function. To illustrate this, Table 8.6 contains the probabilities for the distributions in Figures 8.12 and 8.13. Note that we only display probabilities for values of x up to 30. A geometric random variable can take values greater than 30.

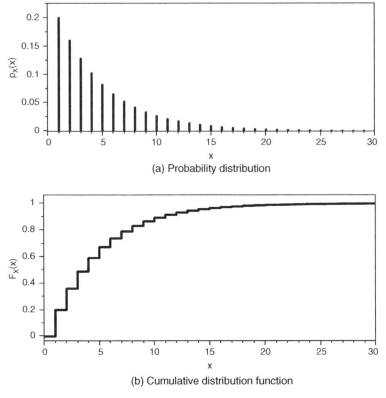

(a) Probability distribution

(b) Cumulative distribution function

Figure 8.12   Geometric distribution with parameter $\pi = 0.2$.

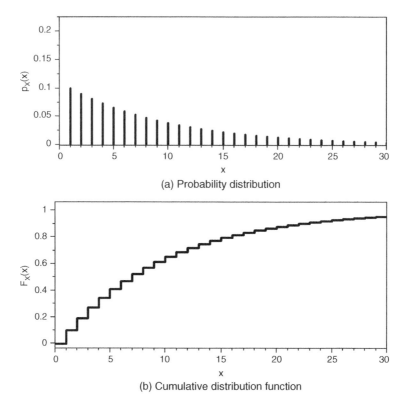

(a) Probability distribution

(b) Cumulative distribution function

Figure 8.13   Geometric distribution with parameter $\pi = 0.1$.

However, if the success probability π is equal to 0.1 or 0.2, this is rather unlikely. For $\pi = 0.2$ for example, $P(X > 30)$ is only 0.001238. Therefore, Figures 8.12 and 8.13 and Table 8.6, do not contain probabilities for $x > 30$.

To demonstrate that the expected value of a geometrically distributed random variable is equal to $1/\pi$, one can use the following series expansion, which states that, for y-values between 0 and 1,

$$\frac{1}{(1-y)^2} = \sum_{i=1}^{+\infty} i y^{i-1}.$$

If we replace the variable y by $1 - \pi$ and i by x in this equation, we obtain

$$\frac{1}{(1 - (1 - \pi))^2} = \frac{1}{\pi^2} = \sum_{x=1}^{+\infty} x(1 - \pi)^{x-1}.$$

The expected value of a geometrically distributed random variable X can then be determined as

$$\mu_X = \sum_{x=1}^{+\infty} x p_X(x; \pi) = \sum_{x=1}^{+\infty} x(1 - \pi)^{x-1} \pi = \pi \sum_{x=1}^{+\infty} x(1 - \pi)^{x-1} = \pi \frac{1}{\pi^2} = \frac{1}{\pi}.$$

Table 8.6 Probabilities for the geometric distributions in Figures 8.12 and 8.13.

x	$\pi = 0.1$		$\pi = 0.2$	
	$p_X(x)$	$F_X(x)$	$p_X(x)$	$F_X(x)$
1	0.1	0.1	0.2	0.2
2	0.09	0.19	0.16	0.36
3	0.081	0.271	0.128	0.488
4	0.0729	0.3439	0.1024	0.5904
5	0.06561	0.40951	0.08192	0.67232
6	0.059049	0.468559	0.05536	0.737856
7	0.0531441	0.5217031	0.0524288	0.7902848
8	0.04782969	0.56953279	0.04194304	0.83222784
9	0.043046721	0.612579511	0.033554432	0.865782272
10	0.038742049	0.651321560	0.026843546	0.892625818
11	0.034867844	0.686189404	0.021474837	0.914100654
12	0.031381060	0.717570464	0.017179869	0.931280523
13	0.028242954	0.745813417	0.013743895	0.945024419
14	0.025418658	0.771232076	0.010995116	0.956019535
15	0.022876793	0.794108868	0.008796093	0.964815628
16	0.020589113	0.814697981	0.007036874	0.971852502
17	0.018530202	0.833228183	0.005629500	0.977482002
18	0.016677182	0.849905365	0.004503600	0.981985602
19	0.015009464	0.864914828	0.003602880	0.985588481
20	0.013508517	0.878423345	0.002882304	0.988470785
21	0.012157666	0.890581011	0.002305843	0.990776628
22	0.010941899	0.901522910	0.001844674	0.992621302
23	0.009847709	0.911370619	0.001475740	0.994097042
24	0.008862938	0.920233557	0.001180592	0.995277634
25	0.007976644	0.928210201	0.000944473	0.996222107
26	0.007178980	0.935389181	0.000755579	0.996977686
27	0.006461082	0.941850263	0.000604463	0.997582148
28	0.005814974	0.947665237	0.000483570	0.998065719
29	0.005233476	0.952898713	0.000386856	0.998452575
30	0.004710129	0.957608842	0.000309485	0.998762060

8.7 The negative binomial distribution

The geometric distribution is a special case of the negative binomial distribution, which is sometimes called the Pascal distribution. The probability distribution is given by

$$p_X(x; \pi, r) = \binom{x-1}{r-1}(1-\pi)^{x-r}\pi^r, \quad x = r, r+1, r+2, \ldots, \quad (8.3)$$

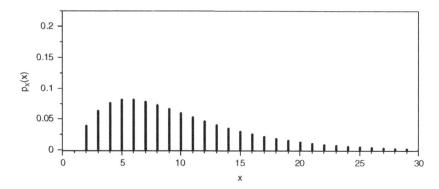

Figure 8.14   Negative binomial distribution with parameters $\pi = 0.2$ and $r = 2$.

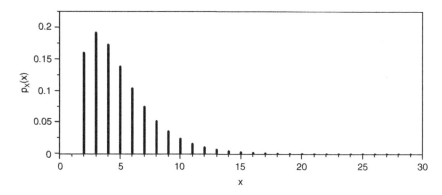

Figure 8.15   Negative binomial distribution with parameters $\pi = 0.4$ and $r = 2$.

where X represents the number of consecutive independent Bernoulli experiments until r successes are obtained. If r is equal to 1, the negative binomial distribution reduces to the geometric distribution.

A negative binomially distributed random variable cannot take any value less than r for the simple reason that one cannot achieve r successes before one has done r attempts. Figures 8.14 and 8.15 show negative binomial distributions with parameters $\pi = 0.2$ and $r = 2$, and with parameters $\pi = 0.4$ and $r = 2$, respectively. The higher success probability in the second figure ensures that there is a higher probability for small values of the random variable. In other words, in the second case, there is a greater chance to obtain two successes with fewer attempts.

Table 8.7 compares the probabilities obtained using the geometric distribution and the negative binomial distribution with $r = 2$ for values of x ranging from 0 to 10, for the same success probability $\pi = 0.2$. Obviously, $P(X = 1) = 0$ for the negative binomial distribution. The reason for this is that it is impossible to obtain two successes in only one Bernoulli experiment.

To a large extent, the derivation of the negative binomial distribution resembles the one of the (ordinary) binomial distribution. Suppose that we are interested in

Table 8.7   Geometric distribution versus negative binomial distribution with $r = 2$, both with success probability $\pi = 0.2$.

x	Neg. Binom.	Geometric	x	Neg. Binom.	Geometric
1	0.0000	0.2000	6	0.0819	0.0655
2	0.0400	0.1600	7	0.0786	0.0524
3	0.0640	0.1280	8	0.0734	0.0419
4	0.0768	0.1024	9	0.0671	0.0336
5	0.0819	0.0819	10	0.0604	0.0268

the probability that we need x independent Bernoulli experiments to achieve r successes. A possible sequence of r successes and $x - r$ failures is, for example, F, S, F, S, F, ..., F, S. This sequence of successes and failures ends at the moment where the r-th success is achieved. To determine the probability of this particular sequence, the multiplication rule (see Section 5.1.2) is used:

$$(1 - \pi)\pi(1 - \pi)\pi(1 - \pi) \ldots (1 - \pi)\pi = \pi^r(1 - \pi)^{x-r}.$$

It is not difficult to see that the exact same probability is obtained for any other sequence involving the outcome S r times and the outcome F $x - r$ times. From combinatorics (see Section 5.2.1), we know that there are

$$\binom{x}{r} = \frac{x!}{r!(x - r)!}$$

such sequences. If, however, we are interested in the number of attempts until we are successful exactly r times, all relevant sequences of successes and failures have to end with a success. Therefore, only the order of the first $x - 1$ outcomes, of which $r - 1$ are successes, can be interchanged for the negative binomial distribution. As a result, there are only

$$\binom{x - 1}{r - 1} = \frac{(x - 1)!}{(r - 1)!(x - 1 - (r - 1))!} = \frac{(x - 1)!}{(r - 1)!(x - r)!}$$

possible orders of r successes and $x - r$ failures that end with the r-th success. If we now apply the summation rule, it follows that the probability that we need x attempts to achieve r successes equals the probability in Equation (8.3).

   Probabilities of the type $P(X = x)$ for negative binomially distributed random variables (and thus also for geometrically distributed random variables) can be obtained in JMP using the function "Neg Binomial Probability($\pi, r, x - r$)". Since, when considering a geometrically distributed random variable, the required number of successes r is equal to 1, probabilities for geometrically distributed random variables can be determined with the function "Neg Binomial Probability($\pi, 1, x - 1$)". Probabilities of the type $P(X \leq x)$ for negative binomially distributed random variables (and thus also for geometrically distributed random variables) can be calculated using the function "Neg Binomial Distribution($\pi, r, x - r$)".

8.8   Probability distributions in JMP

Using JMP, it is not difficult to create probability distributions and cumulative distribution functions. JMP contains numerous functions for this purpose. The use of these functions is demonstrated in this section. JMP can also display probability distributions and cumulative distribution functions graphically.

8.8.1   Tables with probability distributions and cumulative distribution functions

To generate a probability distribution or a cumulative distribution function, we start with a new data table (via the "File" menu, where you successively choose "New" and "Data Table"). We fill the first column with the list of possible values the random variable can take. Suppose that we are interested in a binomially distributed random variable with parameters $n = 9$ and $\pi = 1/3$. For this reason, we list all integer numbers from 0 to 9 in the first column of the data table. These are the possible values of a binomially distributed random variable with $n = 9$. We call this first column x. Changing the name of a variable in a column header can be done by double-clicking on the header.

Next, we create two new columns by right-clicking on the column header next to the first column (see Figure 8.16). We call the two new columns $p_X(x)$ and $F_X(x)$. Then, we need to enter a formula for the two new columns. This can be done by right-clicking on the column header and choosing the option "Formula" (see Figure 8.17).

Now, we only need to enter the appropriate formula. The formulas available in JMP are divided into categories, ranging from formulas on trigonometry (sine, cosine, …),

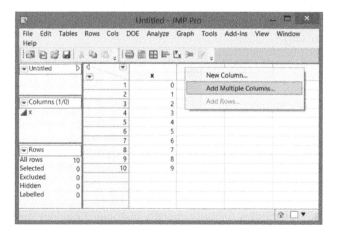

Figure 8.16   Generating a table with a probability distribution and a cumulative distribution function: Step 1.

Figure 8.17   Generating a table with a probability distribution and a cumulative distribution function: Step 2.

transcendental formulas such as logarithms, statistical formulas, and numerical formulas (rounding, ...) to formulas of probability. There are two types of formulas from probability theory, namely formulas for discrete probability distributions and formulas for continuous probability densities. The formulas for discrete probability distributions are contained within the category "Discrete Probability", while the formulas for continuous probability densities can be found in the category "Probability".

Since we want to produce a binomial distribution here, we obviously need the category "Discrete Probability" (see Figure 8.18). For the column $p_X(x)$, we have to select the "Binomial Probability" function, while, for the column $F_X(x)$, the function "Binomial Distribution" is needed. The first function provides probabilities of the type $P(X = x)$, while the latter function provides probabilities of the type $P(X \leq x)$. Both functions have three arguments. The first two arguments are used to specify the two parameters π and n of the binomial distribution. The third argument is used to specify the value of x. Figure 8.19 shows the inputs needed for both the columns $p_X(x)$ and $F_X(x)$.

When entering the two formulas, it is important to click on the name of the column x for the last argument (and not enter a number, letter or name by yourself). JMP then computes the probabilities corresponding to each value of x in the first column of the data table. The final result of all these operations is shown in Figure 8.20.

To calculate probabilities of known families of probability distributions, we can also use the script "JMP Distribution and Probability Calculator". A screen generated by this script is shown in Figure 8.21. The figure shows the probability that a binomially distributed random variable with parameters $n = 9$ and $\pi = 1/3$ is less than or equal to 1. As indicated at the bottom of the screen, that probability is equal to 0.1431.

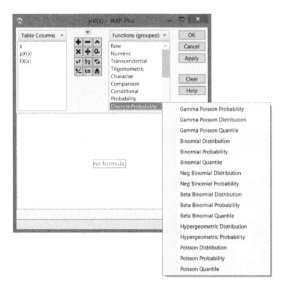

Figure 8.18   Generating a table with a probability distribution and a cumulative distribution function: Step 3.

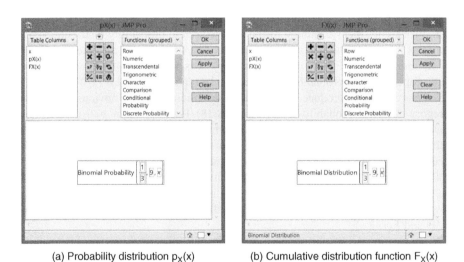

(a) Probability distribution $p_X(x)$ (b) Cumulative distribution function $F_X(x)$

Figure 8.19   Generating a table with a probability distribution and a cumulative distribution function: Step 4.

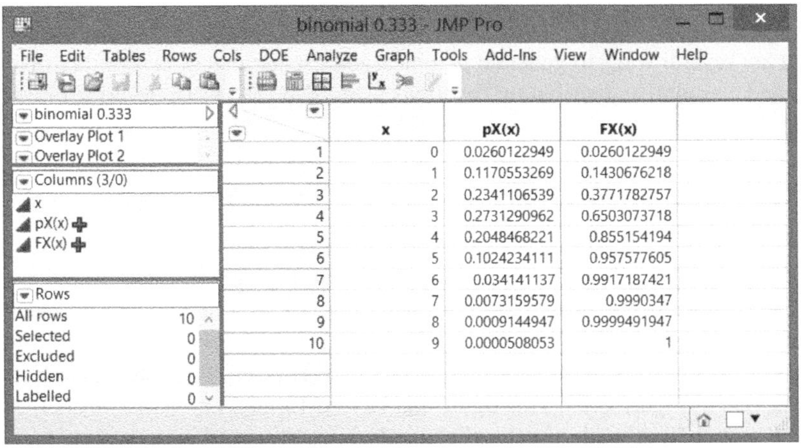

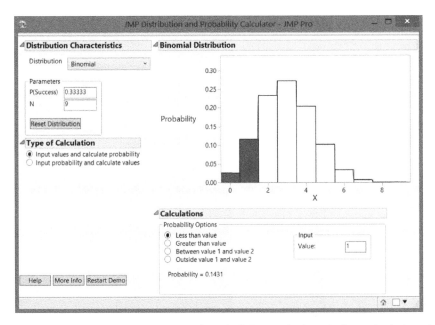

Figure 8.20   Generating a table with a probability distribution and a cumulative distribution function: final result.

Figure 8.21   "JMP Distribution and Probability Calculator" for calculating the probability that a binomially distributed random variable with parameters $n = 9$ and $\pi = 1/3$ is less than or equal to 1.

8.8.2   Graphical representations

There are several ways to generate graphical representations of probability distributions and cumulative distribution functions using JMP. Here, we describe one of the possibilities.

8.8.2.1   Probability distributions

For a picture of a probability distribution of a discrete random variable, a data table like the one in Figure 8.20 is an ideal starting point. If you have created such a table, you can choose the "Overlay Plot" option in the "Graph" menu. You will then get a dialog window where you need to enter the column $p_X(x)$ in the "Y" field and the column x in the "X" field, as shown in Figure 8.22. This will generate an initial graph like the one in Figure 8.23. This figure can be transformed into a needle chart of the desired probability distribution by clicking the hotspot (red triangle) next to the word

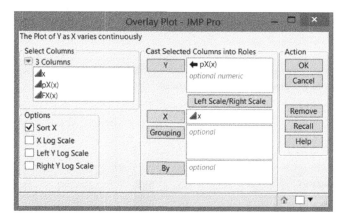

Figure 8.22   Drawing a probability distribution of a discrete random variable: Step 1.

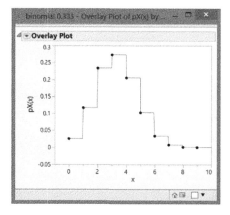

Figure 8.23   Drawing a probability distribution of a discrete random variable: intermediate result.

"Overlay Plot", choosing the option "Needle" and unchecking "Show Points" under "Y Options". This is illustrated in Figure 8.24.

The resulting needle chart is shown in Figure 8.25. This needle chart can still be modified by right-clicking on the horizontal and vertical axis, or the needles of the diagram. In this way, you can make minor aesthetic changes according to your personal taste. To change the color of the needles, use the "Y Options" in the hotspot (red triangle) menu, and choose the "Connect Color" option.

It is also possible to copy a chart from JMP and paste it, for example, in Word or PowerPoint documents. Figure 8.26 shows how this can be done. You can obtain the menu in this figure by right-clicking on the needle diagram.

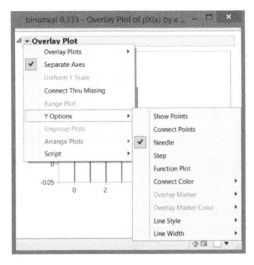

Figure 8.24   Drawing a probability distribution of a discrete random variable: Step 2.

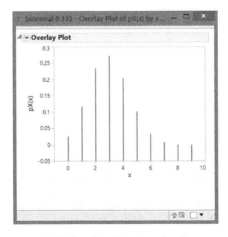

Figure 8.25   Drawing a probability distribution of a discrete random variable: final result.

8.8.2.2   Cumulative distribution functions

To draw a cumulative distribution function of a discrete random variable, the "Overlay Plot" option can also be used, but it requires a little detour. First, a new data table should be created, as in Figure 8.27. In this table, every value of x appears twice. This is necessary to get the desired graph of $F_X(x)$. The first time it appears, every x value is used for the probability $P(X \leq x - 1)$. The second time every x value appears, it is used for the probability $P(X \leq x)$. The fastest way to achieve this is to make use of a

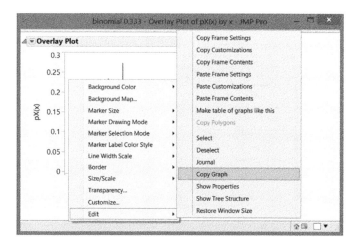

Figure 8.26   Copying a graph from JMP.

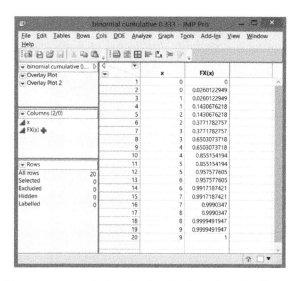

Figure 8.27   Drawing a cumulative distribution function of a discrete random variable: required data table.

special formula, which is shown in Figure 8.28. This formula uses the function "If" from the category "Conditional", the function "Row" from the category "Row", the function "Modulo" from the category "Numeric", and a comparison from the category "Comparison". Finally, the function "Binomial Distribution" from the category "Discrete Probability" is used.

Once the formula in Figure 8.28 has been entered, select the "Overlay Plot" option in the "Graph" menu. Next, enter the column $F_X(x)$ in the "Y" field, and the column x in the "X" field. You then obtain the picture in Figure 8.29. It is important to check either the "Function Plot" or the "Step" option in the "Y Options". Now you can again

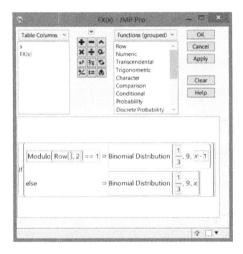

Figure 8.28   Drawing a cumulative distribution function of a discrete random variable: required formula for creating the data table in Figure 8.27.

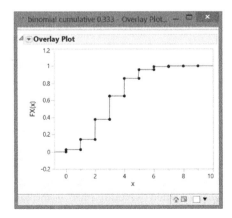

Figure 8.29   Drawing a cumulative distribution function of a discrete random variable: final result.

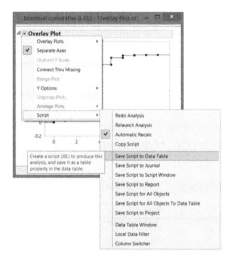

Figure 8.30   Saving the script of a figure in JMP.

Figure 8.31   Reproducing a figure by means of a saved script in JMP.

make aesthetic adjustments by right-clicking on the axes or the figure. Copying the figure can be done in the same manner as in Figure 8.26.

In JMP, it is possible to save a figure together with the data. To do so, click on the hotspot (red triangle) next to the word "Overlay Plot", and choose the option "Save Script to Data Table" under "Script". This is illustrated in Figure 8.30. The result of these operations is a hotspot named "Overlay Plot" in the top left of the data table. The figure can be reproduced at any time by clicking on the hotspot and choosing "Run Script". This is shown in Figure 8.31.

8.9   The simulation of discrete random variables with JMP

In Section 6.6, we already explained how random variables with a predetermined discrete probability distribution can be simulated. For the generation of random variables with a Bernoulli distribution, a binomial distribution, a Poisson distribution or a uniform distribution, it is, however, easier to make use of a number of pre-programmed formulas in JMP.

The first thing you need to do is to create a new data table, as in Figure 8.32. Then, right-click on the header of the second column. Select the option "Formula", as shown in Figure 8.33. In the resulting formula dialog, select "Random". Now, you

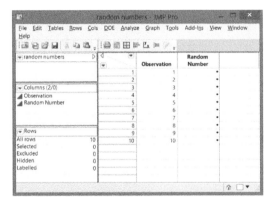

Figure 8.32   Generating pseudo-random numbers of a discrete probability distribution: Step 1.

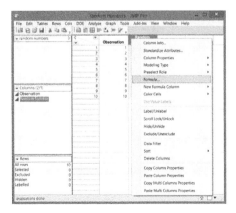

Figure 8.33   Generating pseudo-random numbers of a discrete probability distribution: Step 2.

can choose from a variety of families of probability distributions and probability densities. This is illustrated in Figure 8.34.

If you would like to generate pseudo-random numbers for a discrete uniform distribution with five possible outcomes, you can choose "Random Integer" and the value 5 as an argument. You will then see the screen shown in Figure 8.35. Figure 8.36 contains a possible final result. If you repeat this exercise at home, you

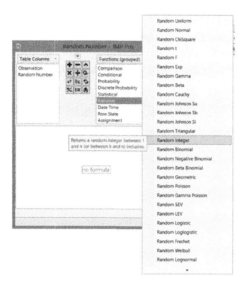

Figure 8.34   Generating pseudo-random numbers of a discrete probability distribution: Step 3.

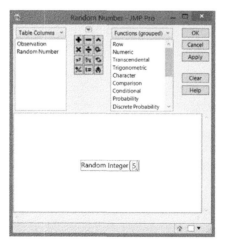

Figure 8.35   Generating pseudo-random numbers of a discrete probability distribution: Step 4.

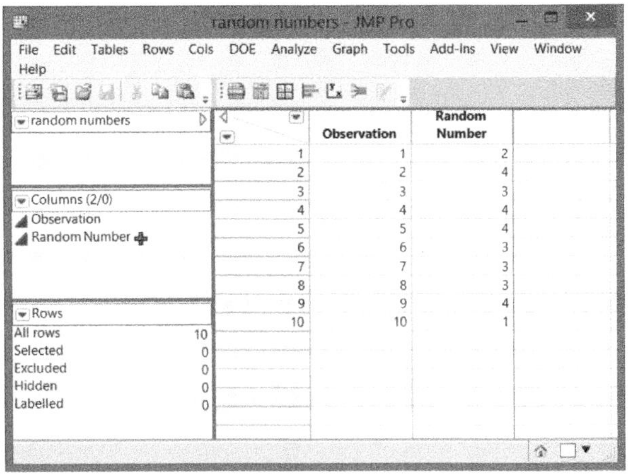

Figure 8.36   Generating pseudo-random numbers of a discrete probability distribution: final result.

will typically obtain other pseudo-random values than those in Figure 8.36. Note that, in Figure 8.36, the outcome 5 does not appear at all. This is not a mistake: just as you may never get a six when you throw a die, it may be that one or more values do not appear when you generate a set of pseudo-random numbers.

9

Important continuous probability densities

I do not pick the place or the time. I use an old math book with a table of random numbers. I have numbered the street index of my Falkplan Berlin mit Cityplan Potsdam. *I combine the two books: The random numbers tell me what street to go to, and at what house number to begin. If you want to play the game of entropy, you have to be as reckless as entropy itself. The random tables also indicate on what day I will appear at a given spot, and how many minutes after sunset this will occur. I love random numbers. There is no pattern to my movements, nobody can predict when and where I will strike. I will never get caught.*

(from *Omega Minor,* Paul Verhaeghen, p. 49)

In Chapter 6, continuous random variables were introduced as random variables that can take any value in an interval. In practice, however, one rarely finds a pure continuous random variable. The lifespan of a light bulb produced by an industrial production process is, in principle, a continuous random variable. But, even with very sophisticated equipment, the lifespan of a light bulb cannot be measured exactly. Each measuring device has a certain precision (for example, the lifespan could be measured to the nearest second), so that in fact we get discrete measurements. Consequently, the set of possible values for the "continuous" random variable is discrete and, strictly speaking, the random variable cannot be treated as a continuous random variable.

Another issue is that in practice, a property is typically studied for a large, but finite population. Despite the finiteness of the population, the property under study is still considered a continuous random variable. Again, the "continuous" random variable can only take a limited number of values. A stock of goods, for example, might

Statistics with JMP: Graphs, Descriptive Statistics, and Probability, First Edition. Peter Goos and David Meintrup.
© 2015 John Wiley & Sons, Ltd. Published 2015 by John Wiley & Sons, Ltd.
Companion Website: wiley.com/go/goosandmeintrup

have a net weight per package that is considered normally distributed. Implicitly, then, the net weight per package is treated as a continuous random variable, even though the stock is finite.

Despite all this, continuous random variables are used in many cases for studying real life problems, generally without any noteworthy trouble. In the following, we will also not worry about these considerations.

In the previous chapter, we discussed discrete random variables that were Bernoulli distributed or binomially distributed. Given that discrete random variables have a probability distribution, this choice of wording makes sense. Continuous random variables are also said to be normally or exponentially distributed, even though, strictly speaking, they do not have a probability distribution but a probability density. This lack of logic in terminology can also be found in many other languages.

The most important probability density is undoubtedly the normal probability density. Therefore, we will discuss this density in a separate chapter, along with the lognormal probability density. This is done in Chapter 10.

9.1 The continuous uniform density

A (continuous) random variable X is uniformly distributed over the interval $[\alpha, \beta]$ if its probability density is given by

$$f_X(x; \alpha, \beta) = \begin{cases} \frac{1}{\beta-\alpha}, & \alpha \leq x \leq \beta, \\ 0, & \text{otherwise}, \end{cases}$$

where the parameters α and β are arbitrary real numbers with $\alpha < \beta$.

Starting from the definitions of the expected value and the variance of continuous random variables in Section 7.1 and Section 7.4, it is easy to verify that

$$\mu_X = E(X) = \frac{\alpha + \beta}{2}$$

and

$$\sigma_X^2 = \text{var}(X) = \frac{(\beta - \alpha)^2}{12}.$$

Applying the definition of the cumulative distribution function of a continuous random variable in Section 6.3, we can derive the cumulative distribution function of a continuous uniformly distributed random variable:

$$F_X(x; \alpha, \beta) = \begin{cases} 0, & x < \alpha, \\ \frac{x-\alpha}{\beta-\alpha}, & \alpha \leq x \leq \beta, \\ 1, & x > \beta. \end{cases}$$

The median of a continuous uniformly distributed random variable is equal to the expected value.

Example 9.1.1 *Figure 9.1 shows the probability density and the corresponding cumulative distribution function of a continuous uniformly distributed random variable with parameters $\alpha = 2$ and $\beta = 7$. The probability density function and the cumulative distribution function are*

$$f_X(x; 2, 7) = \begin{cases} \frac{1}{7-2} = \frac{1}{5}, & 2 \leq x \leq 7, \\ 0, & \text{otherwise,} \end{cases}$$

and

$$F_X(x; 2, 7) = \begin{cases} 0, & x < 2, \\ \frac{x-2}{7-2} = \frac{x-2}{5}, & 2 \leq x \leq 7, \\ 1, & x > 7. \end{cases}$$

The expected value of the random variable is $(2 + 7)/2 = 4.5$, while its variance is equal to $(7 - 2)^2/12 = 25/12$.

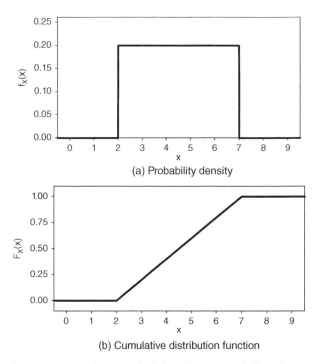

(a) Probability density

(b) Cumulative distribution function

Figure 9.1   Continuous uniform probability density and distribution function for $\alpha = 2$ and $\beta = 7$.

Here, the uniform density was defined on the closed interval $[\alpha, \beta]$. Alternatively, the density can be defined over the open interval $]\alpha, \beta[$, or over a half-open interval. This has no impact on the statistics of the distribution, because the differences only affect one or two x values with zero probability.

9.2   The exponential density

In many applications, we are interested in random variables that can only take positive values. The lifespan of a light bulb or the waiting time for a customer in a bank are two examples of such random variables. Although there are several other suitable probability densities for these random variables, the exponential probability density is by far the best known and most widely used.

9.2.1   Definition and statistics

A random variable X is exponentially distributed with parameter λ if

$$f_X(x; \lambda) = \begin{cases} \lambda e^{-\lambda x}, & x \geq 0, \\ 0, & \text{otherwise.} \end{cases}$$

The parameter λ is an arbitrary real number greater than 0. The exponential density is often used for random variables that indicate waiting times or times between two events.

If X is exponentially distributed with parameter λ, which is often written as

$$X \sim \text{Exp}(\lambda),$$

then

$$\mu_X = E(X) = \frac{1}{\lambda}$$

and

$$\sigma_X^2 = \text{var}(X) = \frac{1}{\lambda^2}.$$

The expected value and the variance can be determined using the method of the moment generating function. This was done in Example 7.6.2.

The cumulative distribution function of an exponentially distributed random variable is given by

$$F_X(x; \lambda) = \begin{cases} 0, & x < 0, \\ 1 - e^{-\lambda x}, & x \geq 0. \end{cases}$$

Using this function, it is not difficult to show that the median of an exponentially distributed random variable with parameter λ is equal to

$$\gamma_{0.5} = \frac{\ln(2)}{\lambda} \approx \frac{0.6931}{\lambda}.$$

Thus, the median is smaller than the expected value. As indicated in Section 3.4, this is typical for right-skewed distributions and densities.

Graphical representations of exponential probability densities with $\lambda = 1/2$ and $\lambda = 1/4$, and the corresponding cumulative distribution functions are given in Figure 9.2. You can verify in Figure 9.2a that an exponential probability density takes the value λ in $x = 0$, and that the densities are skewed to the right.

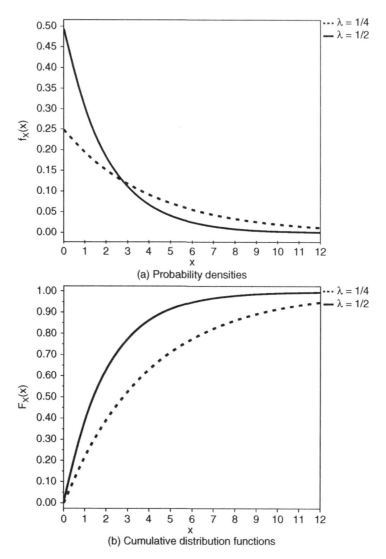

Figure 9.2   Exponential probability densities and cumulative distribution functions for $\lambda = 1/2$ (solid line) and $\lambda = 1/4$ (dotted line).

9.2.2   Some interesting properties

The exponential density has a remarkable property: it is *memoryless*. The following theorem explains the details:

Theorem 9.2.1 *If the random variable X is exponentially distributed, then*

$$P(X > t_1 + t_2 \mid X > t_2) = P(X > t_1) \qquad \text{for any } t_1 > 0 \text{ and } t_2 > 0.$$

Proof:

$$P(X > t_1 + t_2 \mid X > t_2) = \frac{P\{(X > t_1 + t_2) \cap (X > t_2)\}}{P(X > t_2)},$$

$$= \frac{P(X > t_1 + t_2)}{P(X > t_2)},$$

$$= \frac{1 - P(X \le t_1 + t_2)}{1 - P(X \le t_2)},$$

$$= \frac{1 - (1 - e^{-\lambda(t_1 + t_2)})}{1 - (1 - e^{-\lambda t_2})},$$

$$= \frac{e^{-\lambda(t_1 + t_2)}}{e^{-\lambda t_2}},$$

$$= \frac{e^{-\lambda t_1} e^{-\lambda t_2}}{e^{-\lambda t_2}},$$

$$= e^{-\lambda t_1},$$

$$= 1 - (1 - e^{-\lambda t_1}),$$

$$= 1 - P(X \le t_1),$$

$$= P(X > t_1).$$

∎

The practical meaning of this derivation is best illustrated by means of an example.

Example 9.2.1 *Imagine that you are about to visit a post office and that the time T between the arrival of two customers in the office is exponentially distributed. You enter at time zero. The probability that no new customer arrives in the first t_1 minutes is $P(T > t_1)$. At time t_2, you are standing there still waiting. The curious thing about the exponential distribution is that the probability that there is no new customer during the next t_1 minutes still is same as when you came: $P(T > t_1 + t_2 \mid T > t_2) = P(T > t_1)$. After t_2 minutes, the exponential density behaves as if it has forgotten that you have already waited that long. Due to this strange property, the use of the exponential density is not always realistic.*

Most of the other distributions do have a memory. For example, for a uniform density with $\alpha = 0$ and $\beta = 1$ one can verify that $P(X > 0.8 \mid X > 0.5) = P(X > 0.3 + 0.5 \mid X > 0.5) \ne P(X > 0.3)$. However, the geometric distribution is also memoryless: for any value of the success probability π, we have that $P(X > x + c \mid X > c) = P(X > x)$, if X is geometrically distributed. It is a useful exercise to prove this.

The following theorem shows that there is a link between the exponential distribution and the Poisson distribution. We assume, for simplicity, that the dimension in which the exponentially distributed random variable is measured is time.

Theorem 9.2.2 *For a Poisson process with parameter λ, which is the average number of events in a time interval of length 1, the time T between two consecutive events is exponentially distributed with parameter λ. The converse is also true.*

Proof: To prove the theorem, we start from the fact that the probability that we have to wait for more than time t between two consecutive events is equal to the probability that no event occurs in a time interval of length t. In other words,

$$P(T > t) = P(X = 0),$$

where the random variable T represents the time between two events, and X is the number of events in a time interval of length t. If X is Poisson distributed with expected value λt in an interval of length t, then

$$F_T(t) = P(T \leq t),$$

$$= 1 - P(T > t),$$

$$= 1 - P(X = 0),$$

$$= 1 - \frac{e^{-\lambda t}(\lambda t)^0}{0!},$$

$$= 1 - e^{-\lambda t}.$$

Hence, the cumulative distribution function of T is the one of the exponential distribution. ∎

An alternative way to phrase this theorem is as follows: If the number of events in a time interval of length t is Poisson distributed with parameter λt, then the time between successive events is exponentially distributed with parameter λ. This argument is often used in queuing theory, an important field of study within operations research. Queuing theory studies service stations such as counters where customers wait to be served or servers that have to process orders. The purpose of this field is to calculate average waiting times, to determine the occupancy rate of service stations or servers, to assess the effect of additional service stations or servers, and so on. The first applications of queuing theory were intended to address capacity issues of the first telephone lines in the 1920s.

A final interesting feature of the exponential probability density function is related to the minimum of a set of exponentially distributed random variables:

Theorem 9.2.3 *The minimum of k independent exponentially distributed random variables with parameter λ is exponentially distributed with parameter $k\lambda$.*

Proof: Suppose that the random variable X represents the minimum of k independent exponentially distributed random variables $X_1, X_2, \ldots, X_k$, all with parameter λ.

To determine the probability density of the minimum, we start with the cumulative distribution function:

$$F_X(x) = P(X \leq x),$$
$$= P(\min(X_1, X_2, \ldots, X_k) \leq x),$$
$$= 1 - P(\min(X_1, X_2, \ldots, X_k) > x).$$

For the minimum of k random variables to be greater than x, each of them individually must be greater than x. Hence,

$$F_X(x) = 1 - P(X_1 > x \text{ and } X_2 > x \text{ and } \ldots \text{ and } X_k > x).$$

The independence of the random variables $X_1, X_2, \ldots, X_k$ allows us to rewrite this expression as

$$F_X(x) = 1 - P(X_1 > x) \times P(X_2 > x) \times \ldots \times P(X_k > x).$$

As the k random variables $X_1, X_2, \ldots, X_k$ are all exponentially distributed with parameter λ, for any individual X_i, we have that

$$P(X_i > x) = 1 - F_{X_i}(x) = 1 - (1 - e^{-\lambda x}) = e^{-\lambda x}.$$

As a result,

$$F_X(x) = 1 - e^{-\lambda x} e^{-\lambda x} \cdots e^{-\lambda x},$$
$$= 1 - (e^{-\lambda x})^k,$$
$$= 1 - e^{-k\lambda x}.$$

This is the cumulative distribution function of an exponentially distributed random variable with parameter $k\lambda$. We can therefore conclude that the minimum of k independent exponentially distributed random variables is also exponentially distributed.   ■

The exponential density is a special case of the gamma probability density and of the Weibull probability density. This is useful to know in order to perform calculations involving exponentially distributed random variables in JMP because there are no direct functions in JMP for the exponential density. For values of the exponential density function, $f_X(x; \lambda)$, we can use the formula "Gamma Density(x, 1, $1/\lambda$)" in JMP, or the function "Weibull Density(x, 1, $1/\lambda$)". For values of the corresponding cumulative distribution function, $F_X(x; \lambda)$, we can use the function "Gamma Distribution(x, 1, $1/\lambda$)" or "Weibull Distribution(x, 1, $1/\lambda$)". To determine quantiles or percentiles, the functions "Gamma Quantile" or "Weibull Quantile" are available. For example, the median is found with "Gamma Quantile(0.5, 1, $1/\lambda$)" or "Weibull Quantile(0.5, 1, $1/\lambda$)".

If you do not have a computer at your disposal, you can rely on tables (see Appendix D) to calculate probabilities for exponentially distributed random variables. It is a useful exercise to reconstruct one of these tables by yourself, based on the cumulative exponential distribution function.

9.3   The gamma density

The gamma probability density has two strictly positive parameters k and θ. The parameter k is called the shape parameter and the parameter θ is called the scale parameter. The probability density function is given by the following expression for $x \geq 0$:

$$f_X(x; k, \theta) = \frac{1}{\theta^k} \cdot \frac{1}{\Gamma(k)} \cdot x^{k-1} e^{-\frac{x}{\theta}},$$

where $\Gamma()$ is the gamma function[1]. The gamma function is a special function, which is beyond the scope of this book. There is no simple mathematical expression for the cumulative distribution function of a gamma distributed random variable. The expected value and the variance of a gamma distributed random variable are

$$\mu_X = E(X) = k\theta$$

and

$$\sigma_X^2 = \text{var}(X) = k\theta^2.$$

A special case of the gamma probability density is the exponential probability density. Indeed, if $k = 1$ and $\theta = 1/\lambda$, then the gamma probability density simplifies to the exponential probability density. Another special case of the gamma density is the χ^2 distribution (pronounced chi-squared), which will play a prominent role in the book *Statistics with JMP: Hypothesis Tests, ANOVA and Regression*.

The gamma probability density is used for random variables that can only take positive values, such as the magnitude of insurance claims or the amount of rainfall, as well as waiting times and lifespans (as the exponential density). Four different gamma densities are graphically shown in Figure 9.3. All these densities have the same value for the parameter θ, namely the value 2. However, they have different values for the parameter k. The density with $k = 1$ is equal to the exponential probability density with $\lambda = 1/2$ (compare this density from Figure 9.3 to the density represented by the solid line in Figure 9.2a). The gamma density with $k = 0.9$ in Figure 9.3 is very similar to the exponential density, but it has a larger value at $x = 0$ and drops to zero faster. The gamma densities with $k > 1$ differ substantially from the exponential probability density because they are not strictly decreasing functions.

In JMP, values of the gamma density function, $f_X(x; k, \theta)$, can be computed with the formula "Gamma Density(x, k, θ)". For values of the corresponding cumulative

---

[1] The gamma function is an extension of the factorial function for integers, $n! = n \times (n-1) \dots 2 \times 1$. If z is a positive real number, then $\Gamma(z) = (z-1)\Gamma(z-1) = \int_0^{+\infty} e^{-t} t^{z-1} dt$. A special case is $\Gamma(1/2) = \sqrt{\pi}$. In addition, $\Gamma(1) = \Gamma(2) = 1! = 0! = 1$, and $\Gamma(n) = (n-1)!$ for all integers n.

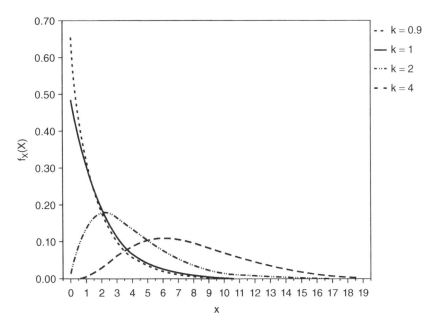

Figure 9.3   Four gamma probability densities with $\theta = 2$ and $k = 0.9$ (dotted line), $k = 1$ (solid line), $k = 2$ (dash-dotted line), and $k = 4$ (dashed line).

distribution function, $F_X(x; k, \theta)$, the formula "Gamma Distribution(x, k, θ)" can be used. To determine quantiles or percentiles, the "Gamma Quantile" function is available. For example, the median is found with the command "Gamma Quantile$(0.5, k, \theta)$".

9.4   The Weibull density

Another continuous probability density is the Weibull[2] density. This probability density is used for modeling lifespans, wind speeds and strength measurements, and for other applications with random variables that can only take values from 0 to $+\infty$. The Weibull probability density has two strictly positive parameters, k and θ:

$$f_X(x; k, \theta) = \frac{k}{\theta} \cdot \left(\frac{x}{\theta} \right)^{k-1} \cdot e^{-(x/\theta)^k}.$$

The corresponding cumulative distribution function is not difficult to derive, and is equal to

$$F_X(x; k, \theta) = 1 - e^{-(x/\theta)^k}.$$

---

[2] Waloddi Weibull (1887–1979) was a Swedish engineer and mathematician, who used the Weibull density for strength measurements and other applications on reliability of materials.

The expected value and the variance of a Weibull distributed random variable are

$$\mu_X = E(X) = \theta \, \Gamma \left(1 + \frac{1}{k} \right)$$

and

$$\sigma_X^2 = \text{var}(X) = \theta^2 \, \Gamma \left(1 + \frac{2}{k} \right) - \mu_X^2,$$

where $\Gamma()$ is again the gamma function. The median is

$$\gamma_{0.5} = \theta \, \{\ln(2)\}^{1/k}.$$

When the parameter k is equal to 1 and the parameter θ is equal to $1/\lambda$, the Weibull probability density simplifies to the exponential density. Four different Weibull densities are graphically shown in Figure 9.4. The density with $k = 1$ and $\theta = 2$ is equivalent to the exponential probability density with $\lambda = 1/2$ (compare this density from Figure 9.4 to the density represented by the solid line in Figure 9.2a). Figure 9.4 clearly shows that the shape of the Weibull probability density changes drastically depending on the parameter values used. The Weibull probability density is strictly decreasing if $k \le 1$.

It is not difficult to show that the minimum of a set of Weibull distributed random variables is also Weibull distributed. Earlier in this chapter, we encountered a similar feature for exponentially distributed random variables. Another interesting exercise is to demonstrate that, if a random variable X is Weibull distributed, the transformed random variable $Y = aX^b$ (with $a > 0$ and $b > 0$) is also Weibull distributed.

In JMP, values of the Weibull density function, $f_X(x; k, \theta)$, can be computed with the formula "Weibull Density(x, k, θ)". For values of the corresponding cumulative distribution function, $F_X(x; k, \theta)$, the formula "Weibull Distribution(x, k, θ)" can be used. To determine quantiles or percentiles, the function "Weibull Quantile" is available. For example, the median is found with the command "Weibull Quantile$(0.5, k, \theta)$".

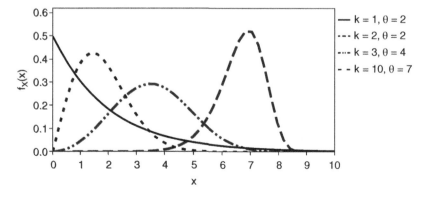

Figure 9.4   Four Weibull probability densities.

9.5   The beta density

The exponential probability density, the gamma probability density and the Weibull probability density all have the interval $[0, +\infty[$ as domain, and thus can serve for continuous random variables that take only positive real values ranging from 0 to $+\infty$ (such as waiting times and lifespans). Other continuous probability densities have the unit interval $[0, 1]$ as domain. An example is the beta density, which has two strictly positive parameters α and β:

$$f_X(x; \alpha, \beta) = \frac{\Gamma(\alpha + \beta)}{\Gamma(\alpha)\Gamma(\beta)} x^{\alpha-1}(1 - x)^{\beta-1},$$

where $\Gamma()$ again represents the gamma function. A beta distributed random variable X has expected value

$$\mu_X = E(X) = \frac{\alpha}{\alpha + \beta}$$

and variance

$$\sigma_X^2 = \text{var}(X) = \frac{\alpha\beta}{(\alpha + \beta)^2(\alpha + \beta + 1)}.$$

The beta probability distribution is suitable for random variables that represent percentages or proportions, as percentages and proportions lie between 0 and 1.

Figure 9.5 shows four beta probability densities with $\alpha = \beta$. In that case, the beta probability density is symmetric and the expected value is equal to $1/2$. If $\alpha = \beta = 1$,

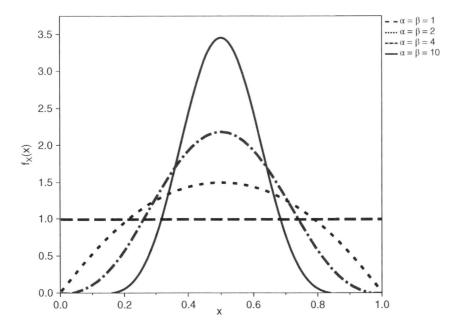

Figure 9.5   Four beta probability densities with $\alpha = \beta$.

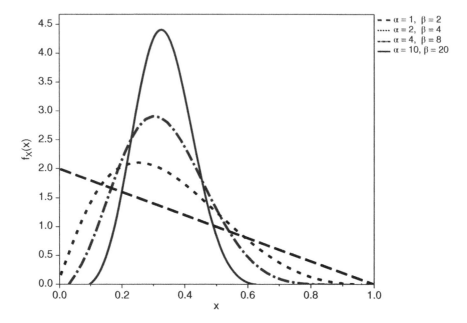

Figure 9.6   Four beta probability densities with $\beta = 2\alpha$.

the beta probability density function is identical to a continuous uniform probability density on the interval $[0, 1]$. The figure clearly shows that the variance of a beta distributed random variable decreases as α and β increase. Figure 9.6 shows four beta probability densities for which β is twice as large as α. In that case, the expected value is equal to $\alpha/(\alpha + \beta) = \alpha/(\alpha + 2\alpha) = 1/3$. Again, the variance decreases as α and β increase. Figure 9.7 shows four beta probability densities for which α is twice as large as β. In that case, the expected value is equal to $\alpha/(\alpha + \beta) = 2\beta/(2\beta + \beta) = 2/3$. Here, too, it is obvious that the variance decreases as α and β increase. Note that the probability densities in Figure 9.7 are mirror images of the probability densities in Figure 9.6. Finally, Figure 9.8 shows four beta probability densities for which at least one of the parameters α or β is smaller than 1.

In JMP, values of the beta density function, $f_X(x; \alpha, \beta)$, can be computed with the formula "Beta Density(x, α, β)". For values of the corresponding cumulative distribution function, $F_X(x; \alpha, \beta)$, the formula "Beta Distribution(x, α, β)" can be used. To determine quantiles or percentiles, the function "Beta Quantile" is available. For example, the median is found with the command "Beta Quantile$(0.5, \alpha, \beta)$".

9.6   Other densities

There are many more densities in the statistical literature. Next to the normal density, the lognormal density, the uniform density, the exponential density, the gamma density, the Weibull density, and the beta density, we mention the logistic,

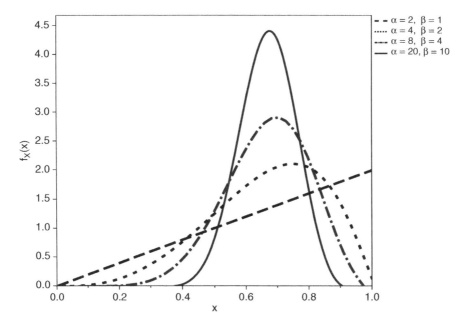

Figure 9.7   Four beta probability densities with $\alpha = 2\beta$.

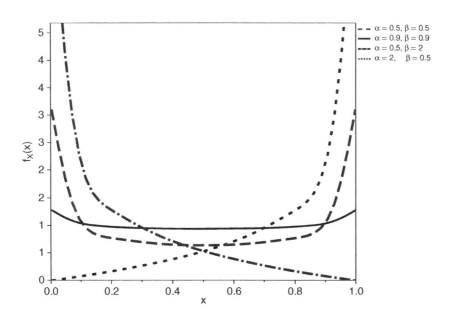

Figure 9.8   Four beta probability densities with at least one parameter α or β smaller than 1.

Gumbel, Pareto, and Cauchy densities. Each of these densities has its specific field of application. For example, the Gumbel density is a so-called extreme value distribution. It is used in areas where researchers are interested in extremely large values, for example when dealing with diamonds (a single large diamond determines the profitability of a diamond mine) or in the reinsurance business (reinsurers need to get an idea of the highest possible claim that they might have to pay sooner or later).

9.7   Graphical representations and probability calculations in JMP

To create a graphical representation of a probability density or cumulative distribution function, your first task is to generate a new data table, as in Figure 9.9. This data table contains a column with a large number of possible values of the random variable X. The more values are listed, the more precise the graph will be. This looks like a lot of work but JMP allows you to create series of values automatically. This is illustrated in Figure 9.10: in the data table, the first two numbers are entered, with the intention to generate a long series of numbers 0.00025, 0.0005, 0.00075, 0.001, 0.00125, ... , all the way up to the number 1. After entering the numbers 0.00025 and 0.0005, select both numbers and right-click on them. From the resulting pop-up menu, choose "Fill", followed by "Continue Sequence to". Then, JMP will ask you to indicate up to which row you want to continue the series. As the final number should

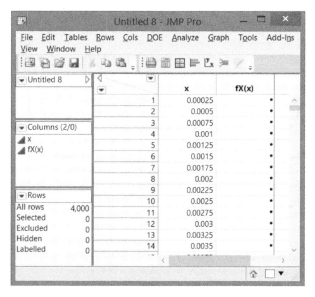

Figure 9.9   Graphical representation of a probability density: Step 1.

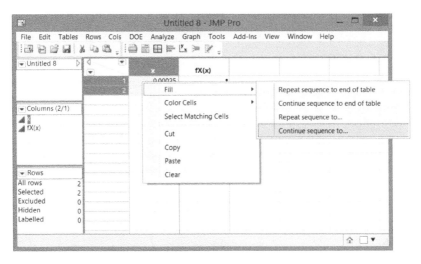

Figure 9.10   Continuing a sequence in JMP.

be 1, we need to specify row number 4000. The resulting data table will then contain 4000 rows.

The next step in the generation of a graph of the probability density is to right-click on the header of the second column. Next, select the "Formula" option in the resulting menu, as shown in Figure 8.33. In the dialog window for building the required formula, you then need to select the category "Probability", after which you can choose one from a variety of families of probability densities. As an example, we choose the "Beta Density" and enter the x column and twice the value 3 as arguments (these values of 3 mean that the beta density with $\alpha = 3$ and $\beta = 3$ will result). This yields the formula in Figure 9.11, and, ultimately, the data table shown in Figure 9.12.

The next step is to select the "Graph Builder" in the "Graph" menu. In the "Graph Builder", you need to drag the column $f_X(x)$ to the "Y" field, and the column x to the "X" field. You will then see the screen shown in Figure 9.13 with the desired graph. Note that the "Smoother" button at the top of the "Graph Builder" has been pressed to ensure that JMP connects the data points using a continuous curve. This button is indicated in Figure 9.13 by the left arrow. Because we have used a formula to determine the values of the probability density function, we can also use the button indicated by the right arrow in Figure 9.13. The latter option usually provides a more accurate graphical representation.

Of course, we can still refine the graph by right-clicking on the axes, and by changing the color or thickness of the curve. The legend can be edited by double-clicking on it.

Creating a graphical representation of a cumulative distribution function for a continuous random variable is done in a similar way to a graph for a probability density.

To calculate probabilities using known families of probability densities, we can also use the script "JMP Distribution and Probability Calculator". A screen generated by the script is shown in Figure 9.14. The figure shows the probability

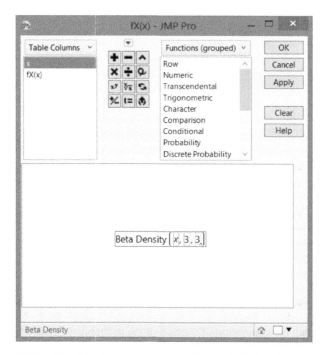

Figure 9.11   Graphical representation of a probability density: Step 2.

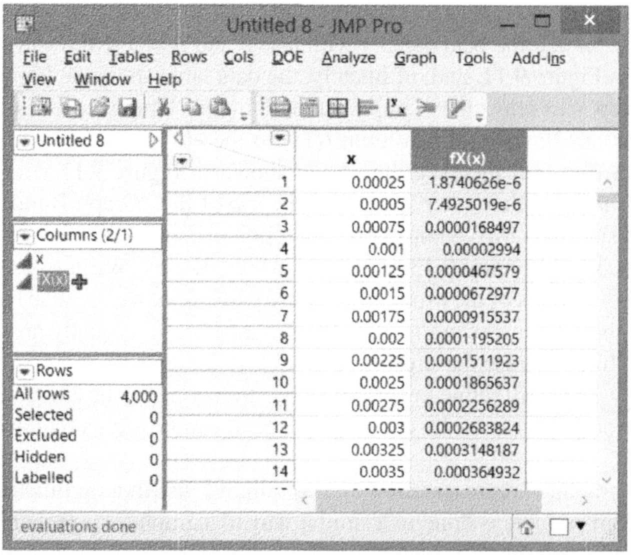

Figure 9.12   Graphical representation of a probability density: Step 3.

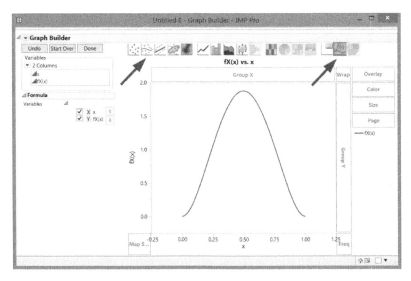

Figure 9.13   Graphical representation of a probability density: Step 4.

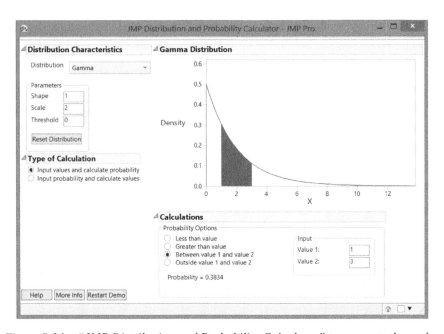

Figure 9.14   "JMP Distribution and Probability Calculator" to compute the probability that an exponentially distributed random variable with parameter $\lambda = 1/2$ (or a gamma distributed random variable with parameters $k = 1$ and $\theta = 2$) takes values between 1 and 3.

that an exponentially distributed random variable with parameter $\lambda = 1/2$, in other words, a gamma distributed random variable with parameters $k = 1$ (the *shape parameter*) and $\theta = 2$ (the *scale parameter*), takes values between 1 and 3.

9.8   Simulating continuous random variables in JMP

We already explained in Section 6.6 how random variables can be simulated for a predetermined continuous probability density. However, to generate random variables with a uniform, exponential, gamma, Weibull, or beta density, it is easier to use existing formulas in JMP.

In order to do so, start with a new data table that has as many rows as you want pseudo-random numbers. An example of such a table is shown in Figure 8.32 in Section 8.9. Then, right-click on the header of the second column, and select the option "Formula" in the resulting pop-up menu, as shown in Figure 8.33. In the dialog window for building the required formula, you then need to select the "Random" option, after which you can choose one from a variety of families of probability distributions and probability densities (see Figure 8.34) to generate pseudo-random numbers from. The functions needed to simulate the random variables discussed in this chapter are "Random Uniform", "Random Exp", "Random Gamma", "Random Weibull", and "Random Beta".

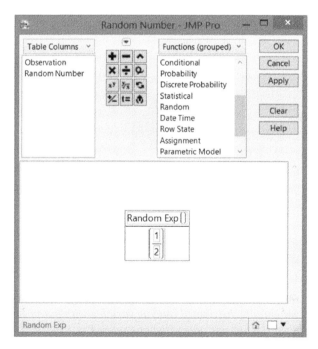

Figure 9.15   Generating pseudo-random numbers from an exponential probability density with parameter $\lambda = 1/2$: required formula.

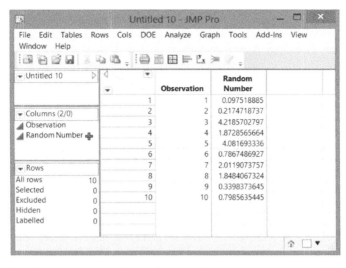

Figure 9.16   Generating pseudo-random numbers from an exponential probability density with parameter $\lambda = 1/2$: final result.

When generating pseudo-random numbers using these functions, the following points are important:

- If you want to generate uniformly distributed pseudo-random numbers on the interval $[\alpha, \beta]$, then you need the formula "Random Uniform(α, β)".

- If you want exponentially distributed pseudo-random numbers with parameter λ, then you need the formula "Random Exp()/λ".

- If you want to generate gamma distributed pseudo-random numbers with parameters k and θ, then you need the formula "θ * Random Gamma(k)".

- If you want to generate Weibull distributed pseudo-random numbers with parameters k and θ, then you need the formula "Random Weibull(k, θ)".

- If you want to generate beta distributed pseudo-random numbers with parameters α and β, then you need the formula "Random Beta(α, β)".

Figure 9.15 shows the formula that is needed to generate pseudo-random numbers from an exponential density with parameter λ equal to $1/2$. Figure 9.16 contains a possible final result. The pseudo-random numbers in Figure 9.16 all lie between 0 and 8, with values closer to 0 being more frequent than values closer to 8. This matches the exponential probability density in Figure 9.14.

10

The normal distribution

Nothing seemed to fit, yet there were plenty of clues to indicate everything was somehow related. I knew the random probability of so many coincidences was zero.

(from *The Eight,* Katherine Neville, p. 120)

The normal density is undoubtedly the most important density in univariate statistics. Some reasons for this are:

- Many practical processes generate data that is (more or less) normally distributed.

- Important functions of normally distributed random variables are again normally distributed.

- Even important functions of random variables that are not normally distributed turn out to be approximately normally distributed.

The last two reasons will be exploited on various occasions in the book *Statistics with JMP: Hypothesis Tests, ANOVA and Regression.*

The normal density was derived for the first time by Abraham de Moivre (1667–1754), a French mathematician who made many contributions to mathematics and probability theory. He obtained the normal density as an approximation of the binomial distribution in 1733. However, due to a historical error, the discovery of the normal density was attributed to the German Johann Carl Friedrich Gauss (1777–1855). The genius Gauss was a nightmare for teachers and professors, so he had many ups and downs on his way to gaining a diploma and doctorate. In 1809, he acquired everlasting fame[1] by briefly mentioning the normal density in a book on astronomy. At that time, the density was not yet referred to as the normal density.

---

[1] The portrait of Gauss with the normal density featured on the German 10 mark banknote for many years (see Figure 10.1).

Statistics with JMP: Graphs, Descriptive Statistics, and Probability, First Edition. Peter Goos and David Meintrup.
© 2015 John Wiley & Sons, Ltd. Published 2015 by John Wiley & Sons, Ltd.
Companion Website: wiley.com/go/goosandmeintrup

Figure 10.1   Tribute to Gauss: former German 10 mark banknote (Deutsche Bundesbank, Frankfurt am Main).

This name came in vogue after an influential Belgian statistician, Adolphe Quetelet[2] (1796–1874), found out that many physiological and behavioral characteristics of humans were distributed around the values of a "normal" person according to this density, which he then logically named normal density. For example, Quetelet showed that the normal distribution provides a good description for the chest sizes of 5738 Scottish soldiers. Later, he taught statistics to Florence Nightingale, who introduced statistical methods in hospitals and, as a result, indirectly saved many people's lives.

10.1   The normal density

A random variable X is normally distributed with parameters μ and σ if its probability density is

$$f_X(x; \mu, \sigma) = \frac{1}{\sigma\sqrt{2\pi}}\, e^{-\frac{(x-\mu)^2}{2\sigma^2}}, \quad -\infty < x < +\infty,$$

where the first parameter μ can be any real number and the second parameter σ can be any positive real number.

To indicate that a random variable X is normally distributed with parameters μ and σ, we use the following notation:

$$X \sim N(\mu, \sigma^2).$$

Some examples of normal densities are shown in Figure 10.2. The figure shows that a change of the parameter μ has no effect on the shape of the curve, but only on the location. The larger the value of μ, the more the curve of the normal probability density shifts to the right. The smaller the value of μ, the more the curve of the normal

---

[2] A portrait of Adolphe Quetelet, who also defined the body mass index, featured on a Belgian stamp published in 1974, the 100th anniversary of his death.

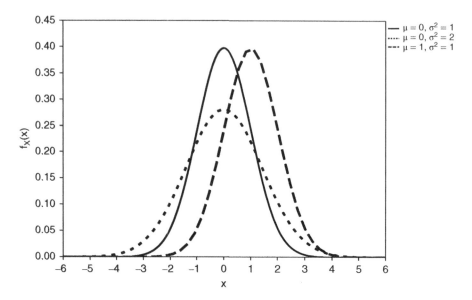

Figure 10.2   Normal densities for different values of μ and σ^2. The solid line indicates a (standard) normal density with $\mu = 0$ and $\sigma^2 = 1$, the dotted line indicates a density with $\mu = 0$ and $\sigma^2 = 2$, and the dashed line indicates a density with $\mu = 1$ and $\sigma^2 = 1$.

probability density shifts to the left. Since μ has no influence on the shape of the curve, this parameter also has no effect on the variance.

Unlike the parameter μ, the parameter σ does drive the spread. The larger the value of σ (or of σ^2), the wider the curve of the normal density. Automatically, the top of the curve is lower for larger values of σ or σ^2. Conversely, the smaller the value of σ (or of σ^2), the narrower the curve of the normal density, and the higher the top.

For a normally distributed random variable X, it can be shown that

$$\mu_X = \mu$$

and

$$\sigma_X^2 = \sigma^2.$$

In other words, the two parameters of a normally distributed random variable, μ and σ, are identical to its expected value and standard deviation.

Other properties of the normal density are the following:

- Any normally distributed random variable can take negative as well as positive values.

- The normal density is bell-shaped and converges asymptotically to zero for $x \to -\infty$ and $x \to +\infty$.

- The density is symmetric around μ. This means that $f_X(\mu + \delta; \mu, \sigma) = f_X(\mu - \delta; \mu, \sigma)$ for all possible values of δ.

- The mode, the median, and the expected value of the normal density coincide.

- The density has inflection points at $x = \mu - \sigma$ and $x = \mu + \sigma$. This can easily be demonstrated by means of the second derivative.

- There is no analytical expression for the cumulative distribution function. The function

$$F_X(x; \mu, \sigma) = P(X \le x) = \int_{-\infty}^{x} \frac{1}{\sigma\sqrt{2\pi}} \, e^{-\frac{(y-\mu)^2}{2\sigma^2}} \, dy,$$

some examples of which are shown in Figure 10.3, can only be calculated by means of numerical methods or using a computer. Seemingly, this implies that one needs tables for all possible values of μ and σ to determine probabilities for normally distributed random variables. However, because a linear transformation of a normally distributed random variable also is a normally distributed random variable (see Theorem 10.1.1 on page 236), we only need the table for the so-called **standard normal density**. The standard normal density is a normal density with $\mu = 0$ and $\sigma = 1$. It is depicted by the solid line in Figure 10.2.

For a standard normal random variable, it is customary to use the letter Z instead of the letter X. The standard normal density is

$$f_Z(z) = \frac{1}{\sqrt{2\pi}} \, e^{-\frac{z^2}{2}}, \quad -\infty < z < +\infty.$$

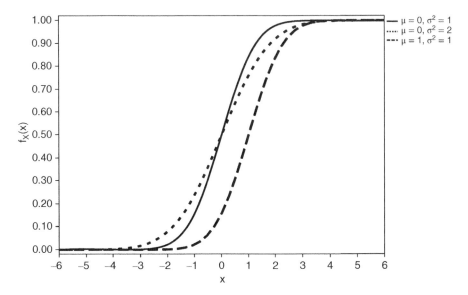

Figure 10.3   Cumulative normal distribution functions corresponding to the densities in Figure 10.2.

The corresponding cumulative distribution function is denoted by

$$\Phi(z) = F_Z(z) = P(Z \le z) = \int_{-\infty}^{z} \frac{1}{\sqrt{2\pi}} e^{-\frac{u^2}{2}} du.$$

As already mentioned previously, to be able to calculate probabilities for normally distributed random variables before the computer age, we had to rely on tables. The following theorem was then indispensable:

Theorem 10.1.1 *A linear transformation* $Y = g(X) = aX + b$ *of a normally distributed random variable* X *with expected value* μ *and variance* σ^2 *is itself a normally distributed random variable with expected value* $E(Y) = a\mu + b$ *and variance* $\text{var}(Y) = a^2\sigma^2$.

Proof: The proof of this theorem is a simple application of Equation (6.1) in Section 6.4.2. Because $x = g^{-1}(y) = (y - b)/a$, we have that

$$\frac{dx}{dy} = \frac{1}{a},$$

and that

$$f_Y(y) = f_X(x) \left| \frac{dx}{dy} \right|,$$

$$= f_X \left(\frac{y - b}{a} \right) \left| \frac{1}{a} \right|,$$

$$= \frac{1}{\sigma\sqrt{2\pi}} e^{-\frac{\left(\frac{y-b}{a} - \mu \right)^2}{2\sigma^2}} \left| \frac{1}{a} \right|,$$

$$= \frac{1}{\sigma\sqrt{2\pi}} e^{-\frac{(y-b-a\mu)^2}{2a^2\sigma^2}} \left| \frac{1}{a} \right|,$$

$$= \frac{1}{|a|\sigma\sqrt{2\pi}} e^{-\frac{(y-(a\mu+b))^2}{2(|a|\sigma)^2}},$$

which corresponds to a normally distributed random variable with mean $a\mu + b$ and variance $a^2\sigma^2$. The standard deviation is $|a|\sigma$. ∎

An immediate consequence of this theorem is that the random variable

$$Z = \frac{X - \mu}{\sigma},$$

where X is normally distributed with expected value μ and variance σ^2, has a standard normal distribution.

10.2   Calculation of probabilities for normally distributed variables

Nowadays, it is easy to determine probabilities for normally distributed random variables with a computer. In this section, we show how to calculate such probabilities without a computer. In addition, we illustrate how probabilities for normally distributed random variables can be calculated in JMP.

10.2.1   The standard normal distribution

The cumulative distribution function for a standard normal random variable is available in tables, which allow the determination of probabilities for variables with a standard normal distribution. An example of such a table is shown in Appendix E. This table contains the so-called exceedance probabilities $P(Z \geq z)$ for $z \geq 0$. For example, we can read from the table that $P(Z \geq 1.54) = 0.06178$. Since Z is a continuous random variable, it is also true that $P(Z > 1.54) = 0.06178$. This probability corresponds to the shaded area under the curve in Figure 10.4.

If we are interested in a probability of the form $P(Z \leq z)$ with $z \geq 0$, then we can still use the table, but it requires a little detour. To do so, we have to remember that $P(Z \leq z) = 1 - P(Z > z) = 1 - P(Z \geq z)$. As a result, we need to search for $P(Z \geq z)$ in the table. For example, $P(Z \leq 1.54) = 1 - P(Z \geq 1.54) = 1 - 0.06178 = 0.93822$. This probability corresponds to the white area under the curve in Figure 10.4.

If $z \leq 0$, then we can get a probability of the form $P(Z \leq z)$ by making use of the symmetry of the standard normal distribution around 0. Due to the symmetry, $P(Z \leq$

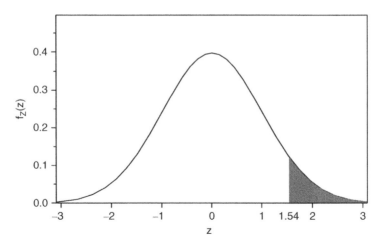

Figure 10.4   Standard normal probability density. The shaded area under the curve represents the probability $P(Z \geq 1.54)$. The white area under the curve represents the probability $P(Z \leq 1.54)$.

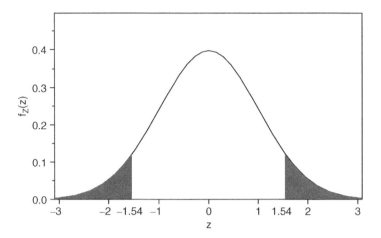

Figure 10.5   Standard normal probability density. The shaded area on the left indicates the probability $P(Z \leq -1.54)$, while the shaded area on the right represents the probability $P(Z \geq 1.54)$. Both probabilities are equal. The white area under the curve represents the probability $P(-1.54 \leq Z \leq 1.54)$.

$z) = P(Z \geq -z)$. For example, Figure 10.5 shows that the probabilities $P(Z \leq -1.54)$ and $P(Z \geq 1.54)$ are the same. The white area under the curve in the figure indicates the probability $P(-1.54 \leq Z \leq 1.54)$. The figure clearly shows that

$$P(-1.54 \leq Z \leq 1.54) = 1 - P(Z \leq -1.54) - P(Z \geq 1.54),$$
$$= 1 - P(Z \geq 1.54) - P(Z \geq 1.54),$$
$$= 1 - 2\,P(Z \geq 1.54),$$
$$= 1 - 2 \times 0.06178,$$
$$= 0.87644.$$

In general, if $z \geq 0$, then $P(-z \leq Z \leq z) = 1 - 2\,P(Z \geq z)$. A probability of the form $P(Z \geq z)$, with $z \leq 0$, can be calculated as

$$P(Z \geq z) = P(Z \leq -z) = 1 - P(Z \geq -z).$$

For example,

$$P(Z \geq -1.54) = P(Z \leq 1.54) = 1 - P(Z \geq 1.54) = 1 - 0.06178 = 0.93822.$$

10.2.2   General normally distributed variables

Theorem 10.1.1 allows us to apply the table in Appendix E to any normally distributed random variable. Indeed, if X is a normally distributed random variable with expected value μ and variance σ^2, then

$$P(a \leq X \leq b) = P(a - \mu \leq X - \mu \leq b - \mu),$$

$$= P\left(\frac{a - \mu}{\sigma} \leq \frac{X - \mu}{\sigma} \leq \frac{b - \mu}{\sigma}\right),$$

$$= P\left(\frac{a - \mu}{\sigma} \leq Z \leq \frac{b - \mu}{\sigma}\right),$$

$$= P\left(Z \geq \frac{a - \mu}{\sigma}\right) - P\left(Z \geq \frac{b - \mu}{\sigma}\right).$$

These last two probabilities can be obtained directly or indirectly from the table in Appendix E, as we explained in Section 10.2.1. For example, assume that the random variable X is normally distributed with expected value 80 and variance 25. The standard deviation of X is thus equal to 5. The probability that X takes a value between 72.3 and 87.7 is

$$P(72.3 \leq X \leq 87.7) = P(72.3 - 80 \leq X - 80 \leq 87.7 - 80),$$

$$= P(-7.7 \leq X - 80 \leq 7.7),$$

$$= P\left(\frac{-7.7}{5} \leq \frac{X - 80}{5} \leq \frac{7.7}{5}\right),$$

$$= P(-1.54 \leq Z \leq 1.54),$$

$$= 1 - 2\,P(Z \geq 1.54),$$

$$= 0.87644.$$

This probability is shown in Figure 10.6 by means of the shaded area.

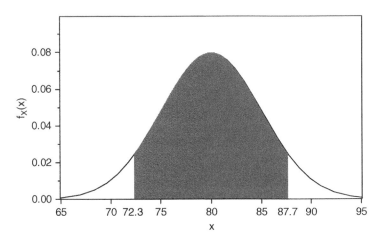

Figure 10.6   Normal density with expected value 80 and variance 25. The shaded area represents the probability $P(72.3 \leq X \leq 87.7)$.

10.2.3   JMP

Obviously, it is easier to make use of available software to calculate probabilities for normally distributed random variables than to use the table in Appendix E. Both the normal densities and the corresponding cumulative distribution functions are available in JMP. The required formulas are "Normal Density(x, μ, σ)" and "Normal Distribution(x, μ, σ)". If you omit the last two arguments of these formulas, by only typing "Normal Density(x)" or "Normal Distribution(x)", then JMP assumes that you want to use the standard normal density with expected value 0 and variance 1.

For the calculation of the probability $P(Z \leq 1.54)$ for a standard normal random variable Z, alternatively, you can enter the JMP formula "Normal Distribution(1.54, 0, 1)" or "Normal Distribution(1.54)". If you want to determine the probability $P(72.3 \leq X \leq 87.7)$ for a normally distributed random variable X with expected value 80 and standard deviation 5, you should use the command "Normal Distribution(87.7, 80, 5) − Normal Distribution(72.3, 80, 5)". Another option is to use the script "Distribution and Probability Calculator". Figure 10.7 shows the probability $P(72.3 \leq X \leq 87.7)$ for the normally distributed random variable X determined by means of the script.

If you want to determine quantiles or percentiles for a normally distributed random variable X, you should use the function "Normal Quantile(p, μ, σ)". If you need quantiles or percentiles of a standard normal random variable, "Normal Quantile(p)"

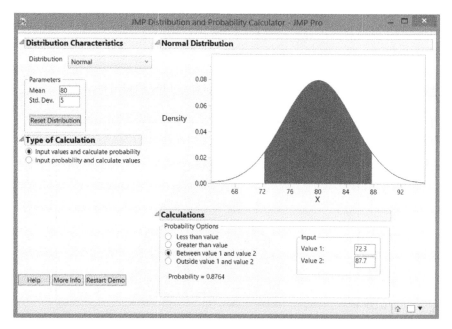

Figure 10.7   Using the "Distribution and Probability Calculator" to compute the probability $P(72.3 \leq X \leq 87.7)$.

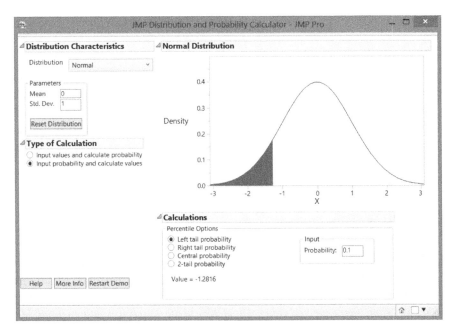

Figure 10.8   Using the "Distribution and Probability Calculator" to compute the 10th quantile or percentile of a standard normal random variable.

is sufficient. The 10th and 90th quantile or percentile of a standard normal variable are -1.28155 and 1.28155, respectively. The required commands for these values are "Normal Quantile(0.1)" and "Normal Quantile(0.9)". Again, an alternative is to make use of the "Distribution and Probability Calculator". This is illustrated in Figure 10.8, where the 10th percentile of a standard normal random variable is computed.

Note that the 10th and the 90th percentiles of a standard normal random variable only differ in their sign. This follows from the fact that the standard normal distribution is symmetric around 0.

10.2.4   Examples

Example 10.2.1 *The following probabilities have some interesting applications:*

$$P(\mu - \sigma \le X \le \mu + \sigma) = P(\mu - \sigma - \mu \le X - \mu \le \mu + \sigma - \mu),$$
$$= P(-\sigma \le X - \mu \le \sigma),$$
$$= P\left(\frac{-\sigma}{\sigma} \le \frac{X - \mu}{\sigma} \le \frac{\sigma}{\sigma}\right),$$
$$= P(-1 \le Z \le 1),$$
$$= P(Z \ge -1) - P(Z \ge 1),$$

$$= 1 - P(Z \geq 1) - P(Z \geq 1),$$

$$= 1 - 2\,P(Z \geq 1),$$

$$= 1 - 2 \times 0.15866,$$

$$= 0.68268;$$

$$P(\mu - 2\sigma \leq X \leq \mu + 2\sigma) = 1 - 2\,P(Z \geq 2),$$

$$= 1 - 2 \times 0.02275,$$

$$= 0.9545;$$

$$P(\mu - 3\sigma \leq X \leq \mu + 3\sigma) = 1 - 2\,P(Z \geq 3),$$

$$= 1 - 2 \times 0.00135,$$

$$= 0.9973.$$

For the last but one step in the three calculations, we used the table in Appendix E. The interpretation of the results in this example is that approximately 68% of the observations from a normally distributed population are less than one standard deviation away from their expected value. About 95% of the data is less than two standard deviations away, and 99.7% of the data is less than three standard deviations away from their expected value.

The calculations in this example offer us the opportunity to counter a common criticism on the use of the normal distribution in practice. Critics point out that the normal distribution is often used for random variables that can only take positive values (lengths, weights, …), despite the fact that the distribution allows values from $-\infty$ to $+\infty$. However, if the value of zero is more than three standard deviations away from the expected value of the normal distribution, then this distribution will only very rarely generate a negative value. Consequently, in these cases, the normal distribution can be safely used.

Example 10.2.2 *Scores on intelligence tests typically follow a normal distribution. The tests can be constructed in such a way that the median or the mean of the distribution is equal to 100 and the standard deviation is equal to 15. This is illustrated in Figure 10.9. Table 10.1 shows how IQ scores are used in the healthcare sector to classify less gifted people. In addition, we have the following facts:*

- *The lower limit for admission to the army of the United States is 70.*

- *Actress Sharon Stone is supposed to have an IQ score of 154.*

- *MENSA is an international organization whose sole criterion for membership is a score at or above the 98th percentile on a standardized IQ test that has been taken under supervision. This means the requirement is a score of 130 or more.*

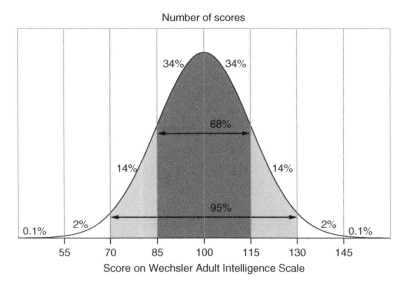

Figure 10.9   IQ scores and the normal density.

Table 10.1   Terminology used in the healthcare sector.

Term	Interval
Normally gifted	IQ ≥ 85
Retarded	70 ≤ IQ < 85
Mildly mentally disabled	50 ≤ IQ < 70
Moderately mentally disabled	35 ≤ IQ < 50
Severely mentally disabled	20 ≤ IQ < 35
Very severely mentally disabled	IQ < 20

Example 10.2.3 *For the construction of confidence intervals (see the book* Statistics with JMP: Hypothesis Tests, ANOVA and Regression*), it is useful to know that*

$$P(\mu - 1.645\sigma \leq X \leq \mu + 1.645\sigma) = 90\%,$$

$$P(\mu - 1.96\sigma \leq X \leq \mu + 1.96\sigma) = 95\%,$$

and

$$P(\mu - 2.576\sigma \leq X \leq \mu + 2.576\sigma) = 99\%.$$

Example 10.2.4 *The label on a package indicates* 16 *grams as minimum net weight. However, the filling machine does not always deposit exactly* 16 *grams in the package: the net weight is normally distributed around the setting point of the filling machine, with a standard deviation of* 0.2 *grams.*

1. *Suppose that the machine is set to a weight of* 16.32 *grams. What percentage of the filled packages will have a weight lower than the* 16 *grams mentioned on the package? If X represents the net weight of a package, then the percentage is given by*

$$P(X < 16) = P\left(\frac{X - 16.32}{0.2} < \frac{16 - 16.32}{0.2}\right),$$

$$= P(Z < -1.6),$$

$$= P(Z > 1.6),$$

$$= P(Z \geq 1.6),$$

$$= 0.0548.$$

Therefore, slightly more than 5% *of all packages will be too light. The last probability can be looked up in the table in Appendix E, or calculated using the formula "1 − Normal Distribution(1.6)" in JMP. An alternative is to calculate the probability P(X < 16) directly by using the formula "Normal Distribution(16,16.32,0.2)" or by using the "Distribution and Probability Calculator".*

2. *To which value of μ should the machine be set to ensure that only* 1% *of all packages have a net weight of less than* 16 *grams? If the random variable X again represents the net weight of an individual package, then the imposed condition is P(X < 16) = 0.01. Now,*

$$P(X < 16) = P\left(\frac{X - \mu}{0.2} < \frac{16 - \mu}{0.2}\right),$$

$$= P\left(Z < \frac{16 - \mu}{0.2}\right).$$

The table in Appendix E tells us that $P(Z \geq 2.33) = 0.0099 \approx 0.01$. *Consequently,* $P(Z < -2.33)$ *is also equal to* $0.0099 \approx 0.01$. *Therefore, to find the desired value of μ, the equation*

$$\frac{16 - \mu}{0.2} = -2.33$$

has to be solved. This results in a value of 16.466 *for μ. An alternative to find the value* −2.33 *is to make use of the formula "Normal Quantile(0.01)" in JMP, or to use the "Distribution and Probability Calculator".*

3. *Increasing the setting value μ of the filling machine is one solution of the problem. However, it is expensive because most of the packages will be overfilled. Another solution might be to buy a new machine that operates more precisely, that is, with a smaller standard deviation. The same effect can be achieved by a better maintenance of the machine, or by replacing parts that are worn out. How small must the standard deviation σ be so that keeping the original setting of* 16.32 *grams yields only* 1% *of packages with insufficient weight? Again, the imposed condition is P(X < 16) = 0.01. However, now we have that*

$$P(X < 16) = P\left(\frac{X - 16.32}{\sigma} < \frac{16 - 16.32}{\sigma}\right),$$

$$= P\left(Z < \frac{16 - 16.32}{\sigma}\right).$$

We already know that $P(Z \geq 2.33) = 0.0099 \approx 0.01$, and that, consequently, $P(Z < -2.33) = 0.0099 \approx 0.01$. To find the required standard deviation σ, we therefore have to solve the equation

$$\frac{16 - 16.32}{\sigma} = -2.33.$$

This returns $\sigma = 0.1373$ as the solution to the problem.

Due to the Japanese quality guru Genichi Taguchi, reducing the variability in products and processes is now one of the key ideas in quality management. Therefore, it is one of the main objectives of the immensely popular quality improvement program **Six Sigma**, as the "Sigma" in the name suggests. This is illustrated in the following example.

Example 10.2.5 *Six Sigma is a brand name of a training program for quality managers. The ultimate goal of the training program is to develop processes that generate less than 3.4 defects per million opportunities (3.4 parts per million, abbreviated as ppm). This is achieved by reducing the variance of the process in such a way that the target value of the process is at least six standard deviations σ (hence the name) away from the specification limits.*

*Specification limits are upper and lower limits that a certain characteristic of a product must satisfy. These limits are typically specified by a customer. For example, a customer who orders elevator cables may want cables with a diameter of 1 cm, and typically defines an interval, such as [0.99 cm, 1.01 cm], of acceptable values for the thickness. The limits of this interval are called specification limits: the lower limit is the **lower specification limit** (LSL) and the upper limit is the **upper specification limit** (USL). The **target** (T) here is 1 cm.*

The intended situation is shown in Figure 10.10: the mean or expected value μ of the production process, which is represented by the normal density, coincides with the target value. The variance of the production process is so small that LSL and USL are 6 standard deviations or 6σ away from the process mean μ. In this situation, the probability of a product being rejected, is equal to

$$P(rejected\ product) = P(product\ characteristic\ is\ out\ of\ specification\ limits),$$

$$= P(X \leq LSL\ or\ X \geq USL),$$

$$= P(X \leq LSL) + P(X \geq USL),$$

$$= P\left(\frac{X - \mu}{\sigma} \leq \frac{LSL - \mu}{\sigma}\right) + P\left(\frac{X - \mu}{\sigma} \geq \frac{USL - \mu}{\sigma}\right),$$

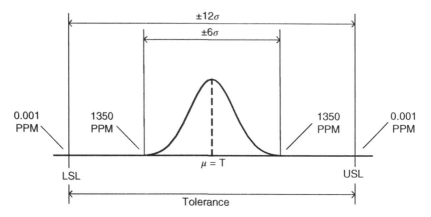

Figure 10.10   Centered 6σ-process.

$$= P\left(\frac{X - \mu}{\sigma} \le \frac{-6\sigma}{\sigma}\right) + P\left(\frac{X - \mu}{\sigma} \ge \frac{6\sigma}{\sigma}\right),$$

$$= P(Z \le -6) + P(Z \ge 6),$$

$$= 2\,P(Z \ge 6),$$

$$= 2 \times 0.000000001,$$

$$= \frac{0.002}{1000000},$$

$$= 0.002\ ppm.$$

We used JMP to compute the probability $P(Z \ge 6)$ in the penultimate step of this calculation because this probability is so small that it is not listed in the table in Appendix E. The result 0.002 ppm means that on average only 0.002 per million produced products will not satisfy the specifications.

A centered process, where the process mean μ is exactly equal to the target T, thus results in very few defects. Now, you might wonder, why the Six Sigma program is associated with an error rate of 3.4 ppm. Indeed, in this centered process we found a defect rate much smaller than 3.4 ppm. The reason for this is that the Six Sigma program assumes that it is extremely difficult to keep the process mean μ exactly equal to the target value T. Therefore, it is part of the Six Sigma philosophy to allow the process mean μ to deviate up to 1.5σ from the target value T. The situation in which the process mean μ differs by 1.5σ from the target T is shown in Figure 10.11. Possible reasons for such deviations include wear of the equipment, or fatigue of the operator. The probability of rejection in the situation depicted in Figure 10.11 is

$$P(rejected\ product) = P(X \le LSL) + P(X \ge USL),$$

$$= P\left(\frac{X - \mu}{\sigma} \le \frac{LSL - \mu}{\sigma}\right) + P\left(\frac{X - \mu}{\sigma} \ge \frac{USL - \mu}{\sigma}\right),$$

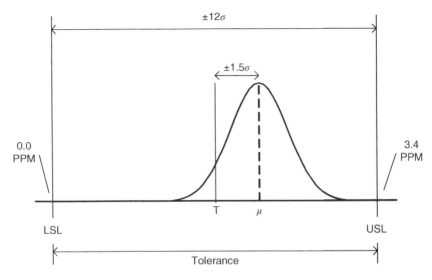

Figure 10.11   Non-centered 6σ-process.

$$= P\left(\frac{X-\mu}{\sigma} \le \frac{-7.5\sigma}{\sigma}\right) + P\left(\frac{X-\mu}{\sigma} \ge \frac{4.5\sigma}{\sigma}\right),$$

$$= P(Z \le -7.5) + P(Z \ge 4.5),$$

$$= 0 + 0.0000034,$$

$$= 3.4 \; ppm.$$

10.3   Lognormal probability density

A normally distributed random variable takes values between $-\infty$ and $+\infty$. In Chapter 9, we have already encountered several probability densities of random variables that only take values between 0 and $+\infty$. The lognormal probability density, which is closely related to the normal density, is another probability density that is suitable for random variables that only take non-negative real values.

A random variable X is lognormally distributed if its probability density is given by

$$f_X(x; \mu, \sigma) = \frac{1}{x\sigma\sqrt{2\pi}} e^{-\frac{(\ln x - \mu)^2}{2\sigma^2}}, \quad 0 < x < +\infty.$$

The expected value of a lognormally distributed random variable is equal to

$$\mu_X = E(X) = e^{\mu + \sigma^2/2},$$

and its variance is equal to

$$\sigma_X^2 = \text{var}(X) = (e^{\sigma^2} - 1)e^{2\mu + \sigma^2}.$$

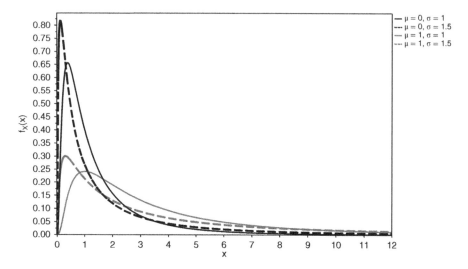

Figure 10.12   Four different lognormal probability densities.

The median is

$$\gamma_{0.5} = e^{\mu},$$

and the mode is $e^{\mu - \sigma^2}$. Some lognormal probability densities, with μ equal to 0 or 1 and σ equal to 1 or 1.5, are shown in Figure 10.12.

The lognormal probability density has many successful applications, including lengths, weights, sizes of cities, the maximum amount of rainfall, income, interest rates, and price indices.

If a random variable X is lognormally distributed with parameters μ and σ, the transformed random variable $Y = \ln(X)$ is normally distributed with expected value μ and variance σ^2. It is a useful exercise to demonstrate this. The proof of this property is analogous to the proof of Theorem 10.1.1. The converse is also true: if a random variable X is normally distributed with expected value μ and variance σ^2, then the random variable $Y = e^X$ is lognormally distributed with parameters μ and σ.

It is not difficult to show some more properties of the lognormal probability density:

- If a random variable X is lognormally distributed with parameters μ and σ, then the transformed random variable $Y_1 = 1/X$ is also lognormally distributed, but with parameters $-\mu$ and σ.

- If a random variable X is lognormally distributed with parameters μ and σ, then the transformed random variable $Y_2 = aX$, where a is a strictly positive constant, is also lognormally distributed, but with parameters $\ln(a) + \mu$ and σ.

- If a random variable X is lognormally distributed with parameters μ and σ, then the transformed random variable $Y_3 = X^a$, where a is a non-zero constant, is also lognormally distributed, but with parameters $a\mu$ and $|a|\sigma$.

- If a random variable X is lognormally distributed with parameters μ and σ, then the transformed random variable $Y_4 = bX^a$, where a is a non-zero constant and b a strictly positive constant, is also lognormally distributed, but with parameters $\ln(b) + a\mu$ and $|a|\sigma$. This follows from combining the results for Y_2 and Y_3.

The starting point for proving these properties is always that $\ln(X)$ is normally distributed with expected value μ and variance σ^2 if X is lognormally distributed with parameters μ and σ. We begin with the random variable $Y_1 = 1/X$. The natural logarithm of Y_1 is

$$\ln(Y_1) = \ln\left(\frac{1}{X}\right) = \ln(1) - \ln(X) = 0 - \ln(X) = -\ln(X).$$

Since $\ln(X)$ is normally distributed with expected value μ and variance σ^2, we know that $-\ln(X)$ is normally distributed with expected value $-\mu$ and variance σ^2. This follows from Theorem 10.1.1. Consequently,

$$Y_1 = e^{\ln(Y_1)} = e^{-\ln(X)}$$

is lognormally distributed with parameters $-\mu$ and σ.

Next, we look at the random variable $Y_2 = aX$. The natural logarithm of Y_2 is

$$\ln(Y_2) = \ln(aX) = \ln(a) + \ln(X).$$

Since $\ln(X)$ is normally distributed with expected value μ and variance σ^2, we know that $\ln(a) + \ln(X)$ is normally distributed with expected value $\ln(a) + \mu$ and variance σ^2. This again follows from Theorem 10.1.1. Hence,

$$Y_2 = e^{\ln(Y_2)} = e^{\ln(a)+\ln(X)}$$

is lognormally distributed with parameters $\ln(a) + \mu$ and σ.

Finally, we consider the random variable $Y_3 = X^a$. The natural logarithm of Y_3 is

$$\ln(Y_3) = \ln(X^a) = a\ln(X).$$

Since $\ln(X)$ is normally distributed with expected value μ and variance σ^2, we know, again from Theorem 10.1.1, that $a\ln(X)$ is normally distributed with expected value $a\mu$ and variance $a^2\sigma^2$. Hence, the standard deviation is $|a|\sigma$ and

$$Y_3 = e^{\ln(Y_3)} = e^{a\ln(X)}$$

is lognormally distributed with parameters $a\mu$ and $|a|\sigma$.

Example 10.3.1 *The yield of an investment fund, expressed in %, is lognormally distributed with parameters $\mu = 0.5$ and $\sigma = 0.75$. Determine the probability that*

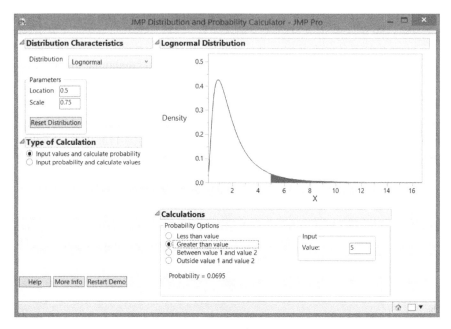

Figure 10.13   Using the "Distribution and Probability Calculator" to compute the probability $P(X \geq 5)$ for a lognormally distributed random variable.

the yield of the investment fund is greater than or equal to 5%. If we denote the yield using the variable X, then the probability we seek is equal to

$$P(X \geq 5) = P(\ln(X) \geq \ln(5)).$$

Since $\ln(X)$ is normally distributed with expected value $\mu = 0.5$ and standard deviation $\sigma = 0.75$, we can determine this probability as follows:

$$P(X \geq 5) = P(\ln(X) - 0.5 \geq \ln(5) - 0.5),$$

$$= P\left(\frac{\ln(X) - 0.5}{0.75} \geq \frac{\ln(5) - 0.5}{0.75}\right),$$

$$= P(Z \geq 1.47925055),$$

$$= 0.06954.$$

Figure 10.13 illustrates how this probability can be determined with the "Distribution and Probability Calculator" in JMP.

Example 10.3.2 *In Example 10.2.1, we learnt that*

$$P(\mu - \sigma \leq X \leq \mu + \sigma) = 0.68268,$$

$$P(\mu - 2\sigma \leq X \leq \mu + 2\sigma) = 0.9545,$$

and

$$P(\mu - 3\sigma \le X \le \mu + 3\sigma) = 0.9973,$$

if X is normally distributed with expected value μ and variance σ^2. It follows that

$$P(e^{\mu-\sigma} \le e^X \le e^{\mu+\sigma}) = 0.68268,$$
$$P(e^{\mu-2\sigma} \le e^X \le e^{\mu+2\sigma}) = 0.9545,$$

and

$$P(e^{\mu-3\sigma} \le e^X \le e^{\mu+3\sigma}) = 0.9973.$$

In these expressions, e^X is lognormally distributed with parameters μ and σ. The expressions can be rewritten as

$$P(e^\mu e^{-\sigma} \le e^X \le e^\mu e^\sigma) = P\left(\frac{e^\mu}{e^\sigma} \le e^X \le e^\mu e^\sigma\right) = 0.68268,$$

$$P(e^\mu e^{-2\sigma} \le e^X \le e^\mu e^{2\sigma}) = P\left(\frac{e^\mu}{e^{2\sigma}} \le e^X \le e^\mu e^{2\sigma}\right),$$

$$= P\left(\frac{e^\mu}{(e^\sigma)^2} \le e^X \le e^\mu (e^\sigma)^2\right),$$

$$= 0.9545,$$

and

$$P(e^\mu e^{-3\sigma} \le e^X \le e^\mu e^{3\sigma}) = P\left(\frac{e^\mu}{e^{3\sigma}} \le e^X \le e^\mu e^{3\sigma}\right),$$

$$= P\left(\frac{e^\mu}{(e^\sigma)^3} \le e^X \le e^\mu (e^\sigma)^3\right),$$

$$= 0.9973.$$

The values e^μ and e^σ are the geometric mean and the geometric standard deviation, respectively, of a lognormally distributed random variable. Just as the geometric mean was an alternative to the arithmetic mean in Chapter 3, the geometric mean and geometric standard deviation are alternatives to the expected value and the ordinary standard deviation here. If we introduce the notation μ^* for the geometric mean and σ^* for the geometric standard deviation, then we can state that a lognormally distributed random variable takes values in the interval $[\mu^*/\sigma^*, \mu^*\sigma^*]$ with a probability of 68.268%, in the interval $[\mu^*/\sigma^{*2}, \mu^*\sigma^{*2}]$ with a probability of 95.45%, and in the interval $[\mu^*/\sigma^{*3}, \mu^*\sigma^{*3}]$ with a probability of 99.73%.

11

Multivariate random variables

Goldfarb was there with his friends from Lowell House. They were sitting in a corner of the room, deeply entrenched in a discussion over a matter of life and death, namely the game-theoretical aspects of blackjack and whether or not it was possible to cheat the bank using purely mathematical principles ... She didn't interrupt her dancing for so much as a nanosecond, but she waved at the boys, or so it seemed, with the sparkling white rubber soles of her flat shoes, thereby drawing even more attention to those marvelous legs. For one long agonizing moment, the four boys each independently considered the possibility that the development of an ultimate probability theory for casino-goers might be just an idle pastime ... On top of that, all their calculations on paper napkins and even the quick simulation study with a real deck of cards had led the group of friends to the final conclusion – unanticipated yet uncontested – that there is after all an untraceable element of chance in blackjack, and this had made them all feel inexplicably blue.

(from *Omega Minor,* Paul Verhaeghen, pp. 167–170)

11.1 Introductory notions

A (univariate) random variable, as introduced in Chapter 6, assigns one real number to each outcome of an experiment. If several real numbers are assigned to each outcome, we have a multivariate random variable. For a k-variate random variable, k numbers are assigned to an outcome ω. We can denote these numbers by $x_1(\omega)$, $x_2(\omega)$, ..., $x_k(\omega)$. This notation emphasizes the fact that a random variable is a function. Usually, however, the k numbers are indicated by x_1, x_2, ..., x_k instead. Those numbers are realizations of the k-variate random variable[1] $\mathbf{X}_k = (X_1, X_2, ..., X_k)$. Again, capital

[1] Sometimes, this variable is also called a probability vector instead of a multivariate random variable.

Statistics with JMP: Graphs, Descriptive Statistics, and Probability, First Edition. Peter Goos and David Meintrup.
© 2015 John Wiley & Sons, Ltd. Published 2015 by John Wiley & Sons, Ltd.
Companion Website: wiley.com/go/goosandmeintrup

letters are used to indicate random variables and lowercase letters are used to indicate real values or realizations of random variables. Bold letters indicate vectors.

Example 11.1.1 *An experiment consists of throwing a red and a blue die. A bivariate random variable can be defined as $X_2 = (X_1, X_2)$, where X_1 represents the number of dots on the red die and X_2 represents the number of dots on the blue die.*

Example 11.1.2 *A cross-hatch test is a test to evaluate the quality of adhesion performed by pulling off an adhesive tape applied to a surface. The adhesion can be "perfect", "excellent", "good", or "insufficient". An experiment consists of carrying out the test on a specific product. A four-variate random variable is defined as*

$$X_1 = 1, X_2 = 0, X_3 = 0, X_4 = 0, \text{ if the adhesion is perfect,}$$

$$X_1 = 0, X_2 = 1, X_3 = 0, X_4 = 0, \text{ if the adhesion is excellent,}$$

$$X_1 = 0, X_2 = 0, X_3 = 1, X_4 = 0, \text{ if the adhesion is good,}$$

$$X_1 = 0, X_2 = 0, X_3 = 0, X_4 = 1, \text{ if the adhesion is insufficient.}$$

This four-variate random variable can be simplified to a three-variate random variable with no loss of information:

$$X = 1, Y = 0, Z = 0, \text{ if the adhesion is perfect,}$$

$$X = 0, Y = 1, Z = 0, \text{ if the adhesion is excellent,}$$

$$X = 0, Y = 0, Z = 1, \text{ if the adhesion is good,}$$

$$X = 0, Y = 0, Z = 0, \text{ if the adhesion is insufficient.}$$

The reason why this simplification can be made is that X_1, X_2, X_3, and X_4 are connected by the constraint $X_1 + X_2 + X_3 + X_4 = 1$. Because of this constraint, the fourth variable can be determined from the other three.

Example 11.1.3 *The surface tension of a plastic material is the sum of a polar and a dispersive component. If an experiment involves measuring the dispersive and polar surface tension, then a bivariate random variable is given by $X_2 = (X_1, X_2)$, where X_1 is the dispersive surface tension of the material, and X_2 is the polar surface tension. The surface tension of materials largely determines how easy or how difficult it is to make a coating stick to a surface.*

Example 11.1.4 *An economist examines household incomes and interviews members of several families in France. The researcher is particularly interested in a bivariate random variable $X_2 = (X_1, X_2)$, where X_1 represents the income of the householder and X_2 the income of his/her partner.*

These examples show that multivariate random variables can, as univariate random variables, be discrete or continuous. In the next sections, we focus on discrete bivariate random variables. Then, we briefly study continuous multivariate and, in more detail, continuous bivariate random variables, which require the use of multiple integrals. An extension to higher dimensions (i.e., to situations with more than two random variables) is conceptually not difficult, but the notation quickly becomes a little confusing.

In the statistical literature, it is common to use the letters X and Y for bivariate random variables instead of the symbols X_1 and X_2. Therefore, in the remainder of this book, we denote the two random variables of a bivariate probability distribution or probability density by X and Y. We call the pair (X, Y) a bivariate random variable. For trivariate probability distributions or densities, one will often find the letters X, Y, and Z for the three random variables involved.

Example 11.1.5 *An experiment consists of randomly drawing a letter that will be delivered by the Royal Mail. With each letter, two real numbers are associated. The first number is a 1 or a 0, depending on whether the letter is a First Class letter or not. The second number is the number of days the Royal Mail needs to deliver the letter. Consequently, (X, Y) is a bivariate random variable, with 0 and 1 as possible values for X, and 0, 1, 2, … as possible values for Y. Randomly selecting a letter yields a realization of the bivariate random variable, leading to an observed or realized value x for X and an observed or realized value y for Y.*

11.2   Joint (discrete) probability distributions

For a bivariate discrete random variable (X, Y), the joint discrete probability distribution is defined as

$$p_{XY}(x, y) = P(X = x, Y = y),$$
$$= P\{(X = x) \cap (Y = y)\}.$$

If D is the set of all possible combinations of values of the bivariate random variable (X, Y), then the probability distribution satisfies the conditions

$$0 \le p_{XY}(x, y) \le 1,$$

and

$$\sum_{(x,y) \in D} p_{XY}(x, y) = 1.$$

Example 11.2.1 *In Table 11.1, a bivariate probability distribution for the bivariate discrete random variable defined in Example 11.1.5 is given. For instance, the probability that a random letter we select is a First Class letter that arrives after one day equals $p_{XY}(1, 1) = P(X = 1, Y = 1) = 0.40$. A graphical representation of the probability distribution is shown in Figure 11.1. To reproduce this figure in JMP, you need to create a data table with three columns (Shipment, Days, and p_{XY}), using the data from*

Table 11.1   Bivariate probability distribution of Example 11.2.1.

			Y			
			1	2	3	4
X	1	First Class	0.400	0.060	0.035	0.005
	0	Second Class	0.180	0.225	0.085	0.010

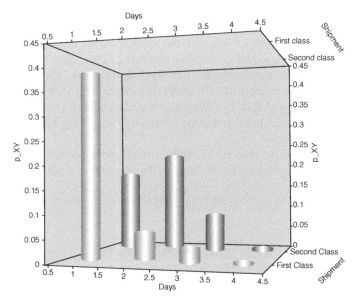

Figure 11.1   Graphical representation of the bivariate probability distribution in Table 11.1.

Table 11.1. Then, you need to select the option "Scatterplot 3D" from the "Graph" menu, and enter the columns "Shipment", "p_{XY}", and "Days", in this order, as "Y" columns. In the resulting output, you then have to click on the hotspot (red triangle) next to the term "Scatterplot 3D" and choose the option "Drop Lines". By adjusting the "Drop Line Thickness" in the same menu, you can change the thickness of the lines according to your taste.

The probability that a randomly selected letter is on the way for at most two days and is not a First Class letter is equal to

$$P\{(X = 0) \cap (Y \le 2)\} = P\{(X = 0) \cap (Y = 1)\} + P\{(X = 0) \cap (Y = 2)\},$$

$$= p_{XY}(0, 1) + p_{XY}(0, 2),$$

$$= 0.18 + 0.225,$$

$$= 0.405.$$

Of course, there is an analogy between the concept of the joint probability distribution and the relative frequency distribution in Table 4.2 in Example 4.2.2. The frequency, however, relates to sample data, while the probability distribution refers to a population or a process.

11.3   Marginal or unconditional (discrete) probability distribution

We again consider a bivariate discrete random variable (X, Y) with joint probability distribution $p_{XY}(x, y)$. A natural question that arises is whether individual univariate probability distributions for X and Y can be derived from the joint probability distribution. The answer to this question is affirmative. These individual probability distributions are called marginal or unconditional distributions. There is essentially no difference between a marginal probability distribution of X or Y and a univariate probability distribution of X or Y, as introduced in Chapter 6. By using the adjective marginal, however, we emphasize that the distribution was derived from a multivariate probability distribution.

If (X, Y) is a bivariate discrete random variable with probability distribution $p_{XY}(x, y)$, then the marginal (or unconditional) distributions of X and Y are

$$p_X(x) = \sum_y p_{XY}(x, y)$$

and

$$p_Y(y) = \sum_x p_{XY}(x, y),$$

respectively. In the expression for $p_X(x)$, we sum over all possible values of Y, while, in the expression for $p_Y(y)$, the sum is over all possible values of X.

Example 11.3.1 *The marginal distributions of X and Y in Example 11.2.1 are shown in Table 11.2. It is easy to verify that, for example, $p_X(1)$ is obtained by summing all probabilities in the first row of Table 11.1. In the same way, $p_Y(2)$ is obtained by adding up all probabilities in the second column of Table 11.1. All row totals form the marginal probability distribution of X, while all column totals form the marginal probability distribution of Y.*

Table 11.2   Bivariate probability distribution and associated marginal probability distributions for Example 11.2.1.

		Y				$p_X(x)$
		1	2	3	4	
X	1	0.400	0.060	0.035	0.005	0.500
	0	0.180	0.225	0.085	0.010	0.500
$p_Y(y)$		0.580	0.285	0.120	0.015	1.000

Although the marginal probability distributions can be derived from a multivariate probability distribution, the converse is usually not the case. Reconstructing a multivariate probability distribution based on the marginal probability distributions is only possible for **independent random variables**.

Definition 11.3.1 *The discrete random variables X and Y are **independent** if their joint probability distribution is the product of the marginal probability distributions for all possible combinations of values of X and Y, that is,*

$$p_{XY}(x, y) = p_X(x)p_Y(y) \text{ for all } (x, y) \in D.$$

If this condition is not met, then X and Y are called **dependent**. The concept of independence here is strongly related to that of independent events in Chapter 4.

Example 11.3.2 *It is easy to verify that the random variables X and Y in Example 11.3.1 are dependent. Indeed,*

$$p_{XY}(1, 1) = 0.4 \neq p_X(1)p_Y(1) = 0.5 \times 0.58 = 0.29.$$

The definition of independent random variables can be extended to a situation involving k random variables instead of two:

Definition 11.3.2 *The discrete random variables $X_1, X_2, \ldots, X_k$ are **independent** if their joint probability distribution is the product of the marginal probability distributions for all possible combinations of values of $X_1, X_2, \ldots, X_k$, that is,*

$$p_{X_1 X_2 \ldots X_k}(x_1, x_2, \ldots, x_k) = p_{X_1}(x_1)p_{X_2}(x_2) \ldots p_{X_k}(x_k)$$

$$\text{for all } (x_1, x_2, \ldots, x_k) \in D.$$

11.4   Conditional (discrete) probability distribution

The concept of a conditional probability distribution is best introduced by means of an example.

Example 11.4.1 *Using the data from Example 11.2.1, we can calculate the probability that a randomly selected letter needs three days to be delivered, given that it is a First Class letter:*

$$P(Y = 3 \mid X = 1) = \frac{P\{(Y = 3) \cap (X = 1)\}}{P(X = 1)} = \frac{p_{XY}(1, 3)}{p_X(1)} = \frac{0.035}{0.500} = 0.070.$$

To calculate this probability, we applied the definition of a conditional probability from Chapter 4 (Definition 4.4.1). Repeating this for all possible values of Y when $X = 1$, we find that $P(Y = 1 \mid X = 1) = 0.800$, $P(Y = 2 \mid X = 1) = 0.120$, and $P(Y = 4 \mid X = 1) = 0.010$. Together, these four probabilities form the conditional probability distribution of Y, given that $X = 1$. Their sum is 1.

In general, a conditional probability distribution of a random variable X given $Y = y$ is defined as

$$p_{X|Y}(x \mid y) = \frac{p_{XY}(x, y)}{p_Y(y)}.$$

The conditional probability distribution of Y given $X = x$ is

$$p_{Y|X}(y \mid x) = \frac{p_{XY}(x, y)}{p_X(x)}.$$

These definitions show that the joint probability distribution $p_{XY}(x, y)$ of (X, Y) cannot be derived from the two marginal distributions, but from the marginal distribution of one random variable and the conditional probability distribution of the other. In general, a joint probability distribution $p_{XY}(x, y)$ can be reconstructed in two ways. The first way is

$$p_{XY}(x, y) = p_{Y|X}(y \mid x)p_X(x),$$

and the second way is

$$p_{XY}(x, y) = p_{X|Y}(x \mid y)p_Y(y).$$

In the special case of independent random variables X and Y, we have that

$$p_{X|Y}(x \mid y) = \frac{p_{XY}(x, y)}{p_Y(y)} = \frac{p_X(x)p_Y(y)}{p_Y(y)} = p_X(x)$$

and

$$p_{Y|X}(y \mid x) = \frac{p_{XY}(x, y)}{p_X(x)} = \frac{p_X(x)p_Y(y)}{p_X(x)} = p_Y(y).$$

In short, for independent random variables, the conditional probability distribution is identical to the marginal probability distribution for each given x or y value.

11.5   Examples of discrete bivariate random variables

Example 11.5.1 *At an international football tournament, the Netherlands play the quarter-final against Germany. After 120 minutes, both teams tie. To determine a winner, each team takes five penalty kicks. For years, the Dutch players have been bad at taking penalty kicks. The probability that a Dutch football player scores a penalty kick is equal to 0.7, while the probability for a German player to score one is 0.9. The team with the highest number of successful penalties wins the match. Assume that the shoot-out does not end when the winner is known with certainty, but that each team takes five penalties anyway. In addition, assume that the probability of converting a penalty kick remains unchanged for each team during the entire shoot-out and that the number of goals scored by the Dutch team is independent of the number of goals scored by the German team.*

1. *What is the expected number of penalties that the Dutch team will convert? What is the expected number of penalties that the German team will convert?*

2. What is the expected number of penalties that the Dutch team should take until they miss a penalty? What is the expected number of penalties that the German team should take until they miss a penalty?

3. What is the probability that Germany leaves as winner after five penalties? What is the chance of a tie after five penalties?

4. What is the most likely outcome after the series of five penalty kicks for each team?

5. Suppose Germany scored three times. What is the probability that the Netherlands win?

To answer all these questions, we need the binomial and the geometric distribution. We also need the joint probability distribution on the set of the number of goals scored by both teams.

1. The number of goals scored by the Dutch team is binomially distributed with parameters $n = 5$ and $\pi = 0.7$, while the number of goals scored by the German team is binomially distributed with parameters $n = 5$ and $\pi = 0.9$. As the expected value of a binomially distributed random variable is equal to $n\pi$, the expected number of Dutch goals is $5 \times 0.7 = 3.5$, while the expected number of German goals equals $5 \times 0.9 = 4.5$.

2. The number of penalties that the Dutch team has to take until missing one, is geometrically distributed with parameter $\pi = 0.3$, the probability that the Netherlands miss a penalty. The expected value of this geometrically distributed random variable is $1/\pi = 1/0.3 = 10/3 = 3.33$. For Germany, we can follow a similar reasoning. The expected number of penalties that the German team has to take until the first miss is $1/0.1 = 10$.

3. In order to determine the probability that Germany wins, or that the shoot-out ends in a tie, we need to determine the joint probability distribution of the numbers of goals scored by the German team and the Dutch team. Suppose that the random variable X represents the number of goals scored by the Dutch team and Y the number of goals scored by the German team. To determine the joint probability distribution of X and Y, we start from the marginal probability distributions. The marginal probability distribution of both X and Y is a binomial distribution with $n = 5$. The two binomial distributions are shown in Table 11.3. The joint probability distribution of X and Y is easy to determine because X and Y are assumed to be independent. Hence,

$$p_{XY}(XY) = p_X(X)p_Y(Y).$$

The joint probability that the Netherlands score twice and Germany scores three times equals

$$p_{XY}(2, 3) = p_X(2)p_Y(3) = (0.13230)(0.07290) = 0.00964467.$$

In the same way, we can compute the joint probability for each possible combination of values of X and Y. All joint probabilities that we obtain in this manner

Table 11.3   Marginal (binomial) probability distributions for the number of goals scored by the Dutch team (X, with success probability $\pi = 0.7$) and by the German team (Y, with success probability $\pi = 0.9$).

x	Netherlands $p_X(x)$	Germany $p_Y(y)$
0	0.00243	0.00001
1	0.02835	0.00045
2	0.13230	0.00810
3	0.30870	0.07290
4	0.36015	0.32805
5	0.16807	0.59049

are shown in Table 11.4, together with the marginal probability distributions $p_X(X)$ and $p_Y(Y)$.

The probability that Germany wins is the sum of all joint probabilities for which $Y > X$:

$P(\text{Germany wins}) = P(Y > X),$

$= P((X = 0) \cap (Y = 1)) + P((X = 0) \cap (Y = 2)) + \ldots$

$\cdots + P((X = 0) \cap (Y = 5))$

$+ P((X = 1) \cap (Y = 2)) + P((X = 1) \cap (Y = 3)) + \ldots$

$\cdots + P((X = 1) \cap (Y = 5))$

$+ P((X = 2) \cap (Y = 3)) + P((X = 2) \cap (Y = 4)) + P((X = 2) \cap (Y = 5))$

$+ P((X = 3) \cap (Y = 4)) + P((X = 3) \cap (Y = 5))$

$+ P((X = 4) \cap (Y = 5)),$

$= p_{XY}(0, 1) + p_{XY}(0, 2) + \cdots + p_{XY}(0, 5)$

$+ p_{XY}(1, 2) + p_{XY}(1, 3) + \cdots + p_{XY}(1, 5)$

$+ p_{XY}(2, 3) + p_{XY}(2, 4) + p_{XY}(2, 5)$

$+ p_{XY}(3, 4) + p_{XY}(3, 5)$

$+ p_{XY}(4, 5),$

$= 0.00000109 + 0.00001968 + 0.00017715 + 0.00079716 + 0.00143489$

$+ 0.00022964 + 0.00206672 + 0.00930022 + 0.01674039$

$+ 0.00964467 + 0.04340102 + 0.07812183$

$+ 0.10126904 + 0.18228426$

$+ 0.21266497,$

$= 0.65815272.$

Table 11.4   Joint probability distribution $p_{XY}(X, Y)$ for the number of goals scored by the Dutch team (X) and by the German team (Y).

X	Y						$p_X(X)$
	0	1	2	3	4	5	
0	0.00000002	0.00000109	0.00001968	0.00017715	0.00079716	0.00143489	0.00243000
1	0.00000028	0.00001276	0.00022964	0.00206672	0.00930022	0.01674039	0.02835000
2	0.00000132	0.00005954	0.00107163	0.00964467	0.04340102	0.07812183	0.13230000
3	0.00000309	0.00013892	0.00250047	0.02250423	0.10126904	0.18228426	0.30870000
4	0.00000360	0.00016207	0.00291722	0.02625494	0.11814721	0.21266497	0.36015000
5	0.00000168	0.00007563	0.00136137	0.01225230	0.05513536	0.09924365	0.16807000
$p_Y(Y)$	0.00001000	0.00045000	0.00810000	0.07290000	0.32805000	0.59049000	1.00000000

Hence, there is a probability of more than 65% that Germany wins after a series of five penalty kicks. The probability that the penalty shoot-out ends in a tie is

$$P(tie) = P(X = Y),$$
$$= P((X = 0) \cap (Y = 0)) + P((X = 1) \cap (Y = 1)) + \dots$$
$$\cdots + P((X = 5) \cap (Y = 5)),$$
$$= p_{XY}(0, 0) + p_{XY}(1, 1) + \cdots + p_{XY}(5, 5),$$
$$= 0.00000002 + 0.00001276 + 0.00107163$$
$$+ 0.02250423 + 0.11814721 + 0.09924365,$$
$$= 0.24097950.$$

The probability that the Netherlands win is

$$P(Netherlands\ win) = 1 - P(Germany\ wins) - P(tie),$$
$$= 1 - 0.65815272 - 0.24097950,$$
$$= 0.10086778.$$

4. *The most likely outcome of the penalty shoot-out is 4–5 in favor of Germany. This outcome has a probability of* 0.21266497.

5. *Finally, the conditional probability that the Netherlands win, given that Germany scored three times, is*

$$P(Netherlands\ win\ |\ Germany\ scored\ 3\ times) = P(X > 3\ |\ Y = 3),$$
$$= \frac{P((X > 3) \cap (Y = 3))}{P(Y = 3)},$$
$$= \frac{P((X = 4) \cap (Y = 3)) + P((X = 5) \cap (Y = 3))}{P(Y = 3)},$$
$$= \frac{p_{XY}(4, 3) + p_{XY}(5, 3)}{p_Y(3)},$$
$$= \frac{0.02625494 + 0.01225230}{0.07290000},$$
$$= 0.52822.$$

This probability can also be determined in a different way, using the fact that the number of goals scored by the Dutch team is independent of the number of goals scored by the German team. As a result, we have that

$$P(X > 3\ |\ Y = 3) = P(X > 3) = P(X = 4) + P(X = 5),$$
$$= 0.36015000 + 0.16807000 = 0.52822.$$

Example 11.5.2 *An urn contains one yellow ball, two red balls, and three blue balls. A blindfolded person randomly draws two balls from the urn. After selecting the first ball, the ball is put back into the urn. As a result, there are again six balls in the urn for the second draw. Suppose we are interested in the number of red balls and the number of blue balls drawn. We call the number of red balls drawn X and the number of blue balls drawn Y.*

To find the joint probability distribution $p_{XY}(x, y)$ of X and Y, we need to enumerate all possible scenarios. For the draw of the first ball, there are three possibilities. For the draw of the second ball, again there are three possibilities, no matter which ball was drawn first. This gives a total of nine scenarios. Not all of these scenarios are equally likely because the numbers of yellow, blue, and red balls in the urn differ. The nine scenarios are listed in Table 11.5, together with the resulting numbers of red and blue balls, X and Y. Note that the sum of the numbers of red and blue balls drawn is at most 2. In the scenario where two yellow balls are drawn, both the random variables X and Y take the value 0. This is the least likely scenario because there is only one yellow ball in the urn.

Based on Table 11.5, it is not difficult to construct the joint probability distribution $p_{XY}(x, y)$ of X and Y. The joint probability distribution is shown in Table 11.6, together with the marginal probability distributions $p_X(x)$ and $p_Y(y)$.

The expected number of red balls is $E(X) = 24/36 = 2/3$, while the expected number of blue balls is equal to $E(Y) = 36/36 = 1$. These expected values can be determined based on the marginal probability distributions. To find the variances of the numbers of red and blue balls, we first compute $E(X^2)$ and $E(Y^2)$. This yields

$$E(X^2) = \frac{16}{36} \times 0^2 + \frac{16}{36} \times 1^2 + \frac{4}{36} \times 2^2 = \frac{32}{36} = \frac{8}{9},$$

and

$$E(Y^2) = \frac{9}{36} \times 0^2 + \frac{18}{36} \times 1^2 + \frac{9}{36} \times 2^2 = \frac{54}{36} = \frac{3}{2}.$$

Table 11.5   Nine possible scenarios in Example 11.5.2 with an indication of the numbers of red (X) and blue (Y) balls drawn and the associated probabilities.

First draw		Second draw		X	Y	Probability
Color	Probability	Color	Probability			
Yellow	$1/6$	Yellow	$1/6$	0	0	$1/36$
		Red	$2/6$	1	0	$2/36$
		Blue	$3/6$	0	1	$3/36$
Red	$2/6$	Yellow	$1/6$	1	0	$2/36$
		Red	$2/6$	2	0	$4/36$
		Blue	$3/6$	1	1	$6/36$
Blue	$3/6$	Yellow	$1/6$	0	1	$3/36$
		Red	$2/6$	1	1	$6/36$
		Blue	$3/6$	0	2	$9/36$

Table 11.6   Joint probability distribution and marginal probability distributions for the numbers of red (X) and blue (Y) balls drawn in Example 11.5.2.

X	Y			$p_X(x)$
	0	1	2	
0	$^1/_{36}$	$^6/_{36}$	$^9/_{36}$	$^{16}/_{36}$
1	$^4/_{36}$	$^{12}/_{36}$	–	$^{16}/_{36}$
2	$^4/_{36}$	–	–	$^4/_{36}$
$p_Y(y)$	$^9/_{36}$	$^{18}/_{36}$	$^9/_{36}$	1

Hence, the variances of X and Y are

$$\sigma_X^2 = E(X^2) - [E(X)]^2 = \frac{8}{9} - \left(\frac{2}{3}\right)^2 = \frac{4}{9}$$

and

$$\sigma_Y^2 = E(Y^2) - [E(Y)]^2 = \frac{3}{2} - 1^2 = \frac{1}{2}.$$

Example 11.5.3 *In Example 11.5.2, the first ball drawn was put back in the urn before drawing the second ball. Suppose that this is not the case, and that, again, we are interested in the numbers of red and blue balls drawn. We again denote the number of red balls drawn by X and the number of blue balls drawn by Y.*

To find the joint probability distribution $p_{XY}(x, y)$ of X and Y, we again have to enumerate all possible scenarios. However, now, there are only eight possible scenarios instead of nine. The reason for this is that there is only one yellow ball in the urn. Because the first ball drawn is not put back into the urn, the second ball cannot be yellow if the first one is. If the first ball drawn is red or blue, there are still three possibilities for the second draw.

The eight possible scenarios are listed in Table 11.7, along with the resulting numbers of red and blue balls, X and Y. Note that, once again, the sum of the numbers of red and blue balls is at most 2.

Based on Table 11.7, the joint probability distribution $p_{XY}(x, y)$ of X and Y can be derived. The joint probability distribution is shown in Table 11.8, together with the marginal probability distributions $p_X(x)$ and $p_Y(y)$.

The expected number of red balls is $E(X) = 20/30 = 2/3$, while the expected number of blue balls is equal to $E(Y) = 30/30 = 1$. To find the variances of the numbers of red and blue balls, we first calculate $E(X^2)$ and $E(Y^2)$. This yields

$$E(X^2) = \frac{12}{30} \times 0^2 + \frac{16}{30} \times 1^2 + \frac{2}{30} \times 2^2 = \frac{24}{30} = \frac{4}{5}$$

and

$$E(Y^2) = \frac{6}{30} \times 0^2 + \frac{18}{30} \times 1^2 + \frac{6}{30} \times 2^2 = \frac{42}{30} = \frac{7}{5}.$$

Table 11.7   Eight possible scenarios in Example 11.5.3, with an indication of the numbers of red (X) and blue (Y) balls drawn, and the associated probabilities.

First draw		Second draw		X	Y	Probability
Color	Probability	Color	Probability			
Yellow	$1/6$	Red	$2/5$	1	0	$2/30$
		Blue	$3/5$	0	1	$3/30$
Red	$2/6$	Yellow	$1/5$	1	0	$2/30$
		Red	$1/5$	2	0	$2/30$
		Blue	$3/5$	1	1	$6/30$
Blue	$3/6$	Yellow	$1/5$	0	1	$3/30$
		Red	$2/5$	1	1	$6/30$
		Blue	$2/5$	0	2	$6/30$

Table 11.8   Joint probability distribution and marginal probability distributions for the numbers of red (X) and blue (Y) balls drawn in Example 11.5.3.

X	Y			$p_X(x)$
	0	1	2	
0	–	$6/30$	$6/30$	$12/30$
1	$4/30$	$12/30$	–	$16/30$
2	$2/30$	–	–	$2/30$
$p_Y(y)$	$6/30$	$18/30$	$6/30$	1

Hence, the variances of X and Y are

$$\sigma_X^2 = E(X^2) - [E(X)]^2 = \frac{4}{5} - \left(\frac{2}{3}\right)^2 = \frac{36}{45} - \frac{20}{45} = \frac{16}{45}$$

and

$$\sigma_Y^2 = E(Y^2) - [E(Y)]^2 = \frac{7}{5} - 1^2 = \frac{2}{5}.$$

In this example, where the first ball drawn is not put back into the urn, the variances are a little smaller than in Example 11.5.2 where the first ball drawn is put back. Similarly, the variance of a hypergeometrically distributed random variable was slightly smaller than the variance of a binomially distributed random variable (see page 185 in Chapter 8). The expected values of X and Y in this example and in Example 11.5.2 are identical. Again, this is consistent with the results from Chapter 8.

11.6   The multinomial probability distribution

In Section 8.3, we introduced the binomial distribution. This distribution is useful when we are interested in the number of successes obtained from a series of independent random experiments with two possible outcomes, success or failure. Such experiments are called Bernoulli experiments. The probability distribution of the number of successes, which we call X, in n consecutive independent Bernoulli experiments, is the binomial distribution

$$p_X(x; n, \pi) = \frac{n!}{x! \, (n-x)!} \, \pi^x (1-\pi)^{n-x}, \quad x = 0, 1, 2, \ldots, n,$$

where π is the probability of success in a single Bernoulli experiment, and $1 - \pi$ is the probability of failure.

We can use an alternative description for the binomial distribution, if we denote by X_1 the number of successes, by X_2 the number of failures, by π_1 the success probability and by $\pi_2 = 1 - \pi_1$ the failure probability. This alternative expression is equal to

$$p_{X_1 X_2}(x_1, x_2; n, \pi_1, \pi_2) = \frac{n!}{x_1! \, x_2!} \, \pi_1^{x_1} \pi_2^{x_2},$$

$$\text{for } x_1, x_2 = 0, 1, 2, \ldots, n, \text{ where } x_1 + x_2 = n.$$

Some random experiments have more than two possible outcomes. Suppose that we have k possible outcomes in a random experiment and that we carry out this random experiment n times. Let X_1 be the number of times that the first outcome occurs, and π_1 be the probability of the first outcome. In the same way, we define $X_2, X_3, \ldots, X_k$, and the corresponding probabilities $\pi_2, \pi_3, \ldots, \pi_k$. The joint probability distribution of $X_1, X_2, \ldots, X_k$ is called the multinomial probability distribution, and is equal to

$$p_{X_1 X_2 \ldots X_k}(x_1, x_2, \ldots, x_k; n, \pi_1, \pi_2, \ldots, \pi_k) = \frac{n!}{x_1! \, x_2! \, \ldots \, x_k!} \, \pi_1^{x_1} \pi_2^{x_2} \ldots \pi_k^{x_k},$$

$$\text{for } x_1, x_2, \ldots, x_k = 0, 1, 2, \ldots, n, \text{ where } x_1 + x_2 + \ldots + x_k = n. \quad (11.1)$$

An example of a multinomial probability distribution is shown in Table 11.9. It is the probability distribution for a situation where four independent experiments with three possible outcomes are carried out. The probability of outcome 1, π_1, is 0.4, while the probabilities of outcome 2 and outcome 3, π_2 and π_3, are equal to 0.1 and 0.5, respectively. The joint probability that outcome 1 occurs once, outcome 2 occurs twice, and outcome 3 occurs once is equal to

$$p_{X_1 X_2 X_3}(1, 2, 1; 4, 0.4, 0.1, 0.5) = \frac{4!}{1! \, 2! \, 1!} \, (0.4)^1 (0.1)^2 (0.5)^1,$$

$$= \frac{24}{1 \times 2 \times 1} \, (0.4)(0.01)(0.5),$$

$$= 12 \times 0.002,$$

$$= 0.024.$$

Table 11.9   Multinomial probability distribution with three possible outcomes for each individual random experiment (also known as trinomial probability distribution), $n = 4$, $\pi_1 = 0.4$, $\pi_2 = 0.1$, and $\pi_3 = 0.5$.

x_1	x_2	x_3	Probability
4	0	0	0.0256
3	1	0	0.0256
3	0	1	0.1280
2	2	0	0.0096
2	1	1	0.0960
2	0	2	0.2400
1	3	0	0.0016
1	2	1	0.0240
1	1	2	0.1200
1	0	3	0.2000
0	4	0	0.0001
0	3	1	0.0020
0	2	2	0.0150
0	1	3	0.0500
0	0	4	0.0625

A multinomial distribution with $k = 3$ is called a trinomial distribution. The general expression for the probability distribution is

$$p_{X_1 X_2 X_3}(x_1, x_2, x_3; n, \pi_1, \pi_2, \pi_3) = \frac{n!}{x_1! \, x_2! \, x_3!} \, \pi_1^{x_1} \pi_2^{x_2} \pi_3^{x_3},$$

for $x_1, x_2, x_3 = 0, 1, 2, \ldots, n$, where $x_1 + x_2 + x_3 = n$.

In JMP, there is no simple formula for calculating the probabilities of a multinomial distribution. However, you can enter the formula of Equation (11.1) yourself. Figure 11.2 shows what the required formula looks like in JMP. To obtain the full probability distribution, you must first create three columns in which all possible values of X_1, X_2, and X_3 are listed (just as in Table 11.9).

Example 11.6.1 *A fair die is tossed 15 times. The probability that, for example, twice the result is a 1, three times a 2, no 3, five times a 4, twice a 5, and three times a 6 can be determined using the multinomial distribution with $k = 6$, $n = 15$ and $\pi_1 = \pi_2 = \cdots = \pi_6 = 1/6$. Suppose that X_1 represents the number of times that a 1 is obtained, X_2 the number of times that a 2 is obtained, and so on. Then, for the requested probability, we have $x_1 = 2$, $x_2 = 3$, $x_3 = 0$, $x_4 = 5$, $x_5 = 2$, and $x_6 = 3$.*

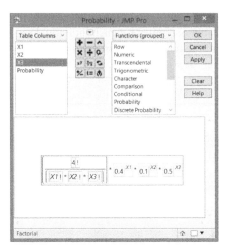

Figure 11.2   Formula required to calculate probabilities for a trinomially distributed random variable in JMP with $n = 4$, $\pi_1 = 0.4$, $\pi_2 = 0.1$ and $\pi_3 = 0.5$.

Notice that $x_1 + x_2 + \cdots + x_6 = 15$. The desired probability is

$$p_{X_1 X_2 X_3 X_4 X_5 X_6}(2, 3, 0, 5, 2, 3; 15, 1/6, 1/6, 1/6, 1/6, 1/6, 1/6)$$

$$= \frac{15!}{2!\,3!\,0!\,5!\,2!\,3!}\,(1/6)^2 (1/6)^3 (1/6)^0 (1/6)^5 (1/6)^2 (1/6)^3,$$

$$= 75675600\,(1/6)^{15} = \frac{350350}{6^{12}} \approx 0.00016.$$

11.7   Joint (continuous) probability density

For a continuous bivariate random variable (X, Y), we denote the joint continuous probability density by $f_{XY}(x, y)$, where X and Y can take values in the domain D. In some cases, the bivariate random variable (X, Y) may take every possible combination of real values. In that case, $D = \mathbb{R}^2$. In other cases, X and Y can both take all possible values in the interval $[0, 1]$. In this case, D a square with vertices $(0,0)$, $(1,0)$, $(0,1)$, and $(1,1)$, which we refer to as $[0, 1]^2$.

In order to be valid, a probability density function $f_{XY}(x, y)$ must satisfy two conditions:

$$f_{XY}(x, y) \geq 0,$$

and

$$\iint_{(x,y) \in D} f_{XY}(x, y)\, dy\, dx = 1.$$

Example 11.7.1 *An example of a bivariate continuous probability density is*

$$f_{XY}(x, y) = \begin{cases} 6x^2y, & 0 \le x \le 1, 0 \le y \le 1, \\ 0, & otherwise. \end{cases}$$

Figure 11.3 shows four graphical representations of this bivariate probability density. The figure was created using the option "Surface Profiler" in JMP. To use this option, you must first choose "Contour Profiler" from the "Graph" menu. Then, you can choose the "Surface Profiler" option. As shown in Figure 11.3, this allows you to view a three-dimensional figure from different angles. Notice that, in this example, the domain D of the bivariate random variable (X, Y) is equal to $[0, 1]^2$.

To check whether the function $f_{XY}(x, y)$ is a valid probability density, we have to verify that the double integral

$$\iint\limits_{(x,y)\in D} f_{XY}(x, y) \, dy \, dx = \int_0^1 \int_0^1 6x^2y \, dy \, dx$$

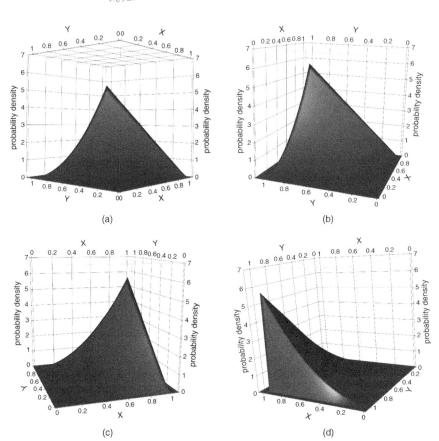

Figure 11.3 Graphical representations of the bivariate probability density of Example 11.7.1.

is equal to 1. This double integral measures the volume under the probability density. Working out the integral in detail, we obtain

$$\int_0^1 \int_0^1 6x^2 y \, dy \, dx = \int_0^1 [3x^2 y^2]_0^1 \, dx,$$

$$= \int_0^1 [3x^2 \times 1^2 - 3x^2 \times 0^2] \, dx,$$

$$= \int_0^1 3x^2 \, dx,$$

$$= [x^3]_0^1,$$

$$= 1^3 - 0^3,$$

$$= 1.$$

In this calculation, we first determined the inner integral, which integrates over the variable y. An alternative method, which can be used here because the limits of the integration domain are not functions of x or y, is as follows:

$$\int_0^1 \int_0^1 6x^2 y \, dy \, dx = \int_0^1 3x^2 \, dx \cdot \int_0^1 2y \, dy,$$

$$= [x^3]_0^1 \cdot [y^2]_0^1,$$

$$= [1^3 - 0^3] \cdot [1^2 - 0^2],$$

$$= 1.$$

Thus, the function $f_{XY}(x, y)$ in this example is a valid probability density.

Example 11.7.2   *Another example of a continuous bivariate probability density is*

$$f_{XY}(x, y) = \begin{cases} \lambda^2 e^{-\lambda y}, & 0 \le x < +\infty, x \le y < +\infty, \\ 0, & otherwise, \end{cases}$$

where λ is a positive real number. Figure 11.4 shows what this bivariate probability density looks like for $\lambda = 1/2$.

To check whether $f_{XY}(x, y)$ is a valid probability density, we have to verify that the double integral

$$\iint_{(x,y) \in D} f_{XY}(x, y) \, dy \, dx = \int_0^{+\infty} \int_x^{+\infty} \lambda^2 e^{-\lambda y} \, dy \, dx$$

is equal to 1. Working out the integral in detail, we obtain

$$\int_0^{+\infty} \int_x^{+\infty} \lambda^2 e^{-\lambda y} \, dy \, dx = \int_0^{+\infty} (-\lambda) \int_x^{+\infty} e^{-\lambda y} \, d(-\lambda y) \, dx,$$

$$= \int_0^{+\infty} (-\lambda)[e^{-\lambda y}]_x^{+\infty} \, dx,$$

$$= \int_0^{+\infty} (-\lambda)[0 - e^{-\lambda x}] \, dx,$$

$$= \int_0^{+\infty} \lambda e^{-\lambda x} \, dx,$$

$$= -\int_0^{+\infty} e^{-\lambda x} \, d(-\lambda x),$$

$$= -[e^{-\lambda x}]_0^{+\infty},$$

$$= -[0 - e^{-\lambda \times 0}],$$

$$= -(-1),$$

$$= 1.$$

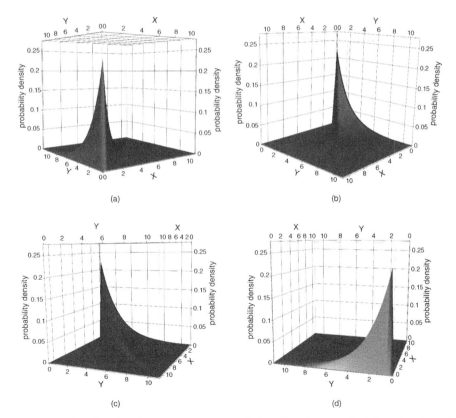

(a)

(b)

(c)

(d)

Figure 11.4   Graphical representations of the bivariate probability density of Example 11.7.2 , in case $\lambda = 1/2$.

Thus, the function $f_{XY}(x, y)$ in this example is a valid probability density. Notice that the integration domain D of the probability density in this example is more complicated than in the previous example, because the values that the random variable Y can take, have the value of the random variable X as lower limit (rather than a constant as lower limit). Therefore, the double integral in this example cannot be split into a product of two single integrals.

Calculating probabilities for bivariate random variables also requires the computation of double integrals. In general, the probability that a bivariate random variable (X, Y) takes values that lie inside a domain G is given by

$$P((X, Y) \in G) = \iint\limits_{(x,y) \in G} f_{XY}(x, y) \, dy \, dx.$$

We illustrate this by means of some examples.

Example 11.7.3 *In this example, we calculate some probabilities for the bivariate random variable (X, Y) defined in Example 11.7.1. We first determine the probability that X and Y are both simultaneously smaller than $1/2$. This is the probability that a realization of the bivariate random variable is located in the area delimited by the inequalities $0 \le x \le 1/2$ and $0 \le y \le 1/2$. The left- and right-hand sides of these inequalities are the lower and upper limits of the double integral we need to calculate*

$$
\begin{aligned}
P(X \le 1/2, Y \le 1/2) &= \int_0^{1/2} \int_0^{1/2} 6x^2 y \, dy \, dx, \\
&= \int_0^{1/2} [3x^2 y^2]_0^{1/2} \, dx, \\
&= \int_0^{1/2} \left[3x^2 \times \left(\frac{1}{2}\right)^2 - 3x^2 \times 0^2 \right] dx, \\
&= \frac{1}{4} \int_0^{1/2} 3x^2 \, dx, \\
&= \frac{1}{4} [x^3]_0^{1/2}, \\
&= \frac{1}{4} \left[\left(\frac{1}{2}\right)^3 - 0^3 \right], \\
&= \frac{1}{32}.
\end{aligned}
$$

Hence, the requested probability is only 1/32. This result means that the volume under the probability density function in Figure 11.3 above the square with vertices $(0, 0)$, $(1/2, 0)$, $(0, 1/2)$, and $(1/2, 1/2)$ is equal to 1/32.

We can also quantify the probability that the random variable X takes a value smaller than the random variable Y. We can write the desired probability as P(X < Y), or, because X and Y are continuous random variables, P(X ≤ Y). Since both X and Y can only take values between 0 and 1, we can rewrite this probability as P(0 ≤ X ≤ Y, 0 ≤ Y ≤ 1). In this way, we get lower and upper bounds for the double integral we have to determine. An alternative expression for the probability P(X ≤ Y) is P(0 ≤ X ≤ 1, X ≤ Y ≤ 1), which gives us different lower and upper limits for the same problem. Note that, in the first notation, the upper limit of X is given by Y and so is not constant. In the second notation, the lower limit of Y is given by X and therefore it is also not a constant. Computing the corresponding double integrals, we must ensure that the integral with two constants as lower and upper limit is the outer integral. If we use the first notation, we obtain

$$P(X < Y) = P(0 \leq X \leq Y, 0 \leq Y \leq 1),$$

$$= \int_0^1 \int_0^y 6x^2 y \, dx \, dy,$$

$$= \int_0^1 [2x^3 y]_0^y \, dy,$$

$$= \int_0^1 [2 \times y^3 \times y - 2 \times 0^3 \times y] \, dy,$$

$$= \int_0^1 2y^4 \, dy,$$

$$= \frac{2}{5}[y^5]_0^1,$$

$$= \frac{2}{5}[1^5 - 0^5],$$

$$= \frac{2}{5}.$$

If we use the second notation, we obtain

$$P(X < Y) = P(0 \leq X \leq 1, X \leq Y \leq 1),$$

$$= \int_0^1 \int_x^1 6x^2 y \, dy \, dx,$$

$$= \int_0^1 [3x^2 y^2]_x^1 \, dx,$$

$$= \int_0^1 [3x^2 \times 1^2 - 3x^2 \times x^2] \, dx,$$

$$= \int_0^1 [3x^2 - 3x^4] \, dx,$$

$$= \left[x^3 - \frac{3}{5}x^5 \right]_0^1,$$

$$= \left[\left(1^3 - \frac{3}{5} \times 1^5 \right) - \left(0^3 - \frac{3}{5} \times 0^5 \right) \right],$$

$$= 1 - \frac{3}{5},$$

$$= \frac{2}{5}.$$

Thus, both methods lead to the same result.

The probability $P(X^2 \geq Y)$ can be calculated in a similar manner:

$$P(X^2 \geq Y) = P(0 \leq X \leq 1, 0 \leq Y \leq X^2),$$

$$= \int_0^1 \int_0^{x^2} 6x^2 y \, dy \, dx,$$

$$= \int_0^1 [3x^2 y^2]_0^{x^2} \, dx,$$

$$= \int_0^1 [3x^2 \times (x^2)^2 - 3x^2 \times 0^2] \, dx,$$

$$= \int_0^1 3x^6 \, dx,$$

$$= \frac{3}{7}[x^7]_0^1,$$

$$= \frac{3}{7}[1^7 - 0^7],$$

$$= \frac{3}{7}.$$

As a final example, we calculate the probability that Y takes values that are greater than or equal to the value of $X/2$:

$$P(Y \geq X/2) = P(0 \leq X \leq 1, X/2 \leq Y \leq 1),$$

$$= \int_0^1 \int_{x/2}^1 6x^2 y \, dy \, dx,$$

$$= \int_0^1 [3x^2 y^2]_{x/2}^1 \, dx,$$

$$= \int_0^1 \left[3x^2 \times 1^2 - 3x^2 \times \left(\frac{x}{2} \right)^2 \right] \, dx,$$

$$= \int_0^1 \left(3x^2 - \frac{3}{4}x^4 \right) \, dx,$$

$$= \left[x^3 - \frac{3}{20}x^5 \right]_0^1 .$$

$$= \left[\left(1^3 - \frac{3}{20} \times 1^5 \right) - \left(0^3 - \frac{3}{20} \times 0^5 \right) \right] ,$$

$$= 1 - \frac{3}{20} ,$$

$$= \frac{17}{20} .$$

The domain of the integral required to compute the probability $P(Y \geq X/2)$ is shaded in gray in Figure 11.5. The calculated probability, 17/20, is the volume above this gray area under the bivariate probability density $f_{XY}(xy)$. The gray area in the figure shows that, for each possible value of X, the values that the random variable Y can take range from $X/2$ to 1. The figure also shows that, as long as $Y \leq 1/2$, the variable X can take values between 0 and 2Y. However, if $Y \geq 1/2$, then the variable X can take any value between 0 and 1. Hence, the probability $P(Y \geq X/2)$ can also be calculated as follows:

$$P(Y \geq X/2) = P(0 \leq Y \leq 1/2, 0 \leq X \leq 2Y) + P(1/2 \leq Y \leq 1, 0 \leq X \leq 1),$$

$$= \int_0^{1/2} \int_0^{2y} 6x^2 y \; dx \; dy + \int_{1/2}^1 \int_0^1 6x^2 y \; dx \; dy,$$

$$= \frac{3}{4} + \frac{1}{10},$$

$$= \frac{17}{20}.$$

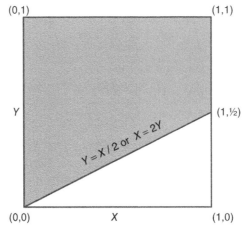

Figure 11.5   Graphical representation of the integration domain for the computation of the probability $P(Y \geq X/2)$ in Example 11.7.3.

11.8 Marginal or unconditional (continuous) probability density

In Section 11.3, we learnt how to convert a probability distribution $p_{XY}(x, y)$ of a discrete bivariate random variable (X, Y) into two separate distributions for X and Y. We have called the separate probability distributions marginal or unconditional probability distributions. In this section, we will see that a continuous bivariate probability density $f_{XY}(x, y)$ can also be turned into two separate, marginal, or unconditional probability densities for X and Y.

If (X, Y) is a bivariate continuous random variable with density $f_{XY}(x, y)$, then the marginal (or unconditional) densities of X and Y, are, respectively,

$$f_X(x) = \int_{-\infty}^{+\infty} f_{XY}(x, y)\, dy$$

and

$$f_Y(y) = \int_{-\infty}^{+\infty} f_{XY}(x, y)\, dx.$$

In the expression for $f_X(x)$, the variable y is eliminated by integrating over all possible values of y, while in the expression for $f_Y(y)$, the variable x is eliminated by integrating over all possible values of x.

Example 11.8.1 *We compute the marginal probability densities of X and Y from Example 11.7.1. For the random variable X, we obtain*

$$
\begin{aligned}
f_X(x) &= \int_{-\infty}^{+\infty} f_{XY}(x, y)\, dy, \\
&= \int_{-\infty}^{0} 0\, dy + \int_{0}^{1} 6x^2 y\, dy + \int_{1}^{+\infty} 0\, dy, \\
&= 0 + [3x^2 y^2]_0^1 + 0, \\
&= [3x^2 \times 1^2 - 3x^2 \times 0^2], \\
&= 3x^2.
\end{aligned}
$$

This marginal probability density is shown in Figure 11.6. Comparing this density to the bivariate probability density in Figure 11.3c, we see that the marginal probability density of X results from the bivariate probability density if we contract the Y-axis. For the random variable Y, we obtain

$$
\begin{aligned}
f_Y(y) &= \int_{-\infty}^{+\infty} f_{XY}(x, y)\, dx, \\
&= \int_{-\infty}^{0} 0\, dx + \int_{0}^{1} 6x^2 y\, dx + \int_{1}^{+\infty} 0\, dx,
\end{aligned}
$$

$$= 0 + [2x^3 y]_0^1 + 0,$$

$$= [2 \times 1^3 \times y - 2 \times 0^3 \times y],$$

$$= 2y.$$

This marginal probability density is shown in Figure 11.7. Comparing this density to the bivariate probability density in Figure 11.3b, we see that the marginal probability density of Y results from the bivariate probability density if we contract the X-axis.

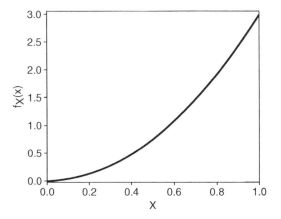

Figure 11.6   Graphical representation of the marginal probability density of X in Example 11.8.1.

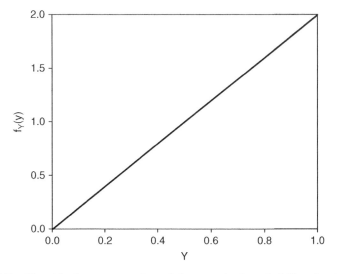

Figure 11.7   Graphical representation of the marginal probability density of Y in Example 11.8.1.

Example 11.8.2   *We compute the marginal distributions of X and Y in Example 11.7.2. For the random variable X, we obtain*

$$f_X(x) = \int_{-\infty}^{+\infty} f_{XY}(x, y)\, dy,$$

$$= \int_{-\infty}^{x} 0\, dy + \int_{x}^{+\infty} \lambda^2 e^{-\lambda y}\, dy,$$

$$= 0 - \lambda \int_{x}^{+\infty} e^{-\lambda y}\, d(-\lambda y),$$

$$= -\lambda[e^{-\lambda y}]_{x}^{+\infty},$$

$$= -\lambda[0 - e^{-\lambda x}],$$

$$= \lambda e^{-\lambda x},$$

for $x \geq 0$. This probability density is the exponential probability density (see Section 9.2).

For the random variable Y, we obtain

$$f_Y(y) = \int_{-\infty}^{+\infty} f_{XY}(x, y)\, dx,$$

$$= \int_{-\infty}^{0} 0\, dx + \int_{0}^{y} \lambda^2 e^{-\lambda y}\, dx + \int_{y}^{+\infty} 0\, dx,$$

$$= 0 + \lambda^2 e^{-\lambda y}[x]_{0}^{y} + 0,$$

$$= \lambda^2 e^{-\lambda y}[y - 0],$$

$$= \lambda^2 y e^{-\lambda y},$$

for $y \geq 0$. This probability density is a gamma probability density with parameters $k = 2$ and $\theta = 1/\lambda$ (see Section 9.3).

Even though the marginal probability densities can be derived from a multivariate probability density, the converse is usually not the case. Reconstructing the multivariate probability density function based on the marginal probability densities can only be achieved for **independent random variables**.

Definition 11.8.1   *The continuous random variables X and Y are **independent** if their joint probability density is the product of the marginal probability densities of X and Y, in other words, if*

$$f_{XY}(x, y) = f_X(x) f_Y(y) \text{ for all } (x, y).$$

If this condition is not satisfied, X and Y are **dependent**. Obviously, the concept of (in)dependence of continuous random variables is essentially the same as that of the (in)dependence of events in Chapter 4.

Example 11.8.3 *It is easy to verify that the random variables X and Y in Examples 11.7.1 and 11.8.1 are independent. Indeed, the joint probability density*

$$f_{XY}(x, y) = 6x^2 y$$

is equal to the product of the marginal probability densities:

$$f_X(x)f_Y(y) = (3x^2)(2y) = 6x^2 y,$$

if both x and y lie between 0 and 1. If x and y do not both lie between 0 and 1, then the joint probability density $f_{XY}(x, y)$ is equal to zero, as is the product of the marginal probability densities $f_X(x)f_Y(y)$.

Example 11.8.4 *The random variables X and Y in Example 11.7.2 are dependent. The values that the random variable Y can take depend on the value of the random variable X. The opposite is also true: the values that the random variable X can take depend on the value of the variable Y. The reason for this is that the probability density in Example 11.7.2 only differs from zero if $0 \le x < +\infty$ and $x \le y < +\infty$.*

The definition of independence can be extended to scenarios with more than two random variables:

Definition 11.8.2 *The continuous random variables $X_1, X_2, \dots, X_k$ are **independent** if their joint probability density is the product of the marginal probability densities of $X_1, X_2, \dots, X_k$, in other words, if*

$$f_{X_1 X_2 \dots X_k}(x_1, x_2, \dots, x_k) = f_{X_1}(x_1)f_{X_2}(x_2) \dots f_{X_k}(x_k)$$

$$\text{for all } (x_1, x_2, \dots, x_k).$$

11.9   Conditional (continuous) probability density

The conditional probability density of a continuous random variable X, given that the random variable Y takes the value y, is defined as

$$f_{X|Y}(x \mid y) = \frac{f_{XY}(x, y)}{f_Y(y)} \text{ if } f_Y(y) > 0.$$

The conditional probability density of Y, given that the random variable X takes the value x, is defined as

$$f_{Y|X}(y \mid x) = \frac{f_{XY}(x, y)}{f_X(x)} \text{ if } f_X(x) > 0.$$

These definitions indicate that, in general, the joint probability density of (X, Y) cannot be deduced from the two marginal densities, but it can be derived from the marginal probability density function of one random variable and the conditional probability density of the other:

$$f_{XY}(x, y) = f_{X|Y}(x \mid y)f_Y(y)$$

or

$$f_{XY}(x, y) = f_{Y|X}(y \mid x)f_X(x).$$

Conditional probability densities are like cross-sectional views of a joint probability density for a specific value of x, or a specific value of y.

In the special case of independent random variables X and Y, we have that

$$f_{X|Y}(x \mid y) = \frac{f_{XY}(x, y)}{f_Y(y)} = \frac{f_X(x)f_Y(y)}{f_Y(y)} = f_X(x)$$

and

$$f_{Y|X}(y \mid x) = \frac{f_{XY}(x, y)}{f_X(x)} = \frac{f_X(x)f_Y(y)}{f_X(x)} = f_Y(y).$$

In short, for independent random variables, the conditional probability densities are identical to the marginal (or unconditional) probability densities.

Example 11.9.1 *If X and Y follow the joint probability density of Example 11.7.2, the conditional probability density of X, given that the random variable Y takes the value y, is*

$$f_{X|Y}(x \mid y) = \frac{f_{XY}(x, y)}{f_Y(y)} = \frac{\lambda^2 e^{-\lambda y}}{\lambda^2 y e^{-\lambda y}} = \frac{1}{y} \text{ for } 0 \le x \le y.$$

Hence, for any value of y, the random variable X has a uniform probability density, with 0 as the smallest possible value and y as the largest possible value.

If X and Y follow the joint probability density of Example 11.7.2, the conditional probability density of Y, given that the random variable X takes the value x, is

$$f_{Y|X}(y \mid x) = \frac{f_{XY}(x, y)}{f_X(x)} = \frac{\lambda^2 e^{-\lambda y}}{\lambda e^{-\lambda x}} = \lambda e^{-\lambda(y-x)} \text{ for } y \ge x.$$

This function is an exponential probability density function, with the variable being $y - x$. The joint probability density $f_{XY}(x, y)$ is positive only if $y \ge x$. The conditional probability density found for Y states that, if the random variable X takes the value x, the difference between Y and x is exponentially distributed.

The conditional probability densities allow us to calculate conditional probabilities for continuous random variables. Suppose, for example, that we want to calculate the conditional probability $P(0 \le X \le 1/2 \mid y = 3/4)$. First, we need to compute the conditional probability density $f_{X|Y}(x \mid 3/4)$. To this end, we replace y by 3/4 in the expression for $f_{X|Y}(x \mid y)$ and obtain

$$f_{X|Y}(x \mid 3/4) = \frac{1}{3/4} = \frac{4}{3} \text{ for } 0 \le x \le 3/4.$$

Then, the desired probability is

$$P(0 \leq X \leq 1/2 \mid y = 3/4) = \int_0^{1/2} f_{X|Y}(x \mid 3/4)\, dx = \int_0^{1/2} \frac{4}{3}\, dx,$$

$$= \left[\frac{4}{3} x\right]_0^{1/2} = \frac{4}{3} \times (1/2 - 0) = \frac{2}{3}.$$

The probability $P(X \geq 1/2 \mid y = 3/4)$ is

$$P(X \geq 1/2 \mid y = 3/4) = P(1/2 \leq X \leq 3/4 \mid y = 3/4) = \int_{1/2}^{3/4} \frac{4}{3}\, dx,$$

$$= \left[\frac{4}{3} x\right]_{1/2}^{3/4} = \frac{4}{3} \times (3/4 - 1/2) = \frac{1}{3}.$$

12

Functions of several random variables

> *"You got a better chance in a plane,"* Wally said.
> *"Yes, a chance,"* Candy said scornfully. *"Why would you want to be anywhere where all you get is a chance?"*
> *"Good question,"* Olive said crossly …
> *"A chance is enough,"* said Homer Wells, who did not immediately recognize the tone in his own voice. *"A chance is all we get, right? In the air, or underwater, or right here, from the minute we're born."*
>
> (from *The Cider House Rules,* John Irving, p. 347)

As in the previous chapter, we discuss discrete random variables as well as continuous random variables. The focus is on bivariate probability distributions and densities, although some more general results are mentioned too. The results in this chapter are generalizations of those in Section 6.4, where functions of one random variable were discussed.

12.1   Functions of several random variables

We are interested in the probability distribution or the probability density of a function $Y = g(X_1, X_2, \ldots, X_k)$ of k random variables $X_1, X_2, \ldots, X_k$. This function Y is also a random variable and takes the value $y = g(x_1, x_2, \ldots, x_k)$ if the realizations $x_1, x_2, \ldots, x_k$ of the random variables $X_1, X_2, \ldots, X_k$ are known.

Example 12.1.1   *In Example 11.1.2, we can look at the number of acceptable products. A product is acceptable if the adhesion is perfect or excellent. Consequently, we*

Statistics with JMP: Graphs, Descriptive Statistics, and Probability, First Edition. Peter Goos and David Meintrup.
© 2015 John Wiley & Sons, Ltd. Published 2015 by John Wiley & Sons, Ltd.
Companion Website: wiley.com/go/goosandmeintrup

consider the random variable $Y = X_1 + X_2$. *This random variable takes the value* 1 *if the adhesion is acceptable, and the value* 0 *if the adhesion is not acceptable.*

Example 12.1.2 *In Example 11.1.3, the total surface tension* $Y = X_1 + X_2$ *is important. This new random variable is a function of the two original random variables that represent the polar and the dispersive surface tension.*

Researchers are usually not only interested in the probability distribution or the probability density of a function of several random variables, but also in the expected value $E(Y) = E[g(X_1, X_2, \ldots, X_k)]$ of this function.

12.2   Expected value of functions of several random variables

Definition 12.2.1 *The **expected value of a function** $W = g(X, Y)$ **of two discrete random variables** with joint probability distribution $p_{XY}(x, y)$, where $(x, y) \in D$, is defined as*

$$E(W) = E[g(X, Y)] = \sum_{(x,y) \in D} g(x, y) p_{XY}(x, y).$$

Definition 12.2.2 *The **expected value of a function** $W = g(X, Y)$ **of two continuous random variables** with joint probability density $f_{XY}(x, y)$ on the area D is defined as*

$$E(W) = E[g(X, Y)] = \iint_{(x,y) \in D} g(x, y) f_{XY}(x, y) \, dy \, dx.$$

These expressions generalize Definitions 7.1.1 and 7.1.2 of the expected value of a univariate discrete and of a continuous random variable.

Example 12.2.1 *Suppose that we want to determine the expected values of the functions $S = X + Y$, $V = Y - X$, $U = XY$ and $Q = X/Y$, in case the joint probability distribution $p_{XY}(x, y)$ of X and Y is given in Table 11.1. Using Definition 12.2.1, we find that*

$$E(S) = (1 + 1)(0.4) + (1 + 2)(0.06) + (1 + 3)(0.035) + (1 + 4)(0.005)$$
$$+ (0 + 1)(0.18) + (0 + 2)(0.225) + (0 + 3)(0.085) + (0 + 4)(0.010) = 2.07,$$

$$E(V) = (1 - 1)(0.4) + (2 - 1)(0.06) + (3 - 1)(0.035) + (4 - 1)(0.005)$$
$$+ (1 - 0)(0.18) + (2 - 0)(0.225) + (3 - 0)(0.085) + (4 - 0)(0.010) = 1.07,$$

$$E(U) = (1)(1)(0.4) + (1)(2)(0.06) + (1)(3)(0.035) + (1)(4)(0.005)$$
$$+ (0)(1)(0.18) + (0)(2)(0.225) + (0)(3)(0.085) + (0)(4)(0.010) = 0.645,$$

and

$$E(Q) = (1/1)(0.4) + (1/2)(0.06) + (1/3)(0.035) + (1/4)(0.005)$$
$$+ (0/1)(0.18) + (0/2)(0.225) + (0/3)(0.085) + (0/4)(0.010) = 0.4429.$$

Using the marginal probability distributions derived in Example 11.3.1, we can also calculate the expected values of X and Y:

$$E(X) = (1)(0.5) + (0)(0.5) = 0.5,$$

and

$$E(Y) = (1)(0.58) + (2)(0.285) + (3)(0.12) + (4)(0.015) = 1.57.$$

These calculations show that $E(S) = E(X + Y) = E(X) + E(Y)$ and $E(V) = E(Y - X) = E(Y) - E(X)$. In contrast, the example shows that $E(U) = E(XY) \neq E(X)E(Y)$ and $E(Q) = E(X/Y) \neq E(X)/E(Y)$.

The fact that $E(X + Y) = E(X) + E(Y)$ and $E(Y - X) = E(Y) - E(X)$ in the example mentioned earlier is not a coincidence. This result is true in general. It is a direct consequence of the following theorem:

Theorem 12.2.1 *For arbitrary constants $a_0, a_1, a_2, \ldots, a_m$ and m functions $Y_i = g_i(X_1, X_2, \ldots, X_k)$ of k random variables $X_1, X_2, \ldots, X_k$, we have*

$$E\left(a_0 + \sum_{i=1}^{m} a_i Y_i \right) = E\left[a_0 + \sum_{i=1}^{m} a_i g_i(X_1, X_2, \ldots, X_k) \right]$$

$$= a_0 + \sum_{i=1}^{m} a_i E[g_i(X_1, X_2, \ldots, X_k)].$$

This theorem, which is valid for discrete as well as continuous random variables, has some interesting consequences:

- The expected value of a linear combination of random variables is the same linear combination of the expected values of the individual random variables. Indeed, if $k = m$ and $g_i(X_1, X_2, \ldots, X_k) = X_i$, then Theorem 12.2.1 states that

$$E\left(a_0 + \sum_{i=1}^{m} a_i X_i \right) = a_0 + \sum_{i=1}^{m} a_i E(X_i).$$

- The expected value of a sum of random variables is the sum of the expected values of the individual random variables. Indeed, if $k = m$, $g_i(X_1, X_2, \ldots, X_k) = X_i$, $a_0 = 0$ and $a_1 = a_2 = \cdots = a_k = 1$, then Theorem 12.2.1 states that

$$E\left(\sum_{i=1}^{m} X_i \right) = \sum_{i=1}^{m} E(X_i).$$

- The expected value of a difference of two random variables is the difference of the expected values. Indeed, if $k = m = 2$, $g_i(X_1, X_2) = X_i$, $a_0 = 0$, $a_1 = 1$, and $a_2 = -1$, then Theorem 12.2.1 states that

$$E(X_1 - X_2) = E(X_1) - E(X_2).$$

An application of the theorem and its consequences can be found in Example 12.2.1. Another example from the financial world is given next.

Example 12.2.2 *An investment project runs over six years and requires an initial investment of €1 million (year 0). The next five years, the returns $X_1, X_2, \ldots, X_5$ of the project are uncertain, but an estimation provides expected values $E(X_1) = 100,000$, $E(X_2) = 300,000$, $E(X_3) = 500,000$, $E(X_4) = 400,000$, and $E(X_5) = 200,000$.*

If interest rates in the coming years are equal to 10%, then the net present value (NPV) of the investment project is equal to

$$NPV = -1,000,000 + \sum_{i=1}^{5} \frac{X_i}{(1.10)^i},$$

which is a linear combination of the random variables $X_1, X_2, \ldots, X_5$. The expected value of the net present value (NPV) is

$$E(NPV) = -1,000,000 + \frac{100,000}{1.1} + \frac{300,000}{(1.1)^2} + \frac{500,000}{(1.1)^3} + \frac{400,000}{(1.1)^4} + \frac{200,000}{(1.1)^5}$$

$$= 111,890,$$

so that the expected return rate of the investment project is $111,890/1,000,000 = 11.189\%$.

In this example, it was assumed that the interest rate is known. In reality, the annual interest rate is uncertain and therefore a random variable. In that case, the calculation of the expected net present value of the project involves expected values of quotients of two random variables (a return divided by a function of the interest rate). Since there is generally no simple way to calculate the expected value of a quotient of random variables, advanced methods are needed.

In general, the expected value of a product of functions of random variables is not equal to the product of the expected values of the functions. This has been demonstrated in Example 12.2.1. If, however, the random variables are independent, then the following theorem can be proven:

Theorem 12.2.2 *If the random variables $X_1, X_2, \ldots, X_k$ are independent, we have*

$$E\left(\prod_{i=1}^{k} g_i(X_i) \right) = \prod_{i=1}^{k} E(g_i(X_i)).$$

A direct consequence of this theorem, which is also valid for discrete as well as continuous multivariate random variables, is that the expected value of the product of independent random variables is equal to the product of the expected values of the individual variables:

$$E(X_1 X_2 \dots X_k) = E(X_1)E(X_2) \dots E(X_k).$$

Hence, for two independent random variables X and Y, we obtain that

$$E(XY) = E(X)E(Y).$$

Moreover,

$$E(X/Y) = E(X)E(1/Y),$$

but this is not equal to $E(X)/E(Y)$. For example, for discrete random variables,

$$E\left(\frac{1}{Y}\right) = \sum_y \frac{1}{y} \, p_Y(y) \neq \frac{1}{E(Y)} = \frac{1}{\sum_y y \, p_Y(y)},$$

where the sum is calculated over all possible values of Y.

The equation $E(XY) = E(X)E(Y)$ for independent random variables is not difficult to prove. For continuous random variables, the proof is as follows:

$$E(XY) = \iint\limits_{(x,y)\in D} xy f_{XY}(x, y) \, dy \, dx,$$

$$= \iint\limits_{(x,y)\in D} xy f_X(x) f_Y(y) \, dy \, dx,$$

$$= \left[\int_x x f_X(x) \, dx\right] \cdot \left[\int_y y f_Y(y) \, dy\right],$$

$$= E(X)E(Y).$$

While the double integral is calculated over all possible values of the pair (x, y), the two simple integrals are calculated over all possible values of x and all possible values of y.

Example 12.2.3 *In Example 11.8.3, we found out that the random variables X and Y from Example 11.7.1 are independent. Therefore, we can calculate the expected value E(XY) for these random variables in two ways. A first method makes use of the joint probability density:*

$$E(XY) = \iint\limits_{(x,y)\in D} xy \, f_{XY}(x, y) \, dy \, dx,$$

$$= \int_0^1 \int_0^1 xy \, 6x^2 y \, dy \, dx,$$

$$= \int_0^1 \int_0^1 6x^3 y^2 \, dy \, dx,$$

$$= \int_0^1 [2x^3 y^3]_0^1 \, dx,$$

$$= \int_0^1 [2x^3 \times 1^3 - 2x^3 \times 0^3] \, dx,$$

$$= \int_0^1 2x^3 \, dx,$$

$$= \left[\frac{2}{4} x^4 \right]_0^1,$$

$$= \left[\frac{2}{4} \times 1^4 - \frac{2}{4} \times 0^4 \right],$$

$$= \frac{2}{4} = \frac{1}{2}.$$

Alternatively, we can determine the expected values $E(X)$ and $E(Y)$ using the marginal probability densities of X and Y derived in Example 11.8.1 and multiply them. The expected values are

$$E(X) = \int_0^1 x \, f_X(x) \, dx,$$

$$= \int_0^1 x \, 3x^2 \, dx,$$

$$= \int_0^1 3x^3 \, dx,$$

$$= \left[\frac{3}{4} x^4 \right]_0^1,$$

$$= \left[\frac{3}{4} \times 1^4 - \frac{3}{4} \times 0^4 \right],$$

$$= \frac{3}{4},$$

and

$$E(Y) = \int_0^1 y \, f_Y(y) \, dy,$$

$$= \int_0^1 y \, 2y \, dy,$$

$$= \int_0^1 2y^2 \, dy,$$

$$= \left[\frac{2}{3} y^3\right]_0^1,$$

$$= \left[\frac{2}{3} \times 1^3 - \frac{2}{3} \times 0^3\right],$$

$$= \frac{2}{3}.$$

The product of these expected values is

$$E(X)E(Y) = \frac{3}{4} \times \frac{2}{3} = \frac{2}{4} = \frac{1}{2},$$

which is equal to E(XY).

12.3   Conditional expected values

The definition of a conditional expected value differs from the definition for an uncon-ditional expected value in only one aspect: instead of the marginal probability distri-bution or probability density, the conditional probability distribution or probability density is used.

Definition 12.3.1 *The **conditional expected value of a discrete random variable** X, given that Y = y, is*

$$E(X|Y = y) = E(X|y) = \sum_x x\, p_{X|Y}(x|y).$$

Analogously, the conditional expected value of a discrete random variable Y, given that X = x, is

$$E(Y|X = x) = E(Y|x) = \sum_y y\, p_{Y|X}(y|x).$$

Example 12.3.1 *For the distribution in Table 11.1, we obtain*

$$E(Y|X = 0) = (1)(0.36) + (2)(0.45) + (3)(0.17) + (4)(0.02) = 1.85.$$

For the computation of the conditional probability distribution that is needed here, we refer to Example 11.4.1. You can also check that

$$E(X|Y = 1) = (1)(0.6897) + (0)(0.3103) = 0.6897.$$

Definition 12.3.2 *The **conditional expected value of a continuous random variable** X, given that Y takes the value y, is*

$$E(X|Y = y) = E(X|y) = \int_x x f_{X|Y}(x|y)\, dx.$$

Analogously, the conditional expected value of a continuous random variable Y, given that X takes the value x, is

$$E(Y|X = x) = E(Y|x) = \int_y y f_{Y|X}(y|x) \, dy.$$

Example 12.3.2 *In Example 11.9.1 we showed that*

$$f_{X|Y}(x|y) = \frac{1}{y} \text{ for } 0 \le x \le y,$$

if, as in Example 11.7.2, the two random variables X and Y have the following bivariate continuous probability density:

$$f_{XY}(x, y) = \begin{cases} \lambda^2 e^{-\lambda y}, & 0 \le x < +\infty, x \le y < +\infty, \\ 0, & \text{otherwise.} \end{cases}$$

Then, the conditional expected value of X given that Y takes the value y, is

$$E(X|Y = y) = \int_0^y x \, \frac{1}{y} \, dx = \left[\frac{x^2}{2} \cdot \frac{1}{y} \right]_0^y = \frac{y^2}{2} \cdot \frac{1}{y} - \frac{0^2}{2} \cdot \frac{1}{y} = \frac{y}{2}.$$

This result is not surprising because we found out in Example 11.9.1 that the random variable X is distributed uniformly on the interval [0, y] and the expected value of a uniformly distributed continuous random variable is the midpoint of the interval it covers. The result for the conditional expected value E(X|Y = y) means, that, for example, the expected value of X is equal to 4 if you know that the random variable Y takes the value 8.

12.4 Probability distributions of functions of random variables

12.4.1 Discrete random variables

If the random variables $X_1, X_2, \ldots, X_k$ are discrete and have a finite number of possible values, then it is relatively easy to derive the probability distribution of a function $Y = g(X_1, X_2, \ldots, X_k)$. The required approach is similar to the one used in Section 6.4.1.

Example 12.4.1 *For the bivariate random variable (X, Y) in Table 11.1, we compute the probability distribution of the function $U = g(X, Y) = XY$. It is easy to check that the new random variable U can only take five values, namely 0, 1, 2, 3, and 4. We obtain the value 0 whenever X takes the value 0. Consequently,*

$$P(U = 0) = P[(X = 0) \cap (Y = 1)] + P[(X = 0) \cap (Y = 2)]$$

$$+ P[(X = 0) \cap (Y = 3)] + P[(X = 0) \cap (Y = 4)],$$

$$= 0.18 + 0.225 + 0.085 + 0.01 = 0.5.$$

The random variable U only takes the value 1 if both X and Y take the value 1. Hence,
$P(U = 1) = P[(X = 1) \cap (Y = 1)] = 0.4$. *In a similar manner, the probabilities that*
U takes the values 2, 3, and 4 can be determined. The probability distribution of the
function U is shown in the following table:

u	0	1	2	3	4
$p_U(u)$	0.500	0.400	0.060	0.035	0.005

This distribution can be used to calculate the expected value of the function U:

$$E(U) = \sum_u u\, p_U(u),$$

where the sum is taken over all values that the function U can take. Applied to the
example, we obtain

$$E(U) = 0(0.5) + 1(0.4) + 2(0.06) + 3(0.035) + 4(0.005) = 0.645.$$

This provides us with a second way to determine the expected value of a function of
random variables (see Example 12.2.1). This exercise can of course also be carried
out for the functions S, V, and Q defined in Example 12.2.1.

12.4.2   Continuous random variables

If the random variables $X_1, X_2, \ldots, X_k$ are continuous, then it is not that easy to derive
the probability density of a function $Y = g(X_1, X_2, \ldots, X_k)$. One possible method uses
the moment generating function, while another technique, the so-called transformation method, makes use of a special matrix, the Jacobian.

12.4.2.1   Method of the moment generating function

To determine the probability density of a function $Y = g(X_1, X_2, \ldots, X_k)$, one can try
to find the moment generating function of Y. In some cases, this moment generating function has a form that we recognize. Most known probability densities have
a recognizable moment generating function. For example, the normal density with
expected value μ and variance σ^2 has the following moment generating function:

$$m_X(t) = e^{\mu t + \frac{1}{2}\sigma^2 t^2}. \tag{12.1}$$

If, at any given time, we encounter a function Y with a moment generating function
of that form, we know immediately that Y is normally distributed. The expected value
of the normal density of Y is the coefficient of t in the exponent in Equation (12.1),
while the variance is equal to the coefficient of $\frac{1}{2}t^2$.

The method of the moment generating function for computing the probability density of $Y = g(X_1, X_2, \ldots, X_k)$ is useful if $X_1, X_2, \ldots, X_k$ are independent random variables and if we are interested in a function of the form

$$Y = a_0 + \sum_{i=1}^{k} a_i X_i.$$

Then, the moment generating function of Y is equal to

$$m_Y(t) = e^{a_0 t} \prod_{i=1}^{k} m_{X_i}(a_i t).$$

It is not difficult to demonstrate this, starting from the definition of the moment generating function and using the fact that all the random variables $X_1, X_2, \ldots, X_k$ are independent:

$$
\begin{aligned}
m_Y(t) &= E(e^{tY}), \\
&= E(e^{t(a_0 + a_1 X_1 + a_2 X_2 + \cdots + a_k X_k)}), \\
&= E(e^{ta_0} e^{ta_1 X_1} e^{ta_2 X_2} \cdots e^{ta_k X_k}), \\
&= E(e^{ta_0}) E(e^{ta_1 X_1}) E(e^{ta_2 X_2}) \cdots E(e^{ta_k X_k}), \\
&= e^{ta_0} E(e^{ta_1 X_1}) E(e^{ta_2 X_2}) \cdots E(e^{ta_k X_k}), \\
&= e^{ta_0} m_{X_1}(ta_1) \, m_{X_2}(ta_2) \, \cdots \, m_{X_k}(ta_k), \\
&= e^{a_0 t} \prod_{i=1}^{k} m_{X_i}(a_i t).
\end{aligned}
$$

In the special case that $a_0 = 0$ and $a_1 = a_2 = \cdots = a_k = 1$, the function

$$Y = \sum_{i=1}^{k} X_i$$

is a simple sum and the moment generating function of this sum is the product of the individual moment generating functions:

$$m_Y(t) = \prod_{i=1}^{k} m_{X_i}(t).$$

Now, suppose that X_1, X_2, and X_3 are all normally distributed with expected values equal to μ_1, μ_2, and μ_3, and variances equal to σ_1^2, σ_2^2, and σ_3^2. The moment generating function of the sum

$$Y = X_1 + X_2 + X_3$$

is the product of the individual moment generating functions:

$$m_Y(t) = m_{X_1}(t)m_{X_2}(t)m_{X_3}(t),$$

$$= e^{\mu_1 t + \frac{1}{2}\sigma_1^2 t^2} \cdot e^{\mu_2 t + \frac{1}{2}\sigma_2^2 t^2} \cdot e^{\mu_3 t + \frac{1}{2}\sigma_3^2 t^2},$$

$$= e^{\mu_1 t + \frac{1}{2}\sigma_1^2 t^2 + \mu_2 t + \frac{1}{2}\sigma_2^2 t^2 + \mu_3 t + \frac{1}{2}\sigma_3^2 t^2},$$

$$= e^{(\mu_1 + \mu_2 + \mu_3)t + \frac{1}{2}(\sigma_1^2 + \sigma_2^2 + \sigma_3^2)t^2}.$$

If we substitute

$$\mu_Y = \mu_1 + \mu_2 + \mu_3$$

and

$$\sigma_Y^2 = \sigma_1^2 + \sigma_2^2 + \sigma_3^2,$$

we obtain the following result:

$$m_Y(t) = e^{\mu_Y t + \frac{1}{2}\sigma_Y^2 t^2}.$$

This moment generating function has exactly the same form as the one in Equation (12.1). Therefore, we can conclude that Y, the sum of three independent normally distributed random variables, is also normally distributed, with expected value $\mu_Y = \mu_1 + \mu_2 + \mu_3$ and variance $\sigma_Y^2 = \sigma_1^2 + \sigma_2^2 + \sigma_3^2$.

Similarly, it can be proven that any linear combination of independent normally distributed random variables is again normally distributed.

12.4.2.2 Transformation method

The transformation method for determining probability densities of functions of several random variables generalizes the second approach in Section 6.4.2, where transformations of a single continuous random variable were studied. We learnt that the computation of a probability density of a strictly increasing or strictly decreasing function $Y = g(X)$ can be based on the equation

$$f_Y(y) = f_X(x)\left|\frac{dx}{dy}\right| = f_X\{g^{-1}(y)\}\left|\frac{dg^{-1}(y)}{dy}\right|, \qquad (12.2)$$

where $g^{-1}()$ is the inverse function of $g()$.

Suppose that two random variables X_1 and X_2 have the joint probability density function $f_{X_1 X_2}(x_1, x_2)$, but that we are interested in the functions $Y_1 = g_1(X_1, X_2)$ and $Y_2 = g_2(X_1, X_2)$. Here, it is important that the functions $g_1()$ and $g_2()$ are bijective. This means that each pair (x_1, x_2) of realizations of X_1 and X_2 corresponds to exactly one pair (y_1, y_2) of realizations of Y_1 and Y_2, and vice versa. In that case, it is possible to express X_1 and X_2 as functions of Y_1 and Y_2:

$$X_1 = h_1(Y_1, Y_2)$$

and

$$X_2 = h_2(Y_1, Y_2).$$

In this case, the joint probability density of Y_1 and Y_2 can be computed as

$$f_{Y_1 Y_2}(y_1, y_2) = f_{X_1 X_2}(x_1, x_2) \left| \det \begin{bmatrix} \dfrac{\partial x_1}{\partial y_1} & \dfrac{\partial x_1}{\partial y_2} \\ \dfrac{\partial x_2}{\partial y_1} & \dfrac{\partial x_2}{\partial y_2} \end{bmatrix} \right|,$$

$$= f_{X_1 X_2}(h_1(y_1, y_2), h_2(y_1, y_2)) \left| \det \begin{bmatrix} \dfrac{\partial x_1}{\partial y_1} & \dfrac{\partial x_1}{\partial y_2} \\ \dfrac{\partial x_2}{\partial y_1} & \dfrac{\partial x_2}{\partial y_2} \end{bmatrix} \right|.$$

The matrix of partial derivatives in this expression is called the Jacobian or Jacobian matrix, after the German mathematician Carl Gustav Jacobi. We need the determinant of this matrix.

It is not difficult to extend the transformation method to more than two random variables. If we are working with three random variables instead of two, the Jacobian is a three-dimensional square matrix instead of a two-dimensional matrix. An important difference between the transformation method and the method of moment generating functions is that the first one can also be used to study dependent random variables.

We illustrate the transformation method for two random variables by means of an example.

Example 12.4.2 *Suppose that X_1 and X_2 are independent exponentially distributed random variables with parameter $\lambda = 1$, so that the joint probability density of these two random variables is equal to*

$$f_{X_1 X_2}(x_1, x_2) = f_{X_1}(x_1) f_{X_2}(x_2) = e^{-x_1} e^{-x_2} = e^{-x_1 - x_2},$$

if x_1 and x_2 are greater than or equal to zero. Suppose that we are rather interested in the random variables

$$Y_1 = X_1 + X_2,$$

and

$$Y_2 = \frac{X_1}{X_1 + X_2}$$

than X_1 and X_2. Hence, we need the probability densities of Y_1 and Y_2.

The first step to determine the probability densities of Y_1 and Y_2 is to find the values that these two random variables can take. Since X_1 and X_2 are exponentially distributed, these variables can both take any non-negative real value: $x_1 \geq 0$ and $x_2 \geq 0$. It follows that Y_1 can also take all non-negative real values, and that Y_2 can take all values between 0 and 1. In mathematical terms, this means that $y_1 \geq 0$ and $0 \leq y_2 \leq 1$.

The next step is to determine an expression for X_1 and X_2 as functions of Y_1 and Y_2. Since $Y_1 = X_1 + X_2$, it is not difficult to see that

$$Y_2 = \frac{X_1}{Y_1}.$$

Hence,

$$X_1 = Y_1 Y_2.$$

Consequently,

$$Y_1 = Y_1 Y_2 + X_2,$$

and

$$X_2 = Y_1 - Y_1 Y_2 = Y_1(1 - Y_2).$$

The partial derivatives that we need for the Jacobian matrix are

$$\frac{\partial x_1}{\partial y_1} = y_2,$$

$$\frac{\partial x_1}{\partial y_2} = y_1,$$

$$\frac{\partial x_2}{\partial y_1} = 1 - y_2,$$

and

$$\frac{\partial x_2}{\partial y_2} = -y_1.$$

Thus, the Jacobian is

$$\begin{bmatrix} y_2 & y_1 \\ 1 - y_2 & -y_1 \end{bmatrix},$$

and the determinant of this matrix is

$$-y_1 y_2 - (1 - y_2)y_1 = -y_1 y_2 - y_1 + y_1 y_2 = -y_1.$$

Hence, the joint probability density of Y_1 and Y_2 is

$$f_{Y_1 Y_2}(y_1, y_2) = f_{X_1 X_2}(x_1, x_2) \cdot |-y_1|,$$
$$= f_{X_1 X_2}(y_1 y_2, y_1(1 - y_2)) \cdot |-y_1|,$$
$$= e^{-y_1 y_2 - y_1(1 - y_2)} \cdot y_1,$$
$$= y_1 e^{-y_1},$$

for $y_1 \geq 0$ and $0 \leq y_2 \leq 1$.
    The marginal probability densities of Y_1 and Y_2 are

$$f_{Y_1}(y_1) = \int_0^1 y_1 e^{-y_1} \, dy_2 = y_1 e^{-y_1},$$

and

$$f_{Y_2}(y_2) = \int_0^{+\infty} y_1 e^{-y_1} \, dy_1 = 1.$$

The conclusion is that Y_1 is gamma distributed with parameters $k = 2$ and $\theta = 1$ (see Section 9.3), and that Y_2 is uniformly distributed over the interval $[0,1]$ (see Section 9.1). Note that the computation of the integral for the marginal probability density of Y_2 is not obvious. The easiest way to see that the integral is equal to 1, is to observe that the integrand, $y_1 e^{-y_1}$, is a gamma density. Hence, the integral

$$\int_0^{+\infty} y_1 e^{-y_1} \, dy_1,$$

the area under the gamma probability density, must be equal to 1. Otherwise, $y_1 e^{-y_1}$ would not be a valid probability density function.

This example illustrates that by combining various random variables (here, exponentially distributed random variables), we can construct new random variables with known probability density functions (here, a gamma probability density and a uniform probability density). This is of great importance in testing hypotheses in statistics, where t-distributed, F-distributed, and χ^2-distributed random variables play a crucial role. The densities of these random variables can be derived from the (standard) normal probability density from Chapter 10. For example, it can be shown that a sum of squared independent standard normally distributed random variables is χ^2-distributed. The t-, F-, and χ^2-distributions are not discussed here, but in the book Statistics with JMP: Hypothesis Tests, ANOVA and Regression, which deals with estimating population parameters and performing hypothesis tests.

12.5 Functions of independent Poisson, normally, and lognormally distributed random variables

In this section, we start with two interesting theorems associated with functions of independent Poisson and normally distributed random variables. For the proof of these theorems, the method of the moment generating functions from Section 12.4.2.1 can be used. Later, we use another approach to prove properties of lognormally distributed random variables.

Theorem 12.5.1 *If $X_1, X_2, \ldots, X_k$ are independent Poisson distributed random variables with parameters equal to $\lambda_1, \lambda_2, \ldots, \lambda_k$, then the sum of these random variables $Y = \sum_{i=1}^k X_i$ is also Poisson distributed, with parameter $\lambda = \sum_{i=1}^k \lambda_i$.*

The proof of this theorem follows the same lines of thought as the proof for the sum of independent normally distributed random variables in Section 12.4.2.1. The only difference is that the theorem is about Poisson distributed random variables instead of normally distributed random variables. The moment generating function of a Poisson distributed random variable X with parameter λ is

$$m_X(t) = e^{\lambda(e^t - 1)}.$$

Example 12.5.1 *An assembly line is fed by three production lines in which identical parts are manufactured. These three production lines produce parts according to three independent Poisson processes: Machine 1 produces an average of $\lambda_1 = 2$ parts per minute, Machine 2 generates an average of $\lambda_2 = 1$ part per minute, and Machine 3 delivers an average of $\lambda_3 = 3.5$ parts per minute. What is the probability that, in a period of 5 minutes, at least 30 parts are supplied to the assembly line?*

The number of parts delivered per minute to the assembly line is a random variable Y, which is the sum of the three independent Poisson distributed random variables. Therefore, the random variable Y is Poisson distributed with parameter $\lambda = 2 + 1 + 3.5 = 6.5$ parts per minute. Hence, every 5 minutes, we have an average number of $5 \times 6.5 = 32.5$ parts. The number of parts that arrive in a time interval of 5 minutes is therefore a Poisson distributed random variable W with parameter 32.5. The desired probability can now be written as

$$P(W \geq 30) = 1 - P(W \leq 29) = 1 - F_W(29),$$

which can be calculated in JMP with the formula "$1 -$ Poisson Distribution(32.5, 29)". This leads to the result 0.6932.

Theorem 12.5.2 *If $X_1, X_2, \ldots, X_k$ are independent normally distributed random variables with expected values $E(X_1) = \mu_1, E(X_2) = \mu_2, \ldots, E(X_k) = \mu_k$ and variances $\text{var}(X_1) = \sigma_1^2, \text{var}(X_2) = \sigma_2^2, \ldots, \text{var}(X_k) = \sigma_k^2$, then the linear function $Y = a_0 + \sum_{i=1}^{k} a_i X_i$ is also normally distributed, with expected value $E(Y) = a_0 + \sum_{i=1}^{k} a_i \mu_i$ and variance $\text{var}(Y) = \sum_{i=1}^{k} a_i^2 \sigma_i^2$.*

This theorem is similar to the central limit theorem in Chapter 14, Section 14.2, which states that the sum of a sufficiently large number of independent random variables *approximately* follows a normal distribution. A direct consequence of Theorem 12.5.2 is that the sum of any number of normally distributed random variables is *exactly* normally distributed. For a proof of Theorem 12.5.2, we can use the method of the moment generating function from Section 12.4.2.1.

If we look at the sum of k independent normally distributed random variables X_i, where all X_i have the same parameters μ and σ^2, then this sum is again normally distributed with expected value $k\mu$ and variance $k\sigma^2$. The random variables' average $\sum_{i=1}^{k} X_i / k$ is also normally distributed, namely with expected value μ and variance σ^2 / k.

The sum $X_1 + X_2$ of two independent normally distributed random variables X_1 (with expected value μ_1 and variance σ_1^2) and X_2 (with expected value μ_2 and variance σ_2^2) is again normally distributed with mean $\mu_1 + \mu_2$ and variance $\sigma_1^2 + \sigma_2^2$. The difference $X_1 - X_2$ is normally distributed with mean $\mu_1 - \mu_2$ and variance $\sigma_1^2 + \sigma_2^2$.

Example 12.5.2 *The filling process of a bottle with a capacity of 1 liter is set to a value of 1.01 l. The process has a standard deviation of 0.02 l and provides a normally distributed content. The bottles are packed in groups of 6. What is the probability that a pack of 6 bottles has a volume that is less than 6 l?*

The content of 6 bottles is normally distributed with expected value $6 \times 1.01 = 6.06 l$ and variance $6 \times (0.02)^2 = 0.0024 \ l^2$ (and thus a standard deviation of $\sqrt{0.0024} = 0.049 \ l$). The desired probability $P(X < 6)$, where X represents the total volume of the 6 bottles, can be calculated using the table of the standard normal distribution in Appendix E, or using the command "Normal Distribution(6, 6.06, 0.049)" in JMP. This gives a probability of 0.1104.

Example 12.5.3 *Assume that the height of a man is normally distributed with expected value $\mu_M = 180$ cm and standard deviation $\sigma_M = 8$ cm, and that the height of a woman is normally distributed with expected value $\mu_F = 170$ cm and standard deviation $\sigma_F = 6$ cm. Also assume that men and women, when choosing a partner, do not take into account each other's height. What is the probability that, in a mixed marriage, the man is taller than the woman?*

To answer this question, we first define the random variable M as the height of a man, and the random variable F as the height of a woman. The probability that the man is taller than the woman is

$$P(M > F) = P(M - F > 0).$$

The difference $M - F$ is a difference of two independent normally distributed random variables. Hence, $M - F$ is itself normally distributed, with expected value

$$\mu_{M-F} = \mu_M - \mu_F = 180 - 170 = 10,$$

and variance

$$\sigma^2_{M-F} = \sigma^2_M + \sigma^2_F = 8^2 + 6^2 = 100.$$

Therefore, the standard deviation of the difference $M - F$, σ_{M-F}, is 10. The desired probability can be calculated as

$$P(M - F > 0) = P\left(\frac{M - F - \mu_{M-F}}{\sigma_{M-F}} > \frac{0 - \mu_{M-F}}{\sigma_{M-F}}\right),$$

$$= P\left(Z > \frac{0 - 10}{10}\right),$$

$$= P(Z > -1),$$

$$= 0.8413.$$

This probability can be found in the table in Appendix E. Alternatively, one can use the formulas "$1 - Normal \ Distribution(-1)$" and "$1 - Normal \ Distribution(0, 10, 10)$" in JMP.

In Belgium, a married couple does not have to consist of a man and a woman. Therefore, let us consider a couple consisting of two men and find the probability that the older one is taller than the younger one. The heights of both the older and the younger man are normally distributed with expected value 180 cm and standard deviation 8 cm. It follows that the difference in height between the two men is normally distributed with expected value 0 and that the probability that the older man is taller than the younger one is exactly equal to 0.5.

Theorem 12.5.3 *If $X_1, X_2, \ldots, X_k$ are independent lognormally distributed random variables with parameters $\mu_1, \mu_2, \ldots, \mu_k$ and $\sigma_1, \sigma_2, \ldots, \sigma_k$, then the product $Y = X_1 X_2 \ldots X_k$ is lognormally distributed with parameters*

$$\mu_Y = \sum_{i=1}^{k} \mu_i \quad and \quad \sigma_Y = \sqrt{\sum_{i=1}^{k} \sigma_i^2}.$$

Proof: For simplicity, we only prove this theorem for two independent lognormally distributed random variables X_1 (with parameters μ_1 and σ_1) and X_2 (with parameters μ_2 and σ_2). In that case, we seek the probability density of a product of two lognormally distributed random variables, $Y = X_1 X_2$. As X_1 and X_2 are lognormally distributed, both $\ln(X_1)$ and $\ln(X_2)$ are normally distributed, with expected values μ_1 and μ_2, and variances σ_1^2 and σ_2^2. Moreover, as X_1 and X_2 are independent, $\ln(X_1)$ and $\ln(X_2)$ are also independent. Since $Y = X_1 X_2$, the natural logarithm of Y equals

$$\ln(Y) = \ln(X_1) + \ln(X_2).$$

In other words, $\ln(Y)$ is the sum of two independent normally distributed random variables with expected values μ_1 and μ_2 and variance σ_1^2 and σ_2^2. Hence, $\ln(Y)$ is normally distributed with expected value $\mu_1 + \mu_2$ and variance $\sigma_1^2 + \sigma_2^2$. As a result,

$$e^{\ln(Y)} = e^{\ln(X_1) + \ln(X_2)} = e^{\ln(X_1 X_2)} = X_1 X_2 = Y$$

is lognormally distributed with parameters $\mu_Y = \mu_1 + \mu_2$ and $\sigma_Y = \sqrt{\sigma_1^2 + \sigma_2^2}$. ∎

Theorem 12.5.4 *If X_1 and X_2 are independent lognormally distributed random variables with parameters μ_1 and σ_1 and μ_2 and σ_2, respectively, then the quotient $Y = X_1/X_2$ is lognormally distributed with parameters*

$$\mu_Y = \mu_1 - \mu_2 \quad and \quad \sigma_Y = \sqrt{\sigma_1^2 + \sigma_2^2}.$$

The proof of this theorem is completely analogous to the proof of Theorem 12.5.3.

Example 12.5.4 *Suppose that X_1, X_2, and X_3 represent annual growth rates of a three-year investment, and that these three growth rates are independent lognormally distributed random variables. The overall growth rate over three years is $Y =$*

$X_1 X_2 X_3$. *The overall growth rate is a product of three independent lognormally distributed random variables, and therefore also lognormally distributed. The parameters of the lognormal probability density of the overall growth rate Y are*

$$\mu_Y = \mu_1 + \mu_2 + \mu_3 \quad and \quad \sigma_Y = \sqrt{\sigma_1^2 + \sigma_2^2 + \sigma_3^2}.$$

Example 12.5.5   *Suppose that X_1 represents the weight of an adult person, expressed in kilograms, and X_2 represents the height of this person, expressed in meters. Suppose that both random variables are independent and lognormally distributed with parameters μ_1 and σ_1 (for X_1) and μ_2 and σ_2 (for X_2). Then, we can wonder what the probability density of the so-called "body mass index" (BMI) is, defined as*

$$BMI = \frac{X_1}{X_2^2}.$$

To answer this question, we first look at the natural logarithm of the BMI:

$$\ln(BMI) = \ln(X_1) - \ln(X_2^2) = \ln(X_1) - 2\,\ln(X_2).$$

The natural logarithm $\ln(BMI)$ is a linear combination of two independent normally distributed random variables, and therefore also normally distributed. More specifically, $\ln(BMI)$ is normally distributed with expected value $\mu_{BMI} = \mu_1 - 2\mu_2$ and variance $\sigma_{BMI}^2 = \sigma_1^2 + 4\sigma_2^2$. As a result,

$$e^{\ln(BMI)} = e^{\ln(X_1) - 2\,\ln(X_2)} = e^{\ln(X_1/X_2^2)} = \frac{X_1}{X_2^2} = BMI$$

is lognormally distributed with parameters

$$\mu_{BMI} = \mu_1 - 2\mu_2 \quad and \quad \sigma_{BMI} = \sqrt{\sigma_1^2 + 4\sigma_2^2}.$$

13

Covariance, correlation, and variance of linear functions

Langdon frowned. Kohler was right. Holy wars were still making headlines. My God is better than your God. It seemed there was always close correlation between true believers and high body counts.

(from *Angels & Demons,* Dan Brown, p. 57)

In Section 3.9.2, we introduced the concepts of covariance and correlation between two quantitative variables as measures of (linear) relationship. Although these concepts were initially defined for sample data, we already briefly indicated that covariances and correlations can be calculated for populations. In this chapter, we focus on the calculation of population correlations and covariances. We also pay attention to the variance, which in fact is a special case of a covariance. The definitions and properties in this section are almost all valid for both continuous and discrete random variables. Only expressions with double summations are only valid for discrete random variables.

13.1 Covariance and correlation

We derive the terms covariance and correlation based on an example, and compute the population covariance for this example, as defined on page 91.

Example 13.1.1 *Assume that we want to investigate the covariance between two random variables X and Y, and that the joint and marginal probability distributions of these variables are given as in Table 13.1. In this example, X represents the size of*

Statistics with JMP: Graphs, Descriptive Statistics, and Probability, First Edition. Peter Goos and David Meintrup.
© 2015 John Wiley & Sons, Ltd. Published 2015 by John Wiley & Sons, Ltd.
Companion Website: wiley.com/go/goosandmeintrup

Table 13.1   Joint probability distribution $p_{XY}(x, y)$ and marginal probability distributions $p_X(x)$ and $p_Y(y)$ of X and Y for Example 13.1.1.

		Y				$p_X(x)$
		1	2	3	4	
	1	0.10	0.10	0.05	0.02	0.27
X	2	0.05	0.10	0.20	0.15	0.50
	3	0.00	0.05	0.10	0.08	0.23
$p_Y(y)$		0.15	0.25	0.35	0.25	1.00

an order. This random variable takes the value 1 for orders of less than €1000, the value 2 for orders between €1000 and 5000, and the value 3 for orders of more than €5000. The random variable Y corresponds to the number of days elapsing between the order and the delivery.

If the population consists of 200 elements, then it follows from the probability distribution that 20 (= 0.10 × 200) of those elements take the value 1 for both variables X and Y, 20 (= 0.10 × 200) elements take the value 1 for X and the value 2 for Y, and so on. Therefore, the population covariance between X and Y can be calculated as

$$\sigma_{XY} = \frac{1}{200}[20(1 - \mu_X)(1 - \mu_Y) + 20(1 - \mu_X)(2 - \mu_Y) + \cdots + 16(3 - \mu_X)(4 - \mu_Y)],$$

$$= \sum_{x_i} \sum_{y_i} p_{XY}(x_i, y_i)(x_i - \mu_X)(y_i - \mu_Y),$$

which shows that the population covariance can be computed based on the joint probability distribution of X and Y directly.

Definition 13.1.1 *For two discrete random variables X and Y with joint probability distribution $p_{XY}(x, y)$, or two continuous random variables X and Y with joint probability density $f_{XY}(x, y)$, the **population covariance** is defined as*

$$\sigma_{XY} = \text{cov}(X, Y) = E[(X - \mu_X)(Y - \mu_Y)].$$

For discrete random variables, the expected value $E[(X - \mu_X)(Y - \mu_Y)]$ can be computed as

$$\sigma_{XY} = \text{cov}(X, Y) = \sum_{x} \sum_{y} (x - \mu_X)(y - \mu_Y)p_{XY}(x, y),$$

where the summation is taken over all possible values of X and Y. This expression is actually an application of the general Definition 12.2.1 of the expected value of a function of several discrete random variables: the covariance between two variables X and Y is the expected value of the function $(X - \mu_X)(Y - \mu_Y)$ of these variables.

For continuous random variables, the expected value can be computed as

$$\sigma_{XY} = \text{cov}(X, Y) = \iint\limits_{(x,y)\in D} (x - \mu_X)(y - \mu_Y)f_{XY}(x, y) \, dy \, dx.$$

Similarly, this expression is an application of the general Definition 12.2.2 of the expected value of a function of several continuous random variables: the covariance between two variables X and Y is the expected value of the function $(X - \mu_X)(Y - \mu_Y)$ of these variables.

The intuition behind these formulas is exactly the same as the one outlined in Chapter 3 on sample data. We will not repeat this explanation here.

A special case of a population covariance is the covariance of a variable X with itself:

$$\sigma_{XX} = E[(X - \mu_X)(X - \mu_X)] = E[(X - \mu_X)^2] = \sigma_X^2,$$

which is the population variance of X. Starting from the population covariance, the population correlation between two random variables can be defined as

$$\rho_{XY} = \text{corr}(X, Y) = \frac{\sigma_{XY}}{\sigma_X \sigma_Y}. \tag{13.1}$$

Here again, we have that $-1 \leq \rho_{XY} \leq 1$. If $Y = aX + b$, with $a \neq 0$, then the correlation will be exactly -1 or $+1$. If $a > 0$, then the correlation is $+1$. In the other case, the correlation is -1. The absolute size of a has no effect on the correlation coefficient. Whether a is 100 or 0.001 does not make a difference: the correlation between X and $Y = aX + b$ will be equal to $+1$. If $a = 0$, then the correlation is not defined (since in that case, $\sigma_{XY} = 0$ and $\sigma_Y = 0$). This means that the correlation of any random variable with a constant is not defined. It is a good exercise to verify all this.

The fact that the correlation can only take values between -1 and $+1$ offers the advantage that the strength of the linear relationship between X and Y is easy to interpret. Covariances can be arbitrarily large or small depending on the units of the measurements, which considerably complicates their interpretation. If, for example, the variable X is expressed in centimeters rather than in meters, this increases the covariance σ_{XY} by a factor of 100. The correlation coefficient is not affected by this rescaling.

Note that, from the definition of the population variance in Equation (13.1), we can derive that

$$\sigma_{XY} = \rho_{XY}\sigma_X\sigma_Y.$$

Example 13.1.2 *It is a useful exercise to verify for the random variables of Example 13.1.1 that $\mu_X = 1.96$, $\mu_Y = 2.7$, $\sigma_{XY} = 0.298$, $\sigma_X = 0.706$, $\sigma_Y = 1.005$, and $\rho_{XY} = 0.42$. This positive correlation indicates that larger orders typically have a longer delivery time than smaller orders.*

The computation of the covariance is usually simplified to

$$\sigma_{XY} = E(XY) - \mu_X \mu_Y.$$

This expression can be found as follows:

$$\begin{aligned}
\sigma_{XY} &= E[(X - \mu_X)(Y - \mu_Y)], \\
&= E[XY - \mu_X Y - X\mu_Y + \mu_X \mu_Y], \\
&= E(XY) - E(\mu_X Y) - E(X\mu_Y) + E(\mu_X \mu_Y), \\
&= E(XY) - \mu_X E(Y) - \mu_Y E(X) + \mu_X \mu_Y, \\
&= E(XY) - \mu_X \mu_Y - \mu_Y \mu_X + \mu_X \mu_Y, \\
&= E(XY) - \mu_X \mu_Y.
\end{aligned}$$

Example 13.1.3   *For the random variables from Example 13.1.1, we have $E(XY) = 5.59$, so that $\sigma_{XY} = 5.59 - (1.96)(2.7) = 0.298$. Indeed, $\mu_X = 1.96$ and $\mu_Y = 2.7$.*

Example 13.1.4   *For the random variables X and Y in Example 11.5.2 (which deals with an urn containing one yellow ball, two red balls, and three blue balls, and where the first ball drawn is put back into the urn), we have that*

$$\begin{aligned}
E(XY) &= \frac{1}{36} \times 0 \times 0 + \frac{6}{36} \times 0 \times 1 + \frac{9}{36} \times 0 \times 2 \\
&\quad + \frac{4}{36} \times 1 \times 0 + \frac{12}{36} \times 1 \times 1 + \frac{4}{36} \times 2 \times 0, \\
&= \frac{12}{36}.
\end{aligned}$$

Hence,

$$\sigma_{XY} = E(XY) - \mu_X \mu_Y = \frac{12}{36} - \frac{2}{3} \times 1 = -\frac{12}{36} = -\frac{1}{3}.$$

Indeed, we calculated in Example 11.5.2 that $\mu_X = E(X) = 2/3$ and $\mu_Y = E(Y) = 1$. We see that there is a negative covariance between the random variables X and Y (the numbers of red and blue balls drawn). This negative relationship is perfectly logical because drawing a red ball implies that no blue ball is drawn, and drawing a blue ball implies that no red ball is drawn.
  The correlation between X and Y is

$$\rho_{XY} = \frac{\sigma_{XY}}{\sigma_X \sigma_Y} = \frac{-1/3}{\sqrt{16/36}\sqrt{1/2}} = -\frac{\sqrt{2}}{2} \approx -0.707,$$

because we found in Example 11.5.2 that $\sigma_X^2 = 16/36$ and $\sigma_Y^2 = 1/2$.

Example 13.1.5   *For the random variables X and Y of Example 11.5.3 (which also deals with an urn containing one yellow ball, two red balls, and three blue balls, but* *where the first ball drawn is not put back into the urn), we have that*

$$E(XY) = \frac{6}{30} \times 0 \times 1 + \frac{6}{30} \times 0 \times 2 + \frac{4}{30} \times 1 \times 0 + \frac{12}{30} \times 1 \times 1 + \frac{2}{30} \times 2 \times 0 = \frac{12}{30}.$$

Hence,

$$\sigma_{XY} = E(XY) - \mu_X\mu_Y = \frac{12}{30} - \frac{2}{3} \times 1 = -\frac{8}{30}.$$

Indeed, we calculated in Example 11.5.3 that $\mu_X = E(X) = 2/3$ and $\mu_Y = E(Y) = 1$. The correlation between X and Y is

$$\rho_{XY} = \frac{\sigma_{XY}}{\sigma_X\sigma_Y} = \frac{-8/30}{\sqrt{16/45}\sqrt{2/5}} = -\frac{1}{\sqrt{2}} = -\frac{\sqrt{2}}{2} \approx -0.707,$$

because we found in Example 11.5.3 that $\sigma_X^2 = 16/45$ and $\sigma_Y^2 = 2/5$.

Theorem 13.1.1 *If X and Y are independent random variables, then $\sigma_{XY} = 0$ and $\rho_{XY} = 0$.*

Proof: This theorem is easy to prove because, for independent random variables, we have that $E(XY) = E(X)E(Y)$. This is a direct consequence of Theorem 12.2.2. Therefore,

$$\sigma_{XY} = E(XY) - \mu_X\mu_Y,$$
$$= E(X)E(Y) - \mu_X\mu_Y,$$
$$= \mu_X\mu_Y - \mu_X\mu_Y,$$
$$= 0.$$

Of course, in that case, the correlation is zero as well, $\rho_{XY} = 0$, because $\rho_{XY} = \sigma_{XY}/(\sigma_X\sigma_Y)$. ∎

The converse of this theorem is not true. It is possible that the correlation between two variables is equal to zero, but that they are not independent. The correlation is a measure of linear relationship between two random variables. Dependence of two variables, however, is a broader concept that includes linear, but also quadratic or logarithmic relationships. This is illustrated in the following example.

Example 13.1.6 *Suppose that the joint probability distribution of the random variables X and Y is given by*

$$p_{XY}(x,y) = \begin{cases} 1/3 & \text{if } x = 0 \text{ and } y = 1, \\ 1/3 & \text{if } x = 1 \text{ and } y = 0, \\ 1/3 & \text{if } x = -1 \text{ and } y = 0. \end{cases}$$

It can be shown that these two random variables are dependent (see Definition 11.3.1). In fact, $Y = 1 - X^2$. However, the covariance and correlation between the variables X and Y are both equal to zero.

13.2   Variance of linear functions of two random variables

In order to determine the variances of linear functions of two random variables, the covariances are also needed. This is shown by the following theorem.

Theorem 13.2.1 *For two random variables X and Y, and arbitrary real constants a, b, and c,*

$$\text{var}(aX + bY + c) = a^2\text{var}(X) + b^2\text{var}(Y) + 2ab \ \text{cov}(X, Y),$$
$$= a^2\sigma_X^2 + b^2\sigma_Y^2 + 2ab \ \sigma_{XY}.$$

Proof:

$$\text{var}(aX + bY + c) = E\{[aX + bY + c - E(aX + bY + c)]^2\},$$
$$= E\{[aX + bY + c - (a\mu_X + b\mu_Y + c)]^2\},$$
$$= E\{[aX + bY + c - a\mu_X - b\mu_Y - c]^2\},$$
$$= E\{[aX + bY - a\mu_X - b\mu_Y]^2\},$$
$$= E\{[a(X - \mu_X) + b(Y - \mu_Y)]^2\},$$
$$= E[a^2(X - \mu_X)^2 + b^2(Y - \mu_Y)^2 + 2ab(X - \mu_X)(Y - \mu_Y)],$$
$$= E[a^2(X - \mu_X)^2] + E[b^2(Y - \mu_Y)^2] + E[2ab(X - \mu_X)(Y - \mu_Y)]$$
$$= a^2E[(X - \mu_X)^2] + b^2E[(Y - \mu_Y)^2] + 2abE[(X - \mu_X)(Y - \mu_Y)]$$
$$= a^2\text{var}(X) + b^2\text{var}(Y) + 2ab \ \text{cov}(X, Y),$$
$$= a^2\sigma_X^2 + b^2\sigma_Y^2 + 2ab \ \sigma_{XY}. \qquad \blacksquare$$

Example 13.2.1 *The previous theorem allows the variance of the sum and difference of the random variables X and Y from Example 13.1.1 to be calculated. For the sum S = X + Y, we have that a = 1, b = 1, and c = 0, while, for the difference V = X − Y, we have that a = 1, b = −1, and c = 0. The variances are*

$$\text{var}(S) = \text{var}(X + Y),$$
$$= \text{var}(X) + \text{var}(Y) + 2 \ \text{cov}(X, Y),$$
$$= (0.706)^2 + (1.005)^2 + 2(0.298),$$
$$= 2.1044,$$

and

$$\text{var}(V) = \text{var}(X - Y),$$

$$= \text{var}(X) + \text{var}(Y) - 2\,\text{cov}(X, Y),$$

$$= (0.706)^2 + (1.005)^2 - 2(0.298),$$

$$= 0.9125.$$

The same result can be found by first computing the probability distribution of S and V (see Section 12.4.1) and then calculating the variance directly. For example, the probability distribution of the variable S is given in the following table:

s	2	3	4	5	6	7
$p_S(s)$	0.10	0.15	0.15	0.27	0.25	0.08

It follows that

$$\mu_S = E(S) = 2(0.10) + 3(0.15) + \cdots + 7(0.08) = 4.66,$$

$$E(S^2) = 2^2(0.10) + 3^2(0.15) + \cdots + 7^2(0.08) = 31.5221,$$

and, by using the formula on page 165,

$$\sigma_S^2 = E(S^2) - \mu_S^2 = 31.5221 - (4.66)^2 = 2.1044.$$

13.3 Variance of linear functions of several random variables

In order to determine the variances of linear functions of several random variables, the concept of a covariance matrix is needed.

Definition 13.3.1 *Let $X_1, X_2, \ldots, X_k$ be random variables with pairwise covariances* $\text{cov}(X_i, X_j) = \sigma_{ij}$. *Then, their **covariance matrix** is given by*

$$C = \begin{bmatrix} \sigma_{11} & \sigma_{12} & \cdots & \sigma_{1k} \\ \sigma_{21} & \sigma_{22} & \cdots & \sigma_{2k} \\ \vdots & \vdots & \ddots & \vdots \\ \sigma_{k1} & \sigma_{k2} & \cdots & \sigma_{kk} \end{bmatrix}.$$

As the covariance of a variable with itself is equal to the variance (and hence $\sigma_{ii} = \sigma_i^2$) and $\text{cov}(X_i, X_j) = \text{cov}(X_j, X_i)$, the covariance matrix can be rewritten as

$$C = \begin{bmatrix} \sigma_1^2 & \sigma_{12} & \cdots & \sigma_{1k} \\ \sigma_{12} & \sigma_2^2 & \cdots & \sigma_{2k} \\ \vdots & \vdots & \ddots & \vdots \\ \sigma_{1k} & \sigma_{2k} & \cdots & \sigma_k^2 \end{bmatrix}.$$

The matrix is symmetric. Using a covariance matrix, Theorem 13.2.1 can be rewritten as follows:

Theorem 13.3.1 *For two random variables X and Y, and arbitrary real constants a, b, and c,*

$$\text{var}(aX + bY + c) = \begin{bmatrix} a & b \end{bmatrix} \begin{bmatrix} \sigma_X^2 & \sigma_{XY} \\ \sigma_{XY} & \sigma_Y^2 \end{bmatrix} \begin{bmatrix} a \\ b \end{bmatrix}.$$

This result can be generalized to more than two random variables:

Theorem 13.3.2 *For k random variables $X_1, X_2, \ldots, X_k$ and arbitrary constants $a_0, a_1, a_2, \ldots, a_k$, we have that*

$$\text{var}\left(a_0 + \sum_{i=1}^{k} a_i X_i\right) = \begin{bmatrix} a_1 & a_2 & \cdots & a_k \end{bmatrix} \begin{bmatrix} \sigma_1^2 & \sigma_{12} & \cdots & \sigma_{1k} \\ \sigma_{12} & \sigma_2^2 & \cdots & \sigma_{2k} \\ \vdots & \vdots & \ddots & \vdots \\ \sigma_{1k} & \sigma_{2k} & \cdots & \sigma_k^2 \end{bmatrix} \begin{bmatrix} a_1 \\ a_2 \\ \vdots \\ a_k \end{bmatrix},$$

$$= \sum_{i=1}^{k} \sum_{j=1}^{k} a_i a_j \sigma_{ij},$$

$$= \sum_{i=1}^{k} a_i^2 \sigma_i^2 + \sum_{i=1}^{k} \sum_{j=i+1}^{k} 2 a_i a_j \sigma_{ij}.$$

The proof of this theorem is not difficult and is left as an exercise.

13.4　Variance of linear functions of independent random variables

13.4.1　Two independent random variables

The variance of a linear function of two independent random variables X and Y is easier to calculate than the variance of a linear function of dependent random variables because the covariance between independent random variables is zero. For independent random variables X and Y, it is true that $\text{cov}(X, Y) = \sigma_{XY} = 0$. It then follows from Theorem 13.2.1 that

$$\text{var}(aX + bY + c) = a^2 \text{var}(X) + b^2 \text{var}(Y) = a^2 \sigma_X^2 + b^2 \sigma_Y^2.$$

The variance of a sum or a difference of two independent random variables X and Y is therefore equal to the sum of the variances of X and Y:

$$\text{var}(X + Y) = \text{var}(X - Y) = \text{var}(X) + \text{var}(Y) = \sigma_X^2 + \sigma_Y^2.$$

Notice that the variance of the difference is not equal to the difference of the variances, but to their sum.

13.4.2   Several pairwise independent random variables

Since, for pairwise independent random variables $X_1, X_2, \ldots, X_k$, all covariances σ_{ij} are equal to zero, the variance of a linear function of several independent random variables $X_1, X_2, \ldots, X_k$ is simply

$$\mathrm{var}\left(a_0 + \sum_{i=1}^{k} a_i X_i\right) = \sum_{i=1}^{k} a_i^2 \mathrm{var}(X_i) = \sum_{i=1}^{k} a_i^2 \sigma_i^2.$$

The variance of a sum of pairwise independent random variables $X_1, X_2, \ldots, X_k$ is thus the sum of the individual variances:

$$\mathrm{var}\left(\sum_{i=1}^{k} X_i\right) = \sum_{i=1}^{k} \mathrm{var}(X_i) = \sum_{i=1}^{k} \sigma_i^2.$$

13.5   Linear functions of normally distributed random variables

Without proof, we mention a theorem on linear functions of normally distributed random variables. This theorem generalizes Theorem 12.5.2, which is only valid for independent normally distributed random variables.

Theorem 13.5.1 *If $X_1, X_2, \ldots, X_k$ are normally distributed random variables with expected values $E(X_1) = \mu_1, E(X_2) = \mu_2, \ldots, E(X_k) = \mu_k$, with variances $\mathrm{var}(X_1) = \sigma_1^2$, $\mathrm{var}(X_2) = \sigma_2^2$, ..., $\mathrm{var}(X_k) = \sigma_k^2$, and with covariances $\mathrm{cov}(X_i, X_j) = \sigma_{ij}$ for every i and j ($i \neq j$), then a linear function $Y = a_0 + \sum_{i=1}^{k} a_i X_i$ is also normally distributed, with expected value*

$$E(Y) = a_0 + \sum_{i=1}^{k} a_i \mu_i$$

and variance

$$\mathrm{var}(Y) = \sum_{i=1}^{k} a_i^2 \sigma_i^2 + \sum_{i=1}^{k} \sum_{j=i+1}^{k} 2 a_i a_j \sigma_{ij}.$$

Example 13.5.1 *Suppose that, as in Example 12.5.3, the height of a man is normally distributed with expected value $\mu_M = 180$ cm and standard deviation $\sigma_M = 8$ cm, and that the height of a woman is normally distributed with expected value $\mu_F = 170$ cm and standard deviation $\sigma_F = 6$ cm. Now, suppose that men and women, when choosing a partner, take into account each other's height: more specifically, that tall* *men/women tend to marry tall women/men. This is reflected by a positive correlation $\rho_{MF} = 51/96$ between the height of a man and a woman in a marriage. What is the probability that, in a mixed marriage, the man is taller than the woman?*

To answer this question, we first define the random variable M as the height of a man, and the random variable F as the height of a woman. The probability that the man in a mixed marriage is taller than the woman is

$$P(M > F) = P(M - F > 0).$$

The difference M − F is a difference of two (dependent) normally distributed random variables. Hence, M − F is itself normally distributed. The expected value of this new normally distributed random variable is

$$\mu_{M-F} = \mu_M - \mu_F = 180 - 170 = 10,$$

while the variance is equal to

$$\sigma^2_{M-F} = \sigma^2_M + \sigma^2_F - 2\sigma_{MF},$$

$$= \sigma^2_M + \sigma^2_F - 2\rho_{MF}\sigma_M\sigma_F,$$

$$= 8^2 + 6^2 - 2 \cdot \frac{51}{96} \cdot 8 \cdot 6,$$

$$= 64 + 36 - \frac{2 \cdot 51 \cdot 8 \cdot 6}{96},$$

$$= 100 - 51,$$

$$= 49.$$

Therefore, the standard deviation of the difference M − F, σ_{M-F}, is 7. The desired probability can be calculated as

$$P(M - F > 0) = P\left(\frac{M - F - \mu_{M-F}}{\sigma_{M-F}} > \frac{0 - \mu_{M-F}}{\sigma_{M-F}}\right),$$

$$= P\left(Z > \frac{0 - 10}{7}\right),$$

$$= P(Z > -1.428571),$$

$$= 0.9234.$$

This probability can be found in the table in Appendix E. Alternatively, one can calculate it with the formulas "1 − Normal Distribution(−1.428571)" and "1 − Normal Distribution(0, 10, 7)" in JMP.

The probability that the man is taller than the woman in a mixed marriage, is bigger if there is a positive correlation between the heights of the two people than if there is no correlation.

13.6   Bivariate and multivariate normal density

In Chapter 10, we introduced the univariate normal probability density. There is also a multivariate version. The multivariate normal density is of great importance in statistics. In regression analysis, the estimators are (approximately) multivariate normally distributed and also many other techniques of multivariate statistics rely on the multivariate normal distribution. These topics are beyond the scope of this book.

13.6.1   Bivariate normal probability density

The simplest multivariate normal density is the bivariate normal density. Two random variables X and Y are bivariate normally distributed if their joint probability density is equal to

$$f_{XY}(x, y) = \frac{1}{2\pi\sigma_X\sigma_Y\sqrt{1 - \rho_{XY}^2}} e^{-\frac{1}{2(1-\rho_{XY}^2)}\left[\frac{(x-\mu_X)^2}{\sigma_X^2} + \frac{(y-\mu_Y)^2}{\sigma_Y^2} - 2\rho_{XY}\frac{(x-\mu_X)(y-\mu_Y)}{\sigma_X\sigma_Y}\right]}, \quad (13.2)$$

where μ_X and σ_X^2 are the expected value and the variance of X, μ_Y and σ_Y^2 are the expected value and the variance of Y, and ρ_{XY} is the correlation between X and Y.

An alternative notation of the bivariate normal probability density uses a vector for the random variables and the covariance matrix

$$\mathbf{C} = \begin{bmatrix} \sigma_X^2 & \sigma_{XY} \\ \sigma_{XY} & \sigma_Y^2 \end{bmatrix} = \begin{bmatrix} \sigma_X^2 & \rho_{XY}\sigma_X\sigma_Y \\ \rho_{XY}\sigma_X\sigma_Y & \sigma_Y^2 \end{bmatrix},$$

where $\sigma_{XY} = \rho_{XY}\sigma_X\sigma_Y$ is the covariance between X and Y. This alternative notation is

$$f_{XY}(x, y) = \frac{1}{2\pi\sqrt{\det(\mathbf{C})}} e^{-\frac{1}{2}\begin{bmatrix} x - \mu_X & y - \mu_Y \end{bmatrix}\mathbf{C}^{-1}\begin{bmatrix} x - \mu_X \\ y - \mu_Y \end{bmatrix}}.$$

Note that

$$\det(\mathbf{C}) = \sigma_X^2\sigma_Y^2 - \sigma_{XY}^2 = \sigma_X^2\sigma_Y^2 - \rho_{XY}^2\,\sigma_X^2\sigma_Y^2 = \sigma_X^2\sigma_Y^2(1 - \rho_{XY}^2),$$

and

$$\mathbf{C}^{-1} = \frac{1}{\sigma_X^2\sigma_Y^2(1 - \rho_{XY}^2)}\begin{bmatrix} \sigma_Y^2 & -\rho_{XY}\sigma_X\sigma_Y \\ -\rho_{XY}\sigma_X\sigma_Y & \sigma_X^2 \end{bmatrix}.$$

13.6.2   Graphical representations

A graphical representation of a bivariate normal density with parameters $\mu_X = \mu_Y = 0$, $\sigma_X^2 = \sigma_Y^2 = 1$ and $\rho_{XY} = 0$ is given in Figure 13.1. This probability density is perfectly bell-shaped and has an equal width in the x- and y-direction. The reason for

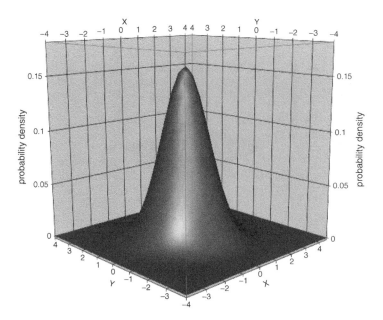

Figure 13.1   Bivariate normal density with parameters $\mu_X = \mu_Y = 0$, $\sigma_X^2 = \sigma_Y^2 = 1$,
and $\rho_{XY} = 0$.

this is that $\sigma_X^2 = \sigma_Y^2$. A characteristic of this probability density function is that it is constant for all pairs (x, y) at the same distance from the point $(\mu_X, \mu_Y) = (0, 0)$. In other words, all the points on a circle around $(\mu_X, \mu_Y) = (0, 0)$ have the same value for $f_{XY}(x, y)$.

Figure 13.2 shows three different bivariate normal probability densities with $\mu_X = 10$, $\mu_Y = 20$, and $\rho_{XY} = 0$. Figure 13.2a shows a density where $\sigma_X^2 = \sigma_Y^2 = 1$. As the one in Figure 13.1, this density is perfectly bell-shaped with the same width in the x- and y-direction. All points on a circle around $(\mu_X, \mu_Y) = (10, 20)$ have the same value for $f_{XY}(x, y)$.

Figure 13.2b shows a density where $\sigma_X^2 = 1$ and $\sigma_Y^2 = 0.25$. This density is not as wide in the x-direction as in the y-direction: because of the smaller variance σ_Y^2, the density is narrower in the y-direction than in the x-direction. It is no longer the case that all points on a circle around $(\mu_X, \mu_Y) = (10, 20)$ have the same value for $f_{XY}(x, y)$. Also note that the probability density in Figure 13.2b is higher than the one in Figure 13.2a. The reason for this is that, because of the smaller variance σ_Y^2, the bell shape in Figure 13.2b is narrower than the one in Figure 13.2a. Therefore, as the volume under the density in Figure 13.2b needs to be equal to 1 (otherwise it is not a valid probability density), the density must be higher.

Finally, Figure 13.2c shows a density where $\sigma_X^2 = 0.25$ and $\sigma_Y^2 = 1$. This density is not as wide in the x-direction as in the y-direction: because of the smaller variance σ_X^2, the density is narrower in the x-direction than in the y-direction. The probability density in Figure 13.2c is higher than the one in Figure 13.2a because its bell shape

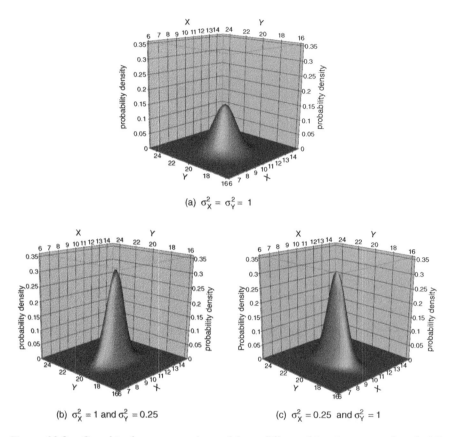

Figure 13.2   Graphical representations of three different bivariate normal probability densities with $\mu_X = 10$, $\mu_Y = 20$, and $\rho_{XY} = 0$.

is narrower. Again, as the volume under the density in Figure 13.2c needs to be 1 (otherwise it is not a valid probability density), the density must be higher.

The three-dimensional graphical representations in Figures 13.1 and 13.2 are called **surface plots**. They can be generated in JMP using the "Contour Profiler" option in the "Graph" menu. Selecting the "Contour Profiler" option will not immediately result in the surface plot. First, you need to click on the hotspot (red triangle icon) next to the word "Profiler" in the header of the output of the "Contour Profiler" and then check the "Surface Profiler" option.

The bivariate probability densities in Figures 13.1 and 13.2 can also be displayed graphically using a so-called **contour plot**. A contour plot is a two-dimensional representation of a function of two variables. Each line in a contour plot corresponds to a set of pairs (x, y) for which the function takes exactly the same value. JMP offers the possibility to generate this type of graphical representation via the "Contour Plot" option in the "Graph" menu. By default, JMP provides a contour plot with white background, but this can be changed by selecting the option "Fill Areas" from the hotspot

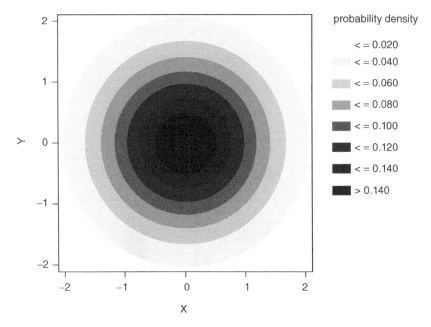

Figure 13.3 Contour plot for the bivariate normal probability density with $\mu_X = \mu_Y = 0$, $\sigma_X^2 = \sigma_Y^2 = 1$ and $\rho_{XY} = 0$ in Figure 13.1.

(red triangle) menu next to the word "Contour Plot". Each colored area indicates a set of combinations of x- and y-values with similar values of the function studied.

The contour plot corresponding to Figure 13.1 is shown in Figure 13.3. All lines in this figure form concentric circles, because all the points located at the same distance from the point with coordinates $(0, 0)$ take the same function value for the bivariate probability density $f_{XY}(x, y)$.

Figure 13.4 shows the contour plots that correspond to the three probability densities in Figure 13.2. The contour plot in Figure 13.4a contains concentric circles because all points at the same distance from the point $(10, 20)$ have the same function value for the probability density. In general, a contour plot involves concentric circles for a bivariate normal probability density if $\sigma_X^2 = \sigma_Y^2$ and $\rho_{XY} = 0$. As soon as $\sigma_X^2 \neq \sigma_Y^2$ and/or $\rho_{XY} \neq 0$, the lines in the contour plot of a bivariate normal probability density become concentric ellipses. This is illustrated in Figures 13.4b and 13.4c. A feature of the ellipses in these two figures is that their symmetry axes are parallel to the horizontal and vertical axes. This is typical for bivariate normal densities with $\rho_{XY} = 0$.

Note that Figure 13.4a does not contain very dark colored areas (unlike Figures 13.4b and 13.4c) because the bivariate normal probability density with $\sigma_X^2 = \sigma_Y^2 = 1$ does not take very large values. This was also demonstrated in Figure 13.2a, which shows that the maximum of the bivariate normal probability density with $\sigma_X^2 = \sigma_Y^2 = 1$ is much lower than when σ_X^2 or σ_Y^2 is equal to 0.25.

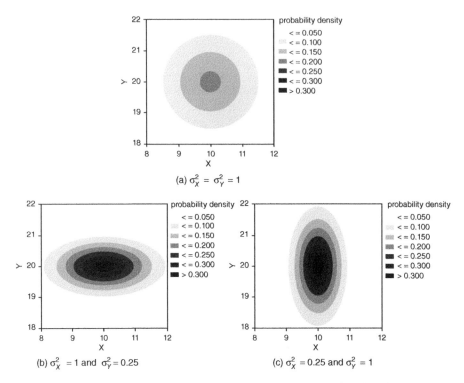

Figure 13.4   Contour plots for the three bivariate normal probability densities with $\mu_X = 10$, $\mu_Y = 20$, *and* $\rho_{XY} = 0$ *in Figure 13.2.*

13.6.3   Independence, marginal, and conditional densities

Two bivariate normally distributed random variables X and Y are independent if the correlation coefficient ρ_{XY} is equal to zero. The converse is also true: if two multivariate normally distributed random variables are uncorrelated, then they are also independent. Consequently, the terms uncorrelated and independent are synonymous for normally distributed random variables. As explained in Section 13.1, this is not true in general. Two random variables that are independent automatically have a correlation of zero. However, two uncorrelated random variables can be dependent. This is the case if there is a relationship between the two random variables that is not linear.

Figure 13.5 contains the graphs of two bivariate normal probability densities with a non-zero correlation ρ_{XY}. Figure 13.5a shows a density with a positive correlation, while Figure 13.5b shows a density with a negative correlation. The corresponding contour plots are shown in Figure 13.6. A striking feature of these contour plots is that their symmetry axes are no longer horizontal and vertical. This applies to all bivariate normal probability densities with $\rho_{XY} \neq 0$.

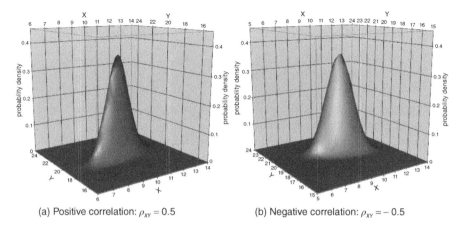

(a) Positive correlation: $\rho_{XY} = 0.5$ (b) Negative correlation: $\rho_{XY} = -0.5$

Figure 13.5   Graphical representations of two bivariate normal probability densities with $\mu_X = 10$, $\mu_Y = 20$, $\sigma_X^2 = 1$, $\sigma_Y^2 = 0.25$, and $\rho_{XY} \neq 0$.

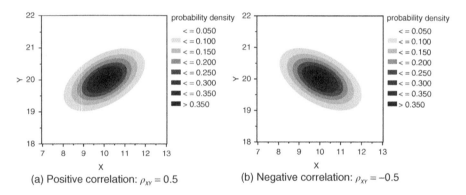

(a) Positive correlation: $\rho_{XY} = 0.5$ (b) Negative correlation: $\rho_{XY} = -0.5$

Figure 13.6   Contour plots for the bivariate normal probability densities with $\mu_X = 10$, $\mu_Y = 20$, $\sigma_X^2 = 1$, $\sigma_Y^2 = 0.25$, and $\rho_{XY} \neq 0$ from Figure 13.5.

The positive correlation between X and Y in Figures 13.5a and 13.6a means that the random variable Y has a tendency to take a large value if X takes a large value (or vice versa), and a tendency to take a small value if X takes a small value (or vice versa). The negative correlation between X and Y in Figures 13.5b and 13.6b means that the random variable Y has a tendency to take a small value if X takes a large value (or vice versa) and a tendency to take a large value if X takes a small value (or vice versa).

The unconditional or marginal probability density of X is the univariate normal density

$$f_X(x) = \frac{1}{\sigma_X \sqrt{2\pi}} \, e^{-\frac{(x-\mu_X)^2}{2\sigma_X^2}},$$

while the unconditional or marginal probability density of Y is the univariate normal density

$$f_Y(y) = \frac{1}{\sigma_Y \sqrt{2\pi}} e^{-\frac{(y-\mu_Y)^2}{2\sigma_Y^2}}.$$

The expected value of X is μ_X, while the variance is σ_X^2. The expected value and variance of Y are μ_Y and σ_Y^2, respectively.

The conditional probability densities are univariate normal probability densities. The conditional probability density of X, given that Y takes the value y, is the univariate normal density with expected value

$$\mu_{X|Y} = \mu_X + \frac{\sigma_X}{\sigma_Y}\rho_{XY}(y - \mu_Y)$$

and variance

$$\sigma_{X|Y}^2 = (1 - \rho_{XY}^2)\sigma_X^2.$$

The conditional probability density of Y, given that X takes the value x, is the univariate normal density with expected value

$$\mu_{Y|X} = \mu_Y + \frac{\sigma_Y}{\sigma_X}\rho_{XY}(x - \mu_X)$$

and variance

$$\sigma_{Y|X}^2 = (1 - \rho_{XY}^2)\sigma_Y^2.$$

It is a good exercise to derive these conditional probability densities, using the definitions $f_{X|Y}(x \mid y) = f_{XY}(x, y)/f_Y(y)$ and $f_{Y|X}(y \mid x) = f_{XY}(x, y)/f_X(x)$. The fact that the conditional probability densities of X and Y are univariate normal probability densities means, in graphical terms, that each horizontal or vertical cross section of the probability densities in Figures 13.1, 13.2, and 13.5 is again a bell-shaped curve.

It is not difficult to show that X and Y are independent if $\rho_{XY} = 0$: the joint probability density of X and Y in Equation (13.2) then is

$$f_{XY}(x, y) = \frac{1}{2\pi\sigma_X\sigma_Y\sqrt{1 - 0^2}} e^{-\frac{1}{2(1-0^2)}\left[\frac{(x-\mu_X)^2}{\sigma_X^2} + \frac{(y-\mu_Y)^2}{\sigma_Y^2} - 2\times 0\times\frac{(x-\mu_X)(y-\mu_Y)}{\sigma_X\sigma_Y}\right]},$$

$$= \frac{1}{2\pi\sigma_X\sigma_Y} e^{-\frac{1}{2}\left[\frac{(x-\mu_X)^2}{\sigma_X^2} + \frac{(y-\mu_Y)^2}{\sigma_Y^2}\right]},$$

$$= \frac{1}{\sqrt{2\pi}\sqrt{2\pi}\sigma_X\sigma_Y} e^{-\frac{(x-\mu_X)^2}{2\sigma_X^2} - \frac{(y-\mu_Y)^2}{2\sigma_Y^2}},$$

$$= \frac{1}{\sigma_X \sqrt{2\pi}} e^{-\frac{(x-\mu_X)^2}{2\sigma_X^2}} \cdot \frac{1}{\sigma_Y \sqrt{2\pi}} e^{-\frac{(y-\mu_Y)^2}{2\sigma_Y^2}},$$

$$= f_X(x)f_Y(y). \tag{13.3}$$

In other words, if $\rho_{XY} = 0$, then the joint probability density of X and Y is equal to the product of the marginal probability density of X and the marginal probability density of Y.

Example 13.6.1 *Around 1900, the heights of fathers were on average 173 cm, while the average height of sons was 175 cm. The standard deviations of the heights of fathers and sons were both equal to 5 cm. The correlation between these two heights was 0.5. It is also known that the heights of fathers and sons were bivariate normally distributed. If we denote the height of a father with the random variable X and the height of a son with the random variable Y, then the parameters of the bivariate normal distribution are $\mu_X = 173$, $\mu_Y = 175$, $\sigma_X = \sigma_Y = 5$, and $\rho_{XY} = 0.5$.*

We can now wonder what the expected height of a son is, given that his father has a height of 2 m or 200 cm. This conditional expected value is

$$\mu_{Y|200} = 175 + \frac{5}{5}(0.5)(200 - 173) = 175 + 0.5 \times 27 = 188.5 \ cm.$$

The variance of the height of the son, given that his father measures 200 cm, is

$$\sigma_{Y|200}^2 = (1 - (0.5)^2) \times 5^2 = 18.75.$$

The conditional probability that the son is taller than his father, given that his father measures 200 cm, is

$$P\left(\frac{Y^* - 188.5}{\sqrt{18.75}} > \frac{200 - 188.5}{\sqrt{18.75}} \right) = P(Z > 2.6558) = 0.00396,$$

where Y^* represents a normally distributed random variable with expected value $\mu_{Y|200} = 188.5$ and variance $\sigma_{Y|200}^2 = 18.75$.

The (unconditional) probability that a son is taller than his father is

$$P(Y > X) = P(Y - X > 0) = P\left(\frac{Y - X - 2}{\sqrt{25}} > \frac{0 - 2}{\sqrt{25}} \right),$$

$$= P(Z > -0.4) = 0.6554.$$

This calculation uses the fact that $Y - X$ has an expected value of $175 - 173 = 2$ and a variance of

$$\sigma_X^2 + \sigma_Y^2 - 2\rho_{XY}\sigma_X\sigma_Y = 5^2 + 5^2 - 2 \times 0.5 \times 5 \times 5 = 25.$$

Note that such an unconditional probability has already been calculated in Example 13.5.1. The information that a father has a height of 200 cm has a negative impact on the probability that his son is taller than him.

13.6.4   General multivariate normal density

For the general multivariate normal density, we have to use a vector notation. Suppose that we study k multivariate normally distributed random variables $X_1, X_2, \ldots, X_k$ with k-dimensional covariance matrix

$$
\mathbf{C} =
\begin{bmatrix}
\sigma_{11} & \sigma_{12} & \cdots & \sigma_{1k} \\
\sigma_{21} & \sigma_{22} & \cdots & \sigma_{2k} \\
\vdots & \vdots & \ddots & \vdots \\
\sigma_{k1} & \sigma_{k2} & \cdots & \sigma_{kk}
\end{bmatrix}
=
\begin{bmatrix}
\sigma_1^2 & \sigma_{12} & \cdots & \sigma_{1k} \\
\sigma_{12} & \sigma_2^2 & \cdots & \sigma_{2k} \\
\vdots & \cdots & \ddots & \vdots \\
\sigma_{1k} & \sigma_{2k} & \cdots & \sigma_k^2
\end{bmatrix},
$$

where each diagonal element $\sigma_{ii} = \sigma_i^2$ represents the variance of the i-th random variable X_i, and each non-diagonal element $\sigma_{ij} = \sigma_{ji}$ represents the covariance between the i-th random variable X_i and the j-th random variable X_j. Moreover, suppose that the expected values of $X_1, X_2, \ldots, X_k$ equal $\mu_1, \mu_2, \ldots, \mu_k$, respectively.

The general expression for the **k-variate normal density** is

$$
f_{X_1 X_2 \ldots X_k}(x_1, x_2, \ldots, x_k) = \frac{1}{(2\pi)^{k/2} \sqrt{\det(\mathbf{C})}} \; e^{-\frac{1}{2}c},
$$

where

$$
c = \begin{bmatrix} x_1 - \mu_1 & x_2 - \mu_2 & \cdots & x_k - \mu_k \end{bmatrix} \mathbf{C}^{-1}
\begin{bmatrix}
x_1 - \mu_1 \\
x_2 - \mu_2 \\
\vdots \\
x_k - \mu_k
\end{bmatrix}.
$$

A shorter notation of this density is

$$
f_{X_1 X_2 \ldots X_k}(x_1, x_2, \ldots, x_k) = \frac{1}{(2\pi)^{k/2} \sqrt{\det(\mathbf{C})}} \; e^{-\frac{1}{2}(\mathbf{x}-\boldsymbol{\mu})'\mathbf{C}^{-1}(\mathbf{x}-\boldsymbol{\mu})},
$$

where

$$
\mathbf{x} =
\begin{bmatrix}
x_1 \\
x_2 \\
\vdots \\
x_k
\end{bmatrix}
$$

and

$$
\boldsymbol{\mu} =
\begin{bmatrix}
\mu_1 \\
\mu_2 \\
\vdots \\
\mu_k
\end{bmatrix}.
$$

14

The central limit theorem

The odds were that one out of every two hundred ball bearings had something wrong with it; at the end of the day, you turned in the bad ball bearings. If you had a day with no bad ball bearings, the foreman told you that you weren't looking each ball bearing over carefully enough.

(from *The Cider House Rules,* John Irving, p. 332)

The central limit theorem is perhaps the most important theorem in statistics. Usually, the theorem is used to make a statement on the probability density of an arithmetic mean or a sample mean. Earlier, we introduced the notation $\bar{x}$ for an arithmetic or sample mean. In this chapter, however, we use the notation $\bar{X}$ to indicate that the sample mean can be interpreted as a function of random variables $X_1, X_2, \ldots, X_n$, which all have the same probability distribution or density, and therefore can itself be considered as a random variable. The sample mean $\bar{X}$ of $X_1, X_2, \ldots, X_n$ therefore has a probability distribution or probability density, just like any other random variable. This chapter deals with the probability distribution or density of a sample mean.

14.1 Probability density of the sample mean from a normally distributed population

In the case of a normally distributed population, we can use Theorem 12.5.2. In this case, all observations $X_1, X_2, \ldots, X_n$ are normally distributed random variables with the same expected value μ and the same variance σ^2. It follows from Theorem 12.5.2 that the mean of all observations $X_1, X_2, \ldots, X_n$ is also normally distributed with expected value μ and variance σ^2/n. Hence, in this case the arithmetic or sample

Statistics with JMP: Graphs, Descriptive Statistics, and Probability, First Edition. Peter Goos and David Meintrup.
© 2015 John Wiley & Sons, Ltd. Published 2015 by John Wiley & Sons, Ltd.
Companion Website: wiley.com/go/goosandmeintrup

mean $\overline{X}$ is normally distributed with mean μ and variance σ^2/n. This is indicated by the notation

$$\overline{X} \sim N\left(\mu, \frac{\sigma^2}{n}\right).$$

This result is valid for any number of random variables n, no matter how small or large. In this context, the number of random variables n is called the sample size: every random variable corresponds to an observation of an element in a population or from a process.

14.2 Probability distribution and density of the sample mean from a non-normally distributed population

When a non-normally distributed population is being studied, the probability distribution of the sample mean can often not be determined exactly. In that case, having a large sample size is helpful, because the central limit theorem can be used for a large number of observations. One version of this theorem, namely Theorem 14.2.3, states that, for large n, the sample mean is approximately normally distributed with mean μ and variance σ^2/n.

14.2.1 Central limit theorem

The so-called **central limit theorem** is one of the main theorems of statistics. This theorem also explains why the normal density is so crucial in statistics. There are different versions of the theorem.

Theorem 14.2.1 *If $X_1, X_2, \ldots, X_n$ are independent random variables with expected values $E(X_i) = \mu_i$ and variances $\mathrm{var}(X_i) = \sigma_i^2$, then, under very general conditions and for a sufficiently large value of n, the following is true:*

1. *The random variable $Y = \sum_{i=1}^{n} X_i$ is approximately normally distributed with mean $\mu_Y = \sum_{i=1}^{n} \mu_i$ and variance $\sigma_Y^2 = \mathrm{var}(Y) = \sum_{i=1}^{n} \sigma_i^2$.*

2. *As a result, the random variable*

$$\frac{Y - \sum_{i=1}^{n} \mu_i}{\sqrt{\sum_{i=1}^{n} \sigma_i^2}}$$

   is approximately standard normally distributed.

The general conditions mentioned in the theorem refer to the fact that none of the individual variances σ_i^2 provides a dominant contribution to the total variance of Y. In many practical applications of the central limit theorem, all random variables $X_1, X_2, \ldots, X_n$ have the same distribution or density. In that case, $X_1, X_2, \ldots, X_n$

all have the same variance, in which case this requirement is fulfilled. If all random variables $X_1, X_2, \ldots, X_n$ have the same distribution or density, then the central limit theorem can be rewritten as follows:

Theorem 14.2.2 *If* $X_1, X_2, \ldots, X_n$ *are independent random variables with expected value* $E(X_i) = \mu$ *and variance* $\text{var}(X_i) = \sigma^2$, *then, for a sufficiently large value of* n, *the following is true:*

1. *The random variable* $Y = \sum_{i=1}^{n} X_i$ *is approximately normally distributed with mean* $\mu_Y = n\mu$ *and variance* $\sigma_Y^2 = \text{var}(Y) = n\sigma^2$.

2. *As a result, the random variable*

$$\frac{Y - n\mu}{\sigma\sqrt{n}}$$

is approximately standard normally distributed.

The central limit theorem can also be stated in terms of the sample mean $\overline{X} = Y/n$:

Theorem 14.2.3 *If* $X_1, X_2, \ldots, X_n$ *are independent random variables with expected value* $E(X_i) = \mu$ *and variance* $\text{var}(X_i) = \sigma^2$, *then, for a sufficiently large value of* n, *the following is true:*

1. *The random variable* $\overline{X} = \dfrac{Y}{n} = \dfrac{\sum_{i=1}^{n} X_i}{n}$ *is approximately normally distributed with mean* μ *and variance* $\dfrac{\sigma^2}{n}$.

2. *As a result, the random variable*

$$\frac{\overline{X} - \mu}{\frac{\sigma}{\sqrt{n}}}$$

is approximately standard normally distributed.

An important practical question is how big the sample size n must be before one can apply the central limit theorem. There is no general answer to this question. The required size of n depends on the distribution or density of the individual random variables X_i:

- If the probability density of X_i is similar to the normal density, $n = 5$ is sufficient.

- If the probability density of X_i does not show any pronounced peaks, such as, for example, the uniform density, then $n = 12$ should be sufficient.

- If the probability distribution or density of X_i shows pronounced peaks, it is difficult to specify a value of n. A value of $n = 100$ will usually suffice.

An example of a distribution with a peak is $P(X = 1) = 0.06$ and $P(X = 10) = 1 - P(X = 1) = 0.94$.

- For continuous variables that appear in practice, typically $n = 30$ is sufficient.

The next section illustrates the third version of the central limit theorem (Theorem 14.2.3) in detail, using simulations.

14.2.2   Illustration of the central limit theorem

Suppose that some students are interested in the value of the Euro Stoxx 50 index, made up of 50 of the largest and most liquid stocks inside the Euro zone. Student 1 will take a sample of n observations of the Euro Stoxx 50 index and calculate the mean, namely $\overline{X}_1$. Student 2 will also take a sample of n observations. Since the Euro Stoxx 50 index changes from minute to minute, Student 2 will obviously observe different values of the Euro Stoxx 50 index (unless by coincidence they observe exactly at the same times). Student 2 also calculates the mean of his sample: $\overline{X}_2$. In this way, all students collect n observations and calculate their sample mean. If there are 200 students, we finally obtain 200 sample means $\overline{X}_1, \overline{X}_2, \ldots, \overline{X}_{200}$.

The third version of the central limit theorem now states that these 200 means have a distribution that is very similar to the normal density. With a histogram of these 200 means, this is easy to verify.

This is exactly what will happen in this section. We will not use real students but we will simulate the scenario outlined here in JMP. So, we will work with hypothetical students. To this end, in JMP, we will create 200 samples of n observations (one for each hypothetical student), calculate the mean for each sample and create a histogram of the 200 sample means. This simulation requires that we specify a probability distribution or probability density in JMP for the generation of the observations.

We start with a normal distribution. Hence we assume that the Euro Stoxx 50 index behaves like a normally distributed random variable. We use $\mu = 3000$ as the mean of this normal distribution (more or less the value of the index when this book project was started in March 2014), and we choose $\sigma = 100$ as the standard deviation. We assume that all observations of the students are independent of each other.

14.2.2.1   Normally distributed X

First, suppose that each student collects a sample of five observations, in other words, that $n = 5$. In that scenario, we need to simulate 5 observations 200 times using JMP. To this end, we create a data table in JMP with 200 rows and 5 columns, filled with pseudo-random numbers from a normal probability density with parameters $\mu = 3000$ and $\sigma = 100$. The formula we use in each of the 5 columns is "Random Normal(3000, 100)". We calculate the mean of the 5 observations in each row, and then display all means in a histogram. If we create a second data table in the same way, this corresponds to a second group of 200 hypothetical students who also collect samples of 5 observations. Two possible histograms obtained in this way are shown in Figure 14.1. The resulting histograms are quite bell-shaped, indicating that the

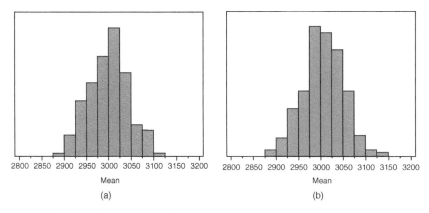

Figure 14.1   Two histograms of 200 *sample means for normally distributed data and samples of* 5 *observations.*

sample means are normally distributed, as Theorem 14.2.3 (and also Theorem 12.5.2 because, here, we assume that the observations are normally distributed) predicts.

The fastest way to generate 200 new samples is to ask JMP to recalculate the formula "Random Normal(3000, 100)". This is done with the command "Rerun Formulas", that appears when you click on the hotspot (red triangle) menu next to the name of the data table. This is illustrated in Figure 14.2.

If students take samples of 20 instead of 5 observations, the histograms have a different shape: they are still bell-shaped but they are significantly narrower. Two histograms for 200 sample means of samples with 20 observations are shown in Figure 14.3. The bell shape tells us that the sample means are still distributed normally. The fact that the histograms are narrower should not come as a surprise since the central limit theorem states that the variance of the sample mean is equal to σ^2/n.

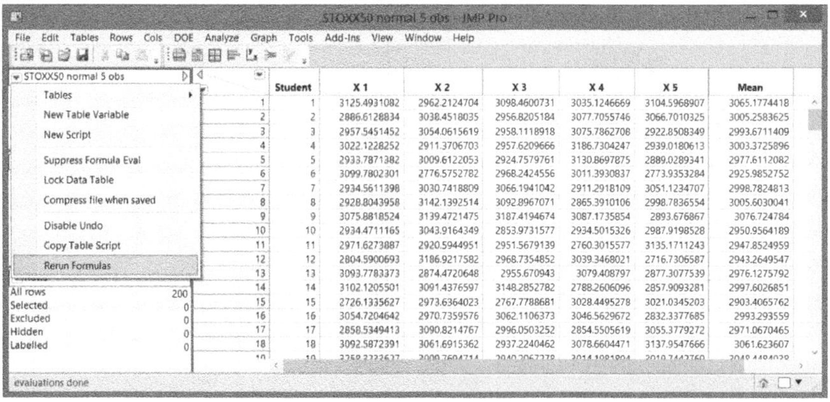

Figure 14.2   Generating new pseudo-random observations in JMP with the option "Rerun Formulas".

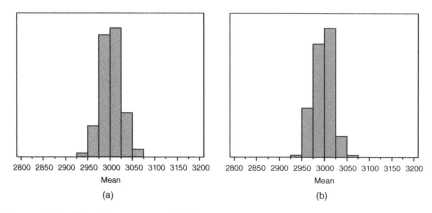

Figure 14.3   Two histograms of 200 sample means for normally distributed data and samples of 20 observations.

As a consequence, sample means of 20 observations have a variance that is four times smaller than the variance of sample means of 5 observations.

14.2.2.2   Uniformly distributed X

Suppose that the value of the Euro Stoxx 50 index is not normally distributed, but uniformly distributed between 2800 and 3200. First, suppose again that each student takes a sample of 5 observations. For this new scenario involving the uniform density, we again simulate 200 samples of 5 observations using JMP. To this end, we need to enter the formula "Random Uniform(2800, 3200)" in five columns of a data table with 200 rows. For each sample of 5 observations, we calculate the mean, and then we display all means in a histogram. Two possible histograms obtained in this way are shown in Figure 14.4. It is striking that, again, the histograms are quite bell-shaped,

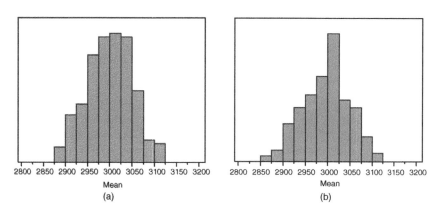

Figure 14.4   Two histograms of 200 sample means for uniformly distributed data and samples of 5 observations.

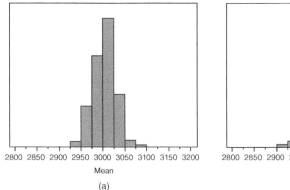
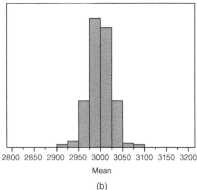

Mean   Mean

(a)   (b)

Figure 14.5   Two histograms of 200 *sample means for uniformly distributed data and samples of* 20 *observations.*

indicating that the sample means are still approximately normally distributed, even though the original data is uniformly distributed.

When the students take samples of 20 instead of 5 observations, the corresponding bell-shaped histograms are significantly narrower. Two histograms for 200 sample means of samples with 20 observations are shown in Figure 14.5.

14.2.2.3   Bernoulli distributed X

Now, suppose that the value of the Euro Stoxx 50 index is Bernoulli distributed, with a 50% chance that the value is 2800, and a 50% chance that the value is 3200. First, suppose again that each student takes a sample of 5 observations. We again need to simulate 200 samples of 5 observations using JMP. This time, we need to enter the formula "2800 + 400 * Random Binomial (1, 0.5)" in 5 columns of a data table with 200 rows. For each sample of 5 observations, we calculate the mean, and display the resulting 200 means in a histogram. Two possible histograms obtained in this way are shown in Figure 14.6. This time, the histograms are not bell-shaped. It is clearly visible that the original data comes from a discrete distribution, namely the Bernoulli distribution. The central limit theorem does not seem to work for the Bernoulli distribution and a sample size of 5 observations.

When, however, the students take samples of 20 instead of 5 observations, the histograms are totally different. Although the histograms still do not show a perfect bell shape, it is not obvious anymore that the original data had a discrete probability distribution. Two possible histograms for 200 sample means of samples with 20 observations are shown in Figure 14.7. In order to obtain an even better bell shape, a slightly larger sample size is required.

This last example demonstrates that the central limit theorem is very powerful. Even probability distributions or probability densities that are quite different from the normal density still lead to distributions of sample means that are approximately normal, provided that the number of observations is sufficiently large.

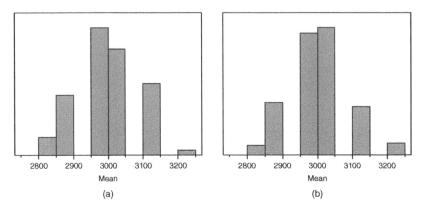

Figure 14.6   Two histograms of 200 *sample means for Bernoulli distributed data and samples of* 5 *observations.*

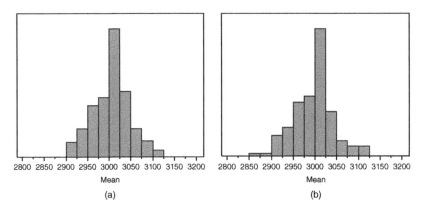

Figure 14.7   Two histograms of 200 *sample means for Bernoulli distributed data and samples of* 20 *observations.*

14.3   Applications

Example 14.3.1   *The label of a coffee package mentions "1 kg net weight". The filling machine is set at an average weight of $\mu = 1.003$ kg. A standard deviation of $\sigma = 10$ g $= 0.01$ kg has to be tolerated for the filling process.*

1. *What is the probability that the average net weight of a batch of* 100 *packages of coffee sent to a customer weighs less than* 1 *kg?*

2. *What is the probability that the total net weight of the batch of* 100 *packages exceeds* 100.5 *kg?*

3. *If the net weight of a filled package is normally distributed, then what is the probability that a randomly picked individual package weighs less than* 1 *kg ?*

For the solution of the first two questions, we need to apply the central limit theorem, whereas this is not necessary for the third question:

1. *Due to the central limit theorem, the average net weight of 100 coffee packages, $\overline{X}$, is approximately normally distributed with expected value $E(\overline{X}) = \mu = 1.003$ kg and variance*

$$\text{var}(\overline{X}) = \frac{\sigma^2}{n} = \frac{(0.01)^2}{100} \text{ kg}^2 = 0.000001 \text{ kg}^2.$$

Thus, the standard deviation of $\overline{X}$ is $\sqrt{0.000001} = 0.001$ kg. The probability that the average net weight of a batch of 100 coffee packages sent to a customer weighs less than 1 kg is

$$P(\overline{X} < 1) = P\left(\frac{\overline{X} - 1.003}{0.001} < \frac{1 - 1.003}{0.001}\right),$$

$$= P\left(Z < \frac{-0.003}{0.001}\right),$$

$$= P(Z < -3),$$

$$= 0.00135.$$

2. *The total net weight of 100 coffee packages is, due to the central limit theorem, approximately normally distributed with expected value*

$$E\left(\sum_{i=1}^{100} X_i\right) = 100\mu = 100.3 \text{ kg}$$

and variance

$$\text{var}\left(\sum_{i=1}^{100} X_i\right) = 100\sigma^2 = 100 \times 0.0001 \text{ kg}^2 = 0.01 \text{ kg}^2.$$

The standard deviation of the total net weight thus is $\sqrt{0.01} = 0.1$ kg. The probability that the total net weight of the batch of 100 packages exceeds 100.5 kg is

$$P\left(\sum_{i=1}^{100} X_i > 100.5\right) = P\left(\frac{\sum_{i=1}^{100} X_i - 100.3}{0.1} > \frac{100.5 - 100.3}{0.1}\right),$$

$$= P\left(Z > \frac{0.2}{0.1}\right),$$

$$= P(Z > 2),$$

$$= 0.02275.$$

3. If the net weight of a filled package, X, is normally distributed, then the probability that a randomly picked individual package weighs less than 1 kg is

$$P(X < 1) = P\left(\frac{X - 1.003}{0.01} < \frac{1 - 1.003}{0.01}\right),$$

$$= P\left(Z < \frac{-0.003}{0.01}\right),$$

$$= P(Z < -0.3),$$

$$= 0.38209.$$

14.4 Normal approximation of the binomial distribution

In 1733, Abraham de Moivre showed that the binomial distribution can be approximated by a normal density. This can easily be demonstrated based on the central limit theorem.

Suppose that $X_1, X_2, \ldots, X_n$ are independent and identically distributed Bernoulli random variables with parameter π, so that $\mu_i = E(X_i) = \pi$ and $\sigma_i^2 = \text{var}(X_i) = \pi(1 - \pi)$. By definition, the sum of these random variables is binomially distributed with parameters n and π. According to the second version of the central limit theorem (Theorem 14.2.2), this sum is also approximately normally distributed with mean $\sum_{i=1}^{n} \mu_i = n\pi$ and variance $\sum_{i=1}^{n} \sigma_i^2 = n\pi(1 - \pi)$.

A good approximation of the binomial distribution by the normal distribution requires that n is large. However, the quality of the approximation also depends on π. Ideally, both $n\pi$ and $n(1 - \pi)$ are bigger than 5. Then, the approximation is considered to be good.

Example 14.4.1 *An exam consists of 60 multiple choice questions with four possible answers per question. What is the probability that a student who randomly answered the 60 questions got at most 15 correct answers?*

If the random variable X represents the number of questions answered correctly, then X is binomially distributed with n = 60 and π = 0.25. The exact answer to our question is

$$P(X \le 15) = \sum_{x=0}^{15} \binom{60}{x} (0.25)^x (0.75)^{60-x}.$$

With the formula "Binomial Distribution(0.25, 60, 15)" in JMP, we obtain the value 0.5688 for this probability.

We now try to approximate the probability using the normal distribution. To use an approximation obviously does not make sense if you have access to JMP, but, during, for instance, a statistics exam, it may be that you only have access to a limited set of tables with probabilities for binomially distributed random variables. Such a set of tables is given in Appendix B. The tables, however, do not include information *concerning binomially distributed random variables with parameter values $n = 60$ and $\pi = 0.25$. In that case, you have to find a solution with an approximate method. We first try the normal approximation with $\mu = n\pi = 15$ and variance $\sigma^2 = n\pi(1 - \pi) = 11.25$. Since $n\pi$ and $n(1 - \pi)$ are both greater than 5, this approximation should indeed be good. However, we find that*

$$P(X \le 15) = P\left(\frac{X - 15}{\sqrt{11.25}} \le \frac{15 - 15}{\sqrt{11.25}} \right) = P(Z \le 0) = 0.5,$$

*which is quite different from the exact value 0.5688. This big difference is due to the fact that a continuous density is used to approximate a discrete distribution. The best illustration of the difference between both is that, according to the binomial distribution, $P(X = 15) = 0.1182$, while the normal approximation states that $P(X = 15) = 0$. In such cases, a **continuity correction** is applied. This means that, instead of computing the probability $P(a \le X \le b)$, the probability $P(a - \frac{1}{2} \le X \le b + \frac{1}{2})$ is calculated. The probability $P(X = a) = P(a \le X \le a)$ is then approximated by $P(a - \frac{1}{2} \le X \le a + \frac{1}{2})$. Applying this type of correction to the example, we obtain*

$$P(X \le 15.5) = P\left(\frac{X - 15}{\sqrt{11.25}} \le \frac{15.5 - 15}{\sqrt{11.25}} \right) = P(Z \le 0.149) = 0.559,$$

which indeed is closer to the correct value. In addition,

$$P(X = 15) = P\left(\frac{14.5 - 15}{\sqrt{11.25}} \le \frac{X - 15}{\sqrt{11.25}} \le \frac{15.5 - 15}{\sqrt{11.25}} \right),$$
$$= 1 - 2\,P(Z \ge 0.149) = 0.1185$$

is a good approximation of the exact value we found using the binomial distribution.

The approximation of the binomial distribution by the normal density is used in the so-called sign test (see the book *Statistics with JMP: Hypothesis Tests, ANOVA and Regression*).

Appendix A

The Greek alphabet

In mathematics and statistics, Greek letters are often used to denote variables, parameters, or functions. Below is a list of all the Greek letters.

Name	Uppercase letter	Lowercase letter
alpha	A	α
beta	B	β
gamma	Γ	γ
delta	Δ	δ
epsilon	E	ϵ
zeta	Z	ζ
eta	H	η
theta	Θ	θ
iota	I	ι
kappa	K	κ
lambda	Λ	λ
mu	M	μ
nu	N	ν
xi	Ξ	ξ
omicron	O	o
pi	Π	π
rho	P	ρ
sigma	Σ	σ
tau	T	τ
upsilon	Y	υ
phi	Φ	ϕ
chi	X	χ
psi	Ψ	ψ
omega	Ω	ω

Statistics with JMP: Graphs, Descriptive Statistics, and Probability, First Edition. Peter Goos and David Meintrup.
© 2015 John Wiley & Sons, Ltd. Published 2015 by John Wiley & Sons, Ltd.
Companion Website: wiley.com/go/goosandmeintrup

Appendix B

Binomial distribution

This table contains exceedance probabilities for the binomial distribution. For example, if $n = 4$ and $\pi = 0.20$, then

$$P(X \geq 2) = 0.1808.$$

n	x	$\pi = 0.05$	0.10	0.15	0.20	0.25	0.30	0.35	0.40	0.45	0.50
2	0	1.0000	1.0000	1.0000	1.0000	1.0000	1.0000	1.0000	1.0000	1.0000	1.0000
	1	0.0975	0.1900	0.2775	0.3600	0.4375	0.5100	0.5775	0.6400	0.6975	0.7500
	2	0.0025	0.0100	0.0225	0.0400	0.0625	0.0900	0.1225	0.1600	0.2025	0.2500

n	x	$\pi = 0.05$	0.10	0.15	0.20	0.25	0.30	0.35	0.40	0.45	0.50
3	0	1.0000	1.0000	1.0000	1.0000	1.0000	1.0000	1.0000	1.0000	1.0000	1.0000
	1	0.1426	0.2710	0.3859	0.4880	0.5781	0.6570	0.7254	0.7840	0.8336	0.8750
	2	0.0073	0.0280	0.0608	0.1040	0.1562	0.2160	0.2818	0.3520	0.4253	0.5000
	3	0.0001	0.0010	0.0034	0.0080	0.0156	0.0270	0.0429	0.0640	0.0911	0.1250

n	x	$\pi = 0.05$	0.10	0.15	0.20	0.25	0.30	0.35	0.40	0.45	0.50
4	0	1.0000	1.0000	1.0000	1.0000	1.0000	1.0000	1.0000	1.0000	1.0000	1.0000
	1	0.1855	0.3439	0.4780	0.5904	0.6836	0.7559	0.8215	0.8704	0.9085	0.9375
	2	0.0140	0.0523	0.1095	0.1808	0.2617	0.3483	0.4370	0.5248	0.6090	0.6875
	3	0.0005	0.0037	0.0120	0.0272	0.0508	0.0837	0.1265	0.1792	0.2415	0.3125
	4		0.0001	0.0005	0.0016	0.0039	0.0081	0.0150	0.0256	0.0410	0.0625

Statistics with JMP: Graphs, Descriptive Statistics, and Probability, First Edition. Peter Goos and David Meintrup.
© 2015 John Wiley & Sons, Ltd. Published 2015 by John Wiley & Sons, Ltd.
Companion Website: wiley.com/go/goosandmeintrup

n	x	π = 0.05	0.10	0.15	0.20	0.25	0.30	0.35	0.40	0.45	0.50
5	0	1.0000	1.0000	1.0000	1.0000	1.0000	1.0000	1.0000	1.0000	1.0000	1.0000
	1	0.2262	0.4095	0.5563	0.6723	0.7627	0.8319	0.8840	0.9222	0.9497	0.9688
	2	0.0226	0.0815	0.1648	0.2627	0.3672	0.4718	0.5716	0.6630	0.7438	0.8125
	3	0.0012	0.0086	0.0266	0.0579	0.1035	0.1631	0.2352	0.3174	0.4069	0.5000
	4		0.0005	0.0022	0.0067	0.0156	0.0308	0.0540	0.0870	0.1312	0.1875
	5			0.0001	0.0003	0.0010	0.0024	0.0053	0.0102	0.0185	0.0313

n	x	π = 0.05	0.10	0.15	0.20	0.25	0.30	0.35	0.40	0.45	0.50
6	0	1.0000	1.0000	1.0000	1.0000	1.0000	1.0000	1.0000	1.0000	1.0000	1.0000
	1	0.2649	0.4686	0.6229	0.7379	0.8220	0.8824	0.9246	0.9533	0.9723	0.9844
	2	0.0328	0.1143	0.2235	0.3446	0.4661	0.5798	0.6809	0.7667	0.8364	0.8906
	3	0.0022	0.0159	0.0473	0.0989	0.1694	0.2557	0.3529	0.4557	0.5585	0.6563
	4	0.0001	0.0013	0.0059	0.0170	0.0376	0.0705	0.1174	0.1792	0.2553	0.3438
	5		0.0001	0.0004	0.0016	0.0046	0.0109	0.0223	0.0410	0.0692	0.1094
	6				0.0001	0.0002	0.0007	0.0018	0.0041	0.0083	0.0156

n	x	π = 0.05	0.10	0.15	0.20	0.25	0.30	0.35	0.40	0.45	0.50
7	0	1.0000	1.0000	1.0000	1.0000	1.0000	1.0000	1.0000	1.0000	1.0000	1.0000
	1	0.3017	0.5217	0.6794	0.7903	0.8665	0.9176	0.9510	0.9720	0.9848	0.9922
	2	0.0444	0.1497	0.2834	0.4233	0.5551	0.6706	0.7662	0.8414	0.8976	0.9375
	3	0.0038	0.0257	0.0738	0.1480	0.2436	0.3529	0.4677	0.5801	0.6836	0.7734
	4	0.0002	0.0027	0.0121	0.0333	0.0706	0.1260	0.1998	0.2898	0.3917	0.5000
	5		0.0002	0.0012	0.0047	0.0129	0.0288	0.0556	0.0963	0.1529	0.2266
	6			0.0001	0.0004	0.0013	0.0038	0.0090	0.0188	0.0357	0.0625
	7					0.0001	0.0002	0.0006	0.0016	0.0037	0.0078

n	x	π = 0.05	0.10	0.15	0.20	0.25	0.30	0.35	0.40	0.45	0.50
8	0	1.0000	1.0000	1.0000	1.0000	1.0000	1.0000	1.0000	1.0000	1.0000	1.0000
	1	0.3366	0.5695	0.7275	0.8322	0.8999	0.9424	0.9681	0.9832	0.9916	0.9961
	2	0.0572	0.1869	0.3428	0.4967	0.6329	0.7447	0.8309	0.8936	0.9368	0.9648
	3	0.0058	0.0381	0.1052	0.2031	0.3215	0.4482	0.5722	0.6846	0.7799	0.8555
	4	0.0004	0.0050	0.0214	0.0563	0.1183	0.1941	0.2936	0.4059	0.5230	0.6367
	5		0.0004	0.0029	0.0104	0.0273	0.0580	0.1061	0.1737	0.2604	0.3633
	6			0.0002	0.0012	0.0042	0.0113	0.0253	0.0498	0.0885	0.1445
	7				0.0001	0.0004	0.0013	0.0036	0.0085	0.0181	0.352
	8						0.0001	0.0002	0.0007	0.0017	0.0039

n	x	π = 0.05	0.10	0.15	0.20	0.25	0.30	0.35	0.40	0.45	0.50
9	0	1.0000	1.0000	1.0000	1.0000	1.0000	1.0000	1.0000	1.0000	1.0000	1.0000
	1	0.3689	0.6126	0.7684	0.8658	0.9249	0.9596	0.9793	0.9899	0.9954	0.9980
	2	0.0712	0.2252	0.4005	0.5638	0.6977	0.8040	0.8789	0.9295	0.9615	0.9805
	3	0.0084	0.0530	0.1409	0.2618	0.3993	0.5372	0.6627	0.7682	0.8505	0.9102
	4	0.0006	0.0083	0.0339	0.0856	0.1657	0.2703	0.3911	0.5174	0.6386	0.7461
	5		0.0009	0.0056	0.0196	0.0489	0.0988	0.1717	0.2666	0.3786	0.5000
	6		0.0001	0.0006	0.0031	0.0100	0.0253	0.0536	0.0994	0.1658	0.2539
	7				0.0003	0.0013	0.0043	0.0112	0.0250	0.0498	0.0898
	8					0.0001	0.0004	0.0014	0.0038	0.0091	0.0195
	9							0.0001	0.0003	0.0008	0.0020

n	x	π = 0.05	0.10	0.15	0.20	0.25	0.30	0.35	0.40	0.45	0.50
10	0	1.0000	1.0000	1.0000	1.0000	1.0000	1.0000	1.0000	1.0000	1.0000	1.0000
	1	0.4013	0.6513	0.8031	0.8926	0.9437	0.9718	0.9865	0.9940	0.9975	0.9990
	2	0.0861	0.2639	0.4557	0.6242	0.7560	0.8507	0.9140	0.9536	0.9767	0.9893
	3	0.0115	0.0702	0.1798	0.3222	0.4744	0.6172	0.7384	0.8327	0.9004	0.9453
	4	0.0010	0.0128	0.0500	0.1209	0.2241	0.3504	0.4862	0.6177	0.7340	0.8281
	5	0.0001	0.0016	0.0099	0.0328	0.0781	0.1503	0.2485	0.3669	0.4956	0.6230
	6		0.0001	0.0014	0.0064	0.0197	0.0473	0.0949	0.1662	0.2616	0.3770
	7			0.0001	0.0009	0.0035	0.0106	0.0260	0.0548	0.1020	0.1719
	8				0.0001	0.0004	0.0016	0.0048	0.0123	0.0274	0.0547
	9						0.0001	0.0005	0.0017	0.0045	0.0107
	10								0.0001	0.0003	0.0010

n	x	π = 0.05	0.10	0.15	0.20	0.25	0.30	0.35	0.40	0.45	0.50
12	0	1.0000	1.0000	1.0000	1.0000	1.0000	1.0000	1.0000	1.0000	1.0000	1.0000
	1	0.4596	0.7176	0.8578	0.9313	0.9683	0.9862	0.9943	0.9978	0.9992	0.9998
	2	0.1184	0.3410	0.5565	0.7251	0.8416	0.9150	0.9576	0.9807	0.9917	0.9968
	3	0.0196	0.1109	0.2642	0.4417	0.6093	0.7472	0.8487	0.9166	0.9579	0.9807
	4	0.0022	0.0256	0.0922	0.2054	0.3512	0.5075	0.6533	0.7747	0.8655	0.9270
	5	0.0002	0.0043	0.0239	0.0726	0.1576	0.2763	0.4167	0.5618	0.6956	0.8062
	6		0.0005	0.0046	0.0197	0.0544	0.1178	0.2127	0.3348	0.4731	0.6128
	7		0.0001	0.0007	0.0039	0.0143	0.0386	0.0846	0.1582	0.2607	0.3872
	8			0.0001	0.0006	0.0028	0.0095	0.0255	0.0573	0.1117	0.1938
	9				0.0001	0.0004	0.0017	0.0056	0.0153	0.0356	0.0730
	10						0.0002	0.0008	0.0028	0.0079	0.0193
	11							0.0001	0.0003	0.0011	0.0032
	12									0.0001	0.0002

n	x	$\pi =0.05$	0.10	0.15	0.20	0.25	0.30	0.35	0.40	0.45	0.50
14	0	1.0000	1.0000	1.0000	1.0000	1.0000	1.0000	1.0000	1.0000	1.0000	1.0000
	1	0.5123	0.7712	0.8972	0.9560	0.9822	0.9932	0.9976	0.9992	0.9998	0.9999
	2	0.1530	0.4154	0.6433	0.8021	0.8990	0.9525	0.9795	0.9919	0.9971	0.9991
	3	0.0301	0.1584	0.3521	0.5519	0.7189	0.8392	0.9161	0.9602	0.9830	0.9935
	4	0.0042	0.0441	0.1465	0.3018	0.4787	0.6448	0.7795	0.8757	0.9368	0.9714
	5	0.0004	0.0092	0.0467	0.1298	0.2585	0.4158	0.5773	0.7207	0.8328	0.9102
	6		0.0015	0.0115	0.0439	0.1117	0.2195	0.3595	0.5141	0.6627	0.7880
	7		0.0002	0.0022	0.0116	0.0383	0.0933	0.1836	0.3075	0.4539	0.6047
	8			0.0003	0.0024	0.0103	0.0315	0.0753	0.1501	0.2586	0.3953
	9				0.0004	0.0022	0.0083	0.0243	0.0583	0.1189	0.2120
	10					0.0003	0.0017	0.0060	0.0175	0.0426	0.0898
	11						0.0002	0.0011	0.0039	0.0114	0.0287
	12							0.0001	0.0006	0.0022	0.0065
	13								0.0001	0.0003	0.0009
	14										0.0001

n	x	$\pi =0.05$	0.10	0.15	0.20	0.25	0.30	0.35	0.40	0.45	0.50
16	0	1.0000	1.0000	1.0000	1.0000	1.0000	1.0000	1.0000	1.0000	1.0000	1.0000
	1	0.5599	0.8147	0.9257	0.9719	0.9900	0.9967	0.9990	0.9997	0.9999	1.0000
	2	0.1892	0.4853	0.7161	0.8593	0.9365	0.9739	0.9902	0.9967	0.9990	0.9997
	3	0.0429	0.2108	0.4386	0.6482	0.8029	0.9006	0.9549	0.9817	0.9934	0.9979
	4	0.0070	0.0684	0.2101	0.4019	0.5950	0.7541	0.8661	0.9349	0.9719	0.9894
	5	0.0009	0.0170	0.0791	0.2018	0.3698	0.5501	0.7108	0.8334	0.9147	0.9616
	6	0.0001	0.0033	0.0245	0.0817	0.1897	0.3402	0.5100	0.6712	0.8024	0.8949
	7		0.0005	0.0056	0.0267	0.0796	0.1753	0.3119	0.4738	0.6340	0.7728
	8		0.0001	0.0011	0.0070	0.0271	0.0744	0.1594	0.2839	0.4371	0.5982
	9			0.0002	0.0015	0.0075	0.0257	0.2671	0.1423	0.2559	0.4018
	10				0.0002	0.0016	0.0071	0.0229	0.0583	0.1241	0.2272
	11					0.0003	0.0016	0.0062	0.0191	0.0486	0.1051
	12						0.0003	0.0013	0.0049	0.0149	0.0384
	13							0.0002	0.0009	0.0035	0.0106
	14								0.0001	0.0006	0.0021
	15									0.0001	0.0003

n	x	$\pi = 0.05$	0.10	0.15	0.20	0.25	0.30	0.35	0.40	0.45	0.50
18	0	1.0000	1.0000	1.0000	1.0000	1.0000	1.0000	1.0000	1.0000	1.0000	1.0000
	1	0.6028	0.8499	0.9464	0.9820	0.9944	0.9984	0.9996	0.9999	1.0000	1.0000
	2	0.2265	0.5497	0.7759	0.9009	0.9605	0.9858	0.9954	0.9987	0.9997	0.9999
	3	0.0581	0.2662	0.5203	0.7287	0.8647	0.9400	0.9764	0.9918	0.9975	0.9993
	4	0.0109	0.0982	0.2798	0.4990	0.6943	0.8354	0.8917	0.9672	0.9880	0.9962
	5	0.0015	0.0282	0.1206	0.2836	0.4813	0.6673	0.8114	0.9058	0.9589	0.9846
	6	0.0002	0.0064	0.0419	0.1329	0.2825	0.4656	0.6450	0.7912	0.8923	0.9519
	7		0.0012	0.0118	0.0513	0.1390	0.2783	0.4509	0.6257	0.7742	0.8811
	8		0.0002	0.0027	0.0163	0.0569	0.1407	0.2717	0.4366	0.6085	0.7597
	9			0.0005	0.0043	0.0193	0.0596	0.1391	0.2632	0.4222	0.5927
	10			0.0001	0.0009	0.0054	0.0210	0.0597	0.1347	0.2527	0.4073
	11				0.0002	0.0012	0.0061	0.0212	0.0576	0.1280	0.2403
	12					0.0002	0.0014	0.0062	0.0203	0.0537	0.1189
	13						0.0003	0.0014	0.0058	0.0183	0.0481
	14							0.0003	0.0013	0.0049	0.0154
	15								0.0002	0.0010	0.0038
	16									0.0001	0.0007
	17										0.0001

n	x	$\pi = 0.05$	0.10	0.15	0.20	0.25	0.30	0.35	0.40	0.45	0.50
20	0	1.0000	1.0000	1.0000	1.0000	1.0000	1.0000	1.0000	1.0000	1.0000	1.0000
	1	0.6415	0.8784	0.9612	0.9885	0.9968	0.9992	0.9998	1.0000	1.0000	1.0000
	2	0.2642	0.6083	0.8244	0.9308	0.9757	0.9924	0.9979	0.9995	0.9999	1.0000
	3	0.0755	0.3231	0.5951	0.7939	0.9087	0.9645	0.9879	0.9964	0.9991	0.9998
	4	0.0159	0.1330	0.3523	0.5886	0.7748	0.8729	0.9556	0.9840	0.9951	0.9987
	5	0.0026	0.0432	0.1702	0.3704	0.5852	0.7625	0.8818	0.9490	0.9811	0.9941
	6	0.0003	0.0113	0.0673	0.1958	0.3838	0.5836	0.7546	0.8744	0.9447	0.9793
	7		0.0024	0.0219	0.0867	0.2142	0.3920	0.5834	0.7500	0.8701	0.9423
	8		0.0004	0.0059	0.0321	0.1018	0.2277	0.3990	0.5841	0.7480	0.8684
	9		0.0001	0.0013	0.0100	0.0409	0.1133	0.2376	0.4044	0.5857	0.7483
	10			0.0002	0.0026	0.0139	0.0480	0.1218	0.2447	0.4086	0.5881
	11				0.0006	0.0039	0.0171	0.0532	0.1275	0.2493	0.4119
	12				0.0001	0.0009	0.0051	0.0196	0.0565	0.1308	0.2517
	13					0.0002	0.0013	0.0060	0.0210	0.0580	0.1316
	14						0.0003	0.0015	0.0065	0.0214	0.0577
	15							0.0003	0.0016	0.0064	0.0207
	16								0.0003	0.0015	0.0059
	17									0.0003	0.0013
	18										0.0002

Appendix C

Poisson distribution

This table contains exceedance probabilities for the Poisson distribution. For example, if $\lambda = 0.6$, then
$$P(X \geq 3) = 0.0231.$$

x	$\lambda = 0.1$	0.2	0.3	0.4	0.5	0.6	0.7	0.8	0.9	1.0
0	1.0000	1.0000	1.0000	1.0000	1.0000	1.0000	1.0000	1.0000	1.0000	1.0000
1	0.0952	0.1813	0.2592	0.3297	0.3935	0.4512	0.5034	0.5507	0.5934	0.6321
2	0.0047	0.0175	0.0369	0.0616	0.0902	0.1219	0.1558	0.1912	0.2275	0.2642
3	0.0002	0.0011	0.0036	0.0079	0.0144	0.0231	0.0341	0.0474	0.0629	0.0803
4		0.0001	0.0003	0.0008	0.0018	0.0034	0.0058	0.0091	0.0135	0.0190
5				0.0001	0.0002	0.0004	0.0008	0.0014	0.0023	0.0037
6							0.0001	0.0002	0.0003	0.0006
7										0.0001

x	$\lambda = 1.1$	1.2	1.3	1.4	1.5	1.6	1.7	1.8	1.9	2.0
0	1.0000	1.0000	1.0000	1.0000	1.0000	1.0000	1.0000	1.0000	1.0000	1.0000
1	0.6671	0.6988	0.7275	0.7534	0.7769	0.7981	0.8173	0.8347	0.8504	0.8647
2	0.3010	0.3374	0.3732	0.4082	0.4422	0.4751	0.5068	0.5372	0.5663	0.5940
3	0.0996	0.1205	0.1249	0.1665	0.1912	0.2166	0.2428	0.2694	0.2963	0.3233
4	0.0257	0.0338	0.0431	0.0537	0.0656	0.0788	0.0932	0.1087	0.1253	0.1429
5	0.0054	0.0077	0.0107	0.0143	0.0186	0.0237	0.0296	0.0364	0.0441	0.0527
6	0.0010	0.0015	0.0022	0.0032	0.0045	0.0060	0.0080	0.0104	0.0132	0.0166
7	0.0001	0.0003	0.0004	0.0006	0.0009	0.0013	0.0019	0.0026	0.0034	0.0045
8			0.0001	0.0001	0.0002	0.0003	0.0004	0.0006	0.0008	0.0011
9							0.0001	0.0001	0.0002	0.0002

Statistics with JMP: Graphs, Descriptive Statistics, and Probability, First Edition. Peter Goos and David Meintrup.
© 2015 John Wiley & Sons, Ltd. Published 2015 by John Wiley & Sons, Ltd.
Companion Website: wiley.com/go/goosandmeintrup

x	$\lambda = 2.1$	2.2	2.3	2.4	2.5	2.6	2.7	2.8	2.9	3.0
0	1.0000	1.0000	1.0000	1.0000	1.0000	1.0000	1.0000	1.0000	1.0000	1.0000
1	0.8875	0.8892	0.8997	0.9093	0.9179	0.9257	0.9328	0.9392	0.9450	0.9502
2	0.6204	0.6454	0.6691	0.6916	0.7127	0.7326	0.7513	0.7689	0.7854	0.8009
3	0.3504	0.3773	0.4040	0.4303	0.4562	0.4816	0.5064	0.5305	0.5540	0.5768
4	0.1614	0.1806	0.2007	0.2213	0.2424	0.2640	0.2859	0.3081	0.3304	0.3528
5	0.0621	0.0725	0.0838	0.0959	0.1088	0.1226	0.1371	0.1523	0.1682	0.1847
6	0.0204	0.0249	0.0300	0.0357	0.0420	0.0490	0.0567	0.0651	0.0742	0.0839
7	0.0059	0.0075	0.0094	0.0116	0.0142	0.0172	0.0206	0.0244	0.0287	0.0335
8	0.0015	0.0020	0.0026	0.0033	0.0042	0.0053	0.0066	0.0081	0.0099	0.0119
9	0.0003	0.0005	0.0006	0.0009	0.0011	0.0015	0.0019	0.0024	0.0031	0.0038
10	0.0001	0.0001	0.0001	0.0002	0.0003	0.0004	0.0005	0.0007	0.0009	0.0011
11					0.0001	0.0001	0.0001	0.0002	0.0002	0.0003
12									0.0001	0.0001

x	$\lambda = 3.2$	3.4	3.6	3.8	4.0	4.2	4.4	4.6	4.8	5.0
0	1.0000	1.0000	1.0000	1.0000	1.0000	1.0000	1.0000	1.0000	1.0000	1.0000
1	0.9592	0.9666	0.9727	0.9776	0.9817	0.9850	0.9877	0.9899	0.9918	0.9933
2	0.8288	0.8532	0.8743	0.8926	0.9084	0.9220	0.9337	0.9437	0.9523	0.9596
3	0.6201	0.6603	0.6973	0.7311	0.7619	0.7898	0.8149	0.8374	0.8575	0.8753
4	0.3975	0.4416	0.4848	0.5265	0.5665	0.6046	0.6406	0.6743	0.7058	0.7350
5	0.2194	0.2558	0.2936	0.3322	0.3712	0.4102	0.4488	0.4868	0.5237	0.5595
6	0.1054	0.1295	0.1559	0.1844	0.2149	0.2469	0.2801	0.3142	0.3490	0.3840
7	0.0446	0.0579	0.0733	0.0909	0.1107	0.1325	0.1564	0.1820	0.2092	0.2378
8	0.0168	0.0231	0.0308	0.0401	0.0511	0.0639	0.0786	0.0951	0.1133	0.1334
9	0.0057	0.0083	0.0117	0.0160	0.0214	0.0279	0.0358	0.0451	0.0558	0.0681
10	0.0018	0.0027	0.0040	0.0058	0.0081	0.0111	0.0149	0.0195	0.0251	0.0318
11	0.0005	0.0008	0.0013	0.0019	0.0028	0.0041	0.0057	0.0078	0.0104	0.0137
12	0.0001	0.0002	0.0004	0.0006	0.0009	0.0014	0.0020	0.0029	0.0040	0.0055
13		0.0001	0.0001	0.0002	0.0003	0.0004	0.0007	0.0010	0.0014	0.0020
14					0.0001	0.0001	0.0002	0.0003	0.0005	0.0007
15							0.0001	0.0001	0.0001	0.0002
16										0.0001

x	λ = 5.5	6.0	6.5	7.0	7.5	8.0	8.5	9.0	9.5	10.0
0	1.0000	1.0000	1.0000	1.0000	1.0000	1.0000	1.0000	1.0000	1.0000	1.0000
1	0.9959	0.9975	0.9985	0.9991	0.9994	0.9997	0.9998	0.9999	0.9999	1.0000
2	0.9744	0.9826	0.9887	0.9927	0.9953	0.9973	0.9981	0.9988	0.9992	0.9995
3	0.9116	0.9380	0.9570	0.9704	0.9797	0.9862	0.9907	0.9938	0.9958	0.9972
4	0.7983	0.8488	0.8882	0.9182	0.9409	0.9576	0.9699	0.9788	0.9851	0.9897
5	0.6425	0.7149	0.7763	0.8270	0.8679	0.9004	0.9256	0.9450	0.9597	0.9707
6	0.4711	0.5543	0.6310	0.6993	0.7586	0.8088	0.8504	0.8843	0.9115	0.9329
7	0.3140	0.3937	0.4735	0.5503	0.6218	0.6866	0.7438	0.7932	0.8351	0.8699
8	0.1905	0.2560	0.3272	0.4013	0.4754	0.5470	0.6144	0.6761	0.7313	0.7798
9	0.1056	0.1528	0.2084	0.2709	0.3380	0.4075	0.4769	0.5443	0.6082	0.6672
10	0.0538	0.0839	0.1226	0.1695	0.2236	0.2834	0.3470	0.4126	0.4782	0.5421
11	0.0253	0.0426	0.0668	0.0985	0.1378	0.1841	0.2366	0.2940	0.3547	0.4170
12	0.0110	0.0201	0.0339	0.0533	0.0792	0.1119	0.1513	0.1970	0.2480	0.3032
13	0.0045	0.0088	0.0160	0.0270	0.0427	0.0638	0.0909	0.1242	0.1636	0.2084
14	0.0017	0.0036	0.0071	0.0128	0.0216	0.0342	0.0514	0.0739	0.1019	0.1355
15	0.0006	0.0014	0.0030	0.0057	0.0103	0.0173	0.0274	0.0415	0.0600	0.0835
16	0.0002	0.0005	0.0012	0.0024	0.0046	0.0082	0.0138	0.0220	0.0335	0.0487
17	0.0001	0.0002	0.0004	0.0010	0.0020	0.0037	0.0066	0.0111	0.0177	0.0270
18		0.0001	0.0002	0.0004	0.0008	0.0016	0.0030	0.0053	0.0089	0.0143
19			0.0001	0.0001	0.0003	0.0007	0.0013	0.0024	0.0043	0.0072
20					0.0001	0.0003	0.0005	0.0011	0.0020	0.0035
21						0.0001	0.0002	0.0004	0.0009	0.0016
22							0.0001	0.0002	0.0004	0.0007
23								0.0001	0.0001	0.0003
24									0.0001	0.0001

Appendix D

Exponential distribution

This table contains exceedance probabilities for the exponential distribution with $\lambda = 1$. For example, if $x = 1.43$, then

$$P(X \geq 1.43) = \int_{1.43}^{+\infty} e^{-x} \, dx = 0.2393.$$

x	0.00	0.01	0.02	0.03	0.04	0.05	0.06	0.07	0.08	0.09
0	1.0000	0.9900	0.9802	0.9704	0.9608	0.9512	0.9418	0.9324	0.9231	0.9139
0.1	0.9048	0.8958	0.8869	0.8781	0.8694	0.8607	0.8521	0.8437	0.8353	0.8270
0.2	0.8187	0.8106	0.8025	0.7945	0.7866	0.7788	0.7711	0.7634	0.7558	0.7483
0.3	0.7408	0.7334	0.7261	0.7189	0.7118	0.7047	0.6977	0.6907	0.6939	0.6771
0.4	0.6703	0.6637	0.6570	0.6505	0.6440	0.6376	0.6313	0.6250	0.6188	0.6126
0.5	0.6065	0.6005	0.5945	0.5886	0.5827	0.5769	0.5712	0.5655	0.5599	0.5543
0.6	0.5488	0.5434	0.5379	0.5326	0.5273	0.5220	0.5169	0.5117	0.5066	0.5016
0.7	0.4966	0.4916	0.4868	0.4819	0.4771	0.4724	0.4677	0.4630	0.4584	0.4538
0.8	0.4493	0.4449	0.4404	0.4360	0.4317	0.4274	0.4232	0.4190	0.4148	0.4107
0.9	0.4066	0.4025	0.3985	0.3946	0.3906	0.3867	0.3829	0.3791	0.3753	0.3716

x	0.00	0.01	0.02	0.03	0.04	0.05	0.06	0.07	0.08	0.09
1.0	0.3679	0.3642	0.3606	0.3570	0.3535	0.3499	0.3465	0.3430	0.3396	0.3362
1.1	0.3329	0.3296	0.3265	0.3230	0.3198	0.3166	0.3135	0.3104	0.3073	0.3042
1.2	0.3012	0.2992	0.2952	0.2923	0.2894	0.2865	0.2837	0.2808	0.2780	0.2753
1.3	0.2725	0.2698	0.2671	0.2645	0.2618	0.2592	0.2567	0.2541	0.2516	0.2491
1.4	0.2466	0.2441	0.2417	0.2393	0.2369	0.2346	0.2322	0.2299	0.2276	0.2254
1.5	0.2231	0.2209	0.2187	0.2165	0.2144	0.2122	0.2101	0.2080	0.2060	0.2039
1.6	0.2019	0.1999	0.1979	0.1959	0.1940	0.1920	0.1901	0.1882	0.1864	0.1845
1.7	0.1827	0.1809	0.1791	0.1773	0.1755	0.1738	0.1720	0.1703	0.1686	0.1670
1.8	0.1653	0.1637	0.1620	0.1604	0.1588	0.1572	0.1557	0.1541	0.1526	0.1511
1.9	0.1496	0.1481	0.1466	0.1451	0.1437	0.1423	0.1409	0.1395	0.1381	0.1367

Statistics with JMP: Graphs, Descriptive Statistics, and Probability, First Edition. Peter Goos and David Meintrup.
© 2015 John Wiley & Sons, Ltd. Published 2015 by John Wiley & Sons, Ltd.
Companion Website: wiley.com/go/goosandmeintrup

x	0.00	0.01	0.02	0.03	0.04	0.05	0.06	0.07	0.08	0.09
2.0	0.1353	0.1340	0.1327	0.1313	0.1300	0.1287	0.1275	0.1262	0.1249	0.1237
2.1	0.1225	0.1212	0.1200	0.1188	0.1177	0.1165	0.1153	0.1142	0.1130	0.1119
2.2	0.1108	0.1097	0.1086	0.1075	0.1065	0.1057	0.1044	0.1035	0.1023	0.1013
2.3	0.1003	0.0993	0.0983	0.0973	0.0963	0.0954	0.0944	0.0935	0.0926	0.0916
2.4	0.0907	0.0898	0.0889	0.0880	0.0872	0.0863	0.0854	0.0846	0.0837	0.0829
2.5	0.0821	0.0813	0.0805	0.0797	0.0789	0.0781	0.0773	0.0765	0.0758	0.0750
2.6	0.0743	0.0735	0.0728	0.0721	0.0714	0.0707	0.0699	0.0693	0.0686	0.0679
2.7	0.0672	0.0665	0.0659	0.0652	0.0646	0.0639	0.0633	0.0627	0.0620	0.0614
2.8	0.0608	0.0602	0.0596	0.0590	0.0584	0.0578	0.0573	0.0567	0.0561	0.0556
2.9	0.0550	0.0545	0.0539	0.0534	0.0529	0.0523	0.0518	0.0513	0.0508	0.0503

x	0.00	0.01	0.02	0.03	0.04	0.05	0.06	0.07	0.08	0.09
3.0	0.0498	0.0493	0.0488	0.0483	0.0478	0.0474	0.0469	0.0464	0.0460	0.0455
3.1	0.0450	0.0446	0.0442	0.0437	0.0433	0.0429	0.0424	0.0420	0.0416	0.0412
3.2	0.0408	0.0404	0.0400	0.0396	0.0392	0.0388	0.0384	0.0380	0.0379	0.0373
3.3	0.0369	0.0365	0.0362	0.0358	0.0354	0.0351	0.0347	0.0344	0.0340	0.0337
3.4	0.0334	0.0330	0.0327	0.0324	0.0321	0.0317	0.0314	0.0311	0.0308	0.0305
3.5	0.0302	0.0299	0.0296	0.0293	0.0290	0.0287	0.0284	0.0282	0.0279	0.0276
3.6	0.0273	0.0271	0.0268	0.0265	0.0263	0.0260	0.0257	0.0255	0.0252	0.0250
3.7	0.0247	0.0245	0.0242	0.0240	0.0238	0.0235	0.0233	0.0231	0.0228	0.0226
3.8	0.0224	0.0221	0.0219	0.0217	0.0215	0.0213	0.0211	0.0209	0.0207	0.0204
3.9	0.0202	0.0200	0.0198	0.0196	0.0194	0.0193	0.0191	0.0189	0.0187	0.0185

x	0.00	0.01	0.02	0.03	0.04	0.05	0.06	0.07	0.08	0.09
4.0	0.0183	0.0181	0.0180	0.0178	0.0176	0.0174	0.0172	0.0171	0.0169	0.0167
4.1	0.0166	0.0164	0.0162	0.0161	0.0159	0.0158	0.0156	0.0155	0.0153	0.0151
4.2	0.0150	0.0148	0.0147	0.0146	0.0144	0.0143	0.0141	0.0140	0.0138	0.0137
4.3	0.0136	0.0134	0.0133	0.0132	0.0130	0.0129	0.0128	0.0127	0.0125	0.0124
4.4	0.0123	0.0122	0.0120	0.0119	0.0118	0.0117	0.0116	0.0114	0.0113	0.0112
4.5	0.0111	0.0110	0.0109	0.0108	0.0107	0.0106	0.0105	0.0104	0.0103	0.0102
4.6	0.0101	0.0100	0.0099	0.0098	0.0097	0.0096	0.0095	0.0094	0.0093	0.0092
4.7	0.0091	0.0090	0.0089	0.0088	0.0087	0.0087	0.0086	0.0085	0.0084	0.0083
4.8	0.0082	0.0081	0.0081	0.0080	0.0079	0.0078	0.0078	0.0077	0.0076	0.0075
4.9	0.0074	0.0074	0.0073	0.0072	0.0072	0.0071	0.0070	0.0069	0.0069	0.0068
5.0	0.0067	0.0067	0.0066	0.0065	0.0065	0.0064	0.0063	0.0063	0.0062	0.0062

Appendix E

Standard normal distribution

This table contains exceedance probabilities of the standard normal distribution. For example, $P(Z \geq 1.96) = 0.02500$.

z	0.00	0.01	0.02	0.03	0.04	0.05	0.06	0.07	0.08	0.09
0.0	0.50000	0.49601	0.49202	0.48803	0.48405	0.48006	0.47608	0.47210	0.46812	0.46414
0.1	0.46017	0.45620	0.45224	0.44828	0.44433	0.44038	0.43644	0.43251	0.42858	0.42465
0.2	0.42074	0.41683	0.41294	0.40905	0.40517	0.40129	0.39743	0.39358	0.38974	0.38591
0.3	0.38209	0.37828	0.37448	0.37070	0.36693	0.36317	0.35942	0.35569	0.35197	0.34827
0.4	0.34458	0.34090	0.33724	0.33360	0.32997	0.32636	0.32276	0.31918	0.31561	0.31207
0.5	0.30854	0.30503	0.30153	0.29806	0.29460	0.29116	0.28774	0.28434	0.28096	0.27760
0.6	0.27425	0.27093	0.26763	0.26435	0.26109	0.25785	0.25463	0.25143	0.24825	0.24510
0.7	0.24196	0.23885	0.23576	0.23270	0.22965	0.22663	0.22363	0.22065	0.21770	0.21476
0.8	0.21186	0.20897	0.20611	0.20327	0.20045	0.19766	0.19489	0.19215	0.18943	0.18673
0.9	0.18406	0.18141	0.17879	0.17619	0.17361	0.17106	0.16853	0.16602	0.16354	0.16109

z	0.00	0.01	0.02	0.03	0.04	0.05	0.06	0.07	0.08	0.09
1.0	0.15866	0.15625	0.15386	0.15151	0.14917	0.14686	0.14457	0.14231	0.14007	0.13786
1.1	0.13567	0.13350	0.13136	0.12924	0.12714	0.12507	0.12302	0.12100	0.11900	0.11702
1.2	0.11507	0.11314	0.11123	0.10935	0.10749	0.10565	0.10383	0.10204	0.10027	0.09853
1.3	0.09680	0.09510	0.09342	0.09176	0.09012	0.08851	0.08692	0.08534	0.08379	0.08226
1.4	0.08076	0.07927	0.07780	0.07636	0.07493	0.07353	0.07214	0.07078	0.06944	0.06811
1.5	0.06681	0.06552	0.06426	0.06301	0.06178	0.06057	0.05938	0.05821	0.05705	0.05592
1.6	0.05480	0.05370	0.05262	0.05155	0.05050	0.04947	0.04846	0.04746	0.04648	0.04551
1.7	0.04457	0.04363	0.04272	0.04182	0.04093	0.04006	0.03920	0.03836	0.03754	0.03673
1.8	0.03593	0.03515	0.03438	0.03362	0.03288	0.03216	0.03144	0.03074	0.03005	0.02938
1.9	0.02872	0.02807	0.02743	0.02680	0.02619	0.02559	0.02500	0.02442	0.02385	0.02330

Statistics with JMP: Graphs, Descriptive Statistics, and Probability, First Edition. Peter Goos and David Meintrup.
© 2015 John Wiley & Sons, Ltd. Published 2015 by John Wiley & Sons, Ltd.
Companion Website: wiley.com/go/goosandmeintrup

z	0.00	0.01	0.02	0.03	0.04	0.05	0.06	0.07	0.08	0.09
2.0	0.02275	0.02222	0.02169	0.02118	0.02068	0.02018	0.01970	0.01923	0.01876	0.01831
2.1	0.01786	0.01743	0.01700	0.01659	0.01618	0.01578	0.01539	0.01500	0.01463	0.01426
2.2	0.01390	0.01355	0.01321	0.01287	0.01254	0.01222	0.01190	0.01160	0.01130	0.01101
2.3	0.01072	0.01044	0.01017	0.00990	0.00964	0.00939	0.00914	0.00889	0.00866	0.00842
2.4	0.00820	0.00798	0.00776	0.00755	0.00734	0.00714	0.00695	0.00676	0.00657	0.00639
2.5	0.00621	0.00604	0.00587	0.00570	0.00554	0.00539	0.00523	0.00509	0.00494	0.00480
2.6	0.00466	0.00453	0.00440	0.00427	0.00415	0.00403	0.00391	0.00379	0.00368	0.00357
2.7	0.00347	0.00336	0.00326	0.00317	0.00307	0.00298	0.00289	0.00280	0.00272	0.00263
2.8	0.00256	0.00248	0.00240	0.00233	0.00226	0.00219	0.00212	0.00205	0.00199	0.00193
2.9	0.00187	0.00181	0.00175	0.00169	0.00164	0.00159	0.00154	0.00149	0.00144	0.00139

z	0.00	0.01	0.02	0.03	0.04	0.05	0.06	0.07	0.08	0.09
3.0	0.00135	0.00131	0.00126	0.00122	0.00118	0.00114	0.00111	0.00107	0.00104	0.00100
3.1	0.00097	0.00094	0.00090	0.00087	0.00085	0.00082	0.00079	0.00076	0.00074	0.00071
3.2	0.00069	0.00066	0.00064	0.00062	0.00060	0.00058	0.00056	0.00054	0.00052	0.00050
3.3	0.00048	0.00047	0.00045	0.00043	0.00042	0.00040	0.00039	0.00038	0.00036	0.00035
3.4	0.00034	0.00032	0.00031	0.00030	0.00029	0.00028	0.00027	0.00026	0.00025	0.00024
3.5	0.00023	0.00022	0.00022	0.00021	0.00020	0.00019	0.00019	0.00018	0.00017	0.00017
3.6	0.00016	0.00015	0.00015	0.00014	0.00014	0.00013	0.00013	0.00012	0.00012	0.00011
3.7	0.00011	0.00010	0.00010	0.00010	0.00009	0.00009	0.00009	0.00008	0.00008	0.00008
3.8	0.00007	0.00007	0.00007	0.00006	0.00006	0.00006	0.00006	0.00005	0.00005	0.00005
3.9	0.00005	0.00005	0.00004	0.00004	0.00004	0.00004	0.00004	0.00004	0.00004	0.00003

z	0.00	0.01	0.02	0.03	0.04	0.05	0.06	0.07	0.08	0.09
4.0	0.00003	0.00003	0.00003	0.00003	0.00003	0.00002	0.00002	0.00002	0.00002	0.00002

Index

Statistics with JMP: Graphs, Descriptive Statistics, and Probability, First Edition. Peter Goos and David Meintrup.
© 2015 John Wiley & Sons, Ltd. Published 2015 by John Wiley & Sons, Ltd.
Companion Website: wiley.com/go/goosandmeintrup

Printed and bound by CPI Group (UK) Ltd, Croydon, CR0 4YY

27/10/2024

14580298-0002